AF689167

Dearest Michael,

At last! thank you
for inspiring the best
in me and for nurturin[g]
it to fruition!

With love + profound r[...]

Rita.

STATE AGENCY AND THE PATENTING OF LIFE IN INTERNATIONAL LAW

Dedicated to my beloved parents,
Homayoun and Ali,
Loving husband, Reza,
and
My precious Daniel,
To Create is Divine.

In Loving Memory of Uncle Aflatoon Adhami

State Agency and the Patenting of Life in International Law

Merchants and Missionaries in a Global Society

BITA AMANI

Queen's University, Canada

ASHGATE

Published by

Ashgate Publishing Limited	Ashgate Publishing Company
Wey Court East	Suite 420
Union Road	101 Cherry Street
Farnham	Burlington, VT 05401-4405
Surrey GU9 7PT	USA
England	

www.ashgate.com

British Library Cataloguing in Publication Data
Amani, Bita.
State agency and the patenting of life in international law. -- (Globalization and law)
1. Life sciences--Patents. 2. Human rights. 3. Patents (International law) 4. International and municipal law.
I. Title II. Series
346'.0486-dc22

Library of Congress Cataloging-in-Publication Data
Amani, Bita.
State agency and the patenting of life in international law : merchants and missionaries in a global society / by Bita Amani.
p. cm. -- (Globalization and law)
Includes bibliographical references and index.
ISBN 978-0-7546-7438-2
1. Biotechnology--Patents. 2. Biotechnology--Law and legislation. 3. International obligations. 4. World Trade Organization. I. Title.

K1519.B54A74 2009
346.04'86--dc22

2009007623

ISBN 978-0-7546-7438-2

Mixed Sources
Product group from well-managed forests and other controlled sources
www.fsc.org Cert no. SGS-COC-2482
© 1996 Forest Stewardship Council
FSC

Printed and bound in Great Britain by
TJ International Ltd, Padstow, Cornwall

Contents

Abbreviations

Berne	Berne Convention
CBAC	Canadian Biotech Advisory Committee
CBD	Convention on Biological Diversity
CGIAR	Consultative Group on International Agricultural Research
CIPO	Canadian Intellectual Property Office
CIPO	Canadian Intellectual Property Office
CP	Contracting Parties (to the GATT 1947)
CPA	Canadian Patent Act
CPO	Canadian Patent Office
CPR	civil and political rights
DNA	deoxyribonucleic acid
DOE	Department of Health (U.S.A)
DSU	Dispute Settlement Understanding
EBD	European Biotechnolody Directive
ECD	Equitable Conduct Defence (also referred to as the Public Interest Defence)
ECOSOC	United Nations Economic and Social Council
ECSR	Economic Cultural and Social Rights
EPC	European Patent Convention
EPO	European Patent Office
EU	European Union
FAO	Food and Agricultural Organization
GA	General Assembly
GATT	General Agreement on Tariff and Trade
GM	genetically modified
GMOs	genetically modified organisms
GNP	Gross National Product
GSP	Generalized System of Preferences
HD	Human Development
HGHRD	Universal Declaration on the Human Genome and Human Rights
HR	human rights
ICCPR	International Covenant for Civil and Political Rights
ICESCR	International Covenant for Economic Social and Cultural Rights
IP	intellectual property
IPP	intellectual property protection
IPRs	intellectual property rights
LMOs	living modified organisms

MDD	Millennium Development Declaration
MDGs	Millennium Development Goals
Members	Members of the WTO
MFA	Multi-Fibre Arrangement
MGL	Myriad Genetic Laboratories Inc.
MNCs	Multinational Corporations
MOPOP	Manual of Patent Office Practice (Canada)
NGOs	Non-governmental organizations
NIH	National Institutes for Health (USA)
NTB	Non-tariff barrier (to trade)
OECD	Organization for Economic Cooperation and Development
Paris	Paris Convention
PAP	Protocol of Provisional Application
PCT	Patent Cooperation Treaty
PGRs	Plant Genetic Resources
PPA	Plant Protection Act
PVPA	Plant Variety Protection Act
RNA	Ribonucleic acid
RRC	Round Up Ready Canola (Monsanto's)
SNPs	single nucleotide polymorphisms
SPLT	Substantive Patent Law Treaty
TK	Traditional Knowledge
TNCs	Transnational Corporations
TRIPS	Trade Related Aspects of Intellectual Property Agreement
UDHR	Universal Declaration of Human Rights
UN	United Nations
UNC	United Nations Charter
UNCTAD	United Nations Commission on Trade and Development
UNDHR	Universal Declaration of Human Rights
UNDP	United Nations Development Program
UNESCO	United Nations Economic Social and Cultural Organization
USA	United States of America
USPTO	United States Patent and Trademark Office
VCLT	Vienna Convention on the Law of Treaties
WIPO	World Intellectual Property Organization
WTO	World Trade Organization

Acknowledgments

I have been blessed with a number of truly fantastic mentors who I admire as respected scholars and as genuinely great persons. I owe an enormous debt to Michael Trebilcock for his clear guidance and patience in supervising my doctoral work, generously supported by the Social Science and Humanities Research Council and the Centre for Innovation Law and Policy, on which this book is based; to Rosemary Coombe for creating many growth opportunities and always guiding their outcome with her sage counsel; and to David Vaver, my original mentor and professor of intellectual property law, for inciting in me a fond appreciation for the study of IP law and then providing sufficient fodder for me to reflect on through his writings. I am tremendously honoured to have David provide the foreword to this book. My grade 10 geography teacher planted the seed many moons ago, and Professor Couckell nurtured it by introducing me to the vast world of genetics in his course, the 'Genetic Revolution'. Trudo Lemmens extended a rare opportunity to do policy work on gene patenting. Many thanks to those who offered technical support, in particular Nancy McCormack and Sharron Sluiter at Queen's University, Pauline Beavers and Alison Kirk at Ashgate, and Pat FitzGerald. Thanks also to Richard Cumyn. I am deeply indebted to the Brocher Foundation in Geneva, Switzerland for their generous support in funding my stay on the beautiful and peaceful banks of Lake Geneva in the summer of 2008, enabling me to work on this book and to share some of its contents through an interdisciplinary workshop and visitorships at the Oxford Intellectual Property Research Centre and the Centre for International Governance at Leeds University.

I cannot thank my personal support network enough, old and new friends, colleagues, and the students I have taught and who have taught me. My particular appreciation extends to my immediate and extended family, for suffering the silence that quiet study and the solitary task of writing bring, and for offering nothing but support in return. My dear parents, Homayoun and Ali, have provided unconditional love and gentle prodding through the years. I thank them and my siblings for their guidance. My father instilled in me a love of books while my mother, to whom I am particularly indebted, emphasized the importance of higher education, provided unrelenting support, and kept me company during many late nights of scholarly pursuits. I am most obliged to my uncle Aflatoon Adhami, for enabling me to call Canada home; his generous spirit will live in the Aflatoon Adhami Foundation. To my love Reza, thank you for your loving support, patience and forbearance over the years. Finally, precious Daniel, thank you simply for being you! One can never appropriately acknowledge or thank all who have contributed to her intellectual development, her sense of social justice or desire for political accountability. These people have shaped my experiences and, therefore, the primacy of the humanist approach I adopt in this text towards creating a life of meaningful difference. They remain unnamed, however, because authorship is a romantically individual enterprise in Western intellectual property law. My mother taught me early that 'experience is a class with a high tuition'. Happily, I continue to pay the price. All errors and omissions remain my own.

Foreword

A lot of nonsense has been written about patents and biotechnology. Polarized opinions and lobbying are often presented as undisputed fact, law or good policy, with the result that much confusion reigns among law makers, the general public, and even legal and scientific experts.

Against this state of affairs comes this important and passionate, yet richly textured and scholarly work by Bita Amani. She views with a detached but sceptical eye the well-nigh universal trend, with all its anomalies and inconsistencies, towards allowing patents over living matter and genetic technologies, at least in the developed world. The process by which this trend occurred was far from inevitable and Dr Amani traces its curious course through patent law and practice. Globalization pushed the trend beyond the developed to the developing world through trade-related instruments such as the World Trade Organization Agreement and other international treaties. While these agreements promised improvements in health, food security and knowledge transfer, the more immediate results were losses of national sovereignty, higher consumer welfare costs, and public disquiet. Difficulties about how to reconcile biopatenting with cultural, religious and ethical norms and with biodiversity imperatives continued unabated.

Dr Amani meticulously traces these developments and argues that none of them was or is necessary. She posits that international trading agreements remain flexible enough for states to adopt a broader range of biotechnology strategies than many have thought possible and that may better accord with a state's level of economic development and other internal concerns. Trading treaties and intellectual property regimes, far from supplanting or contradicting human rights obligations, can be read so as to accommodate such critical rights as access to health, food security and human dignity. Dr Amani argues that WTO dispute resolution processes must embrace such considerations if the TRIPs agreement is to command general support.

Such are some of the themes and lines of argument found in this prodigiously researched and cautionary book. It is bound to stir controversy and debate. Dr Amani deserves our thanks for raising the issues and their possible resolution so clearly and crisply. For anyone interested in health and science policy, and how law may be shaped to fulfil the promise of widespread access to the benefits that genetic technologies may offer, this work is essential reading, for experts and non-experts alike.

David Vaver
Oxford
May 2009

Preface

This book explores state agency in determining the legality and impact of biopatenting. The state's role in resolving the patenting life debate within its own jurisdiction is consistent with the historical territorial and statutory nature of patents and is integral to whether international human rights commitments at the intersection of patent rights are domestically realized. Patenting life and its genetic components has proven controversial in various jurisdictions because of the social, economic, religious, ethical and developmental issues engaged. Patents create a trade-off between long term public interest promoting gains of greater innovation and short term costs of legislatively created monopolies which limit the price and access to such innovations. Biopatenting can affect the right to health, food, and farmer's rights to save seeds. Patents are essentially anti-competitive and substantively antithetical to the multilateral trading regime in which they are now located as part of the overall composite package of obligations to which all WTO member countries must adhere. While the Trade Related Aspects to Intellectual Property Agreement (TRIPS) does require compliance by all Member states with universal minimum prescriptions for patent protection, the agreement has sufficient provision for regulatory diversity and autonomy. Since many states are both signatory to the UN Covenant of Social, Economic and Cultural Rights as well as Members of the WTO, they may find themselves faced with competing international obligations under trade and human rights regimes. My objective in writing this book is to provide a framework for appropriate state agency in an age of legal fragmentation through an examination of how states can exercise the right to self-determination to order domestic law and policy preferences at the intersection of biopatenting to better accord with welfare and competition-promoting objectives under trade law.

The interplay of the two bodies of international law is evident from their post-war historical emergence, stated goals, and various textual references. The ensuing analyses demonstrate how the emergence of separate systems for facilitating trade and preserving human rights, both arguably to promote welfare, the public interest, dignity and development, has led to an artificial divisibility of the world's governance institutions, and therefore our dually legally constructed identities, as though these identities are mutually exclusive despite the indivisibility of their goals. At any given moment, a state is perceived to be acting as agent for either 'merchants' or 'missionaries', never both at the same time, creating a moral divide with regrettable results that potentially undermine the synergistic gains of a reconciled 'identity'. This dualism is based on a false dichotomy that pits the liberal self empowered with an individual human right against state authorized traders and actors operating on a capitalist imperative, also strategically framed as a 'right' (IPRs). But the dichotomy is illusory for reasons accounted for in this text. To maintain any coherence in our international institutions for governance, I argue, we must reconcile human rights with trade obligations and eliminate binary interpretations of the instruments arising from the twin institutions (UN and WTO) for global governance. Trade and human rights must be read in a complementary and coherent manner, as a dialogue with each state about how to devise a better world where respect for *dignity* and *human development* is made paramount in a self-determined manner. Such an approach to international obligations is both desirable and congruent with the domestic and international prescriptions offered. While the 'publics' constituted by domestic governance and defined by increasingly imagined territorial boundaries are not juridical parties before the WTO, it is in their 'interest' that the WTO has been constituted with promises to improve social welfare by lowering consumption costs facilitated by reduced trade barriers. TRIPS, however, reveals that trade liberalization in this area is not motivated by concern over people (consumers in market terms) but corporate greed linked to the push for minimal substantive IP standards based on economic motivations that transcend state boundaries for articulation while relying on state agency for enforcement. The failure of representative governments to improve domestic law and policy in a manner consistent with the human rights of its citizens is a direct consequence of the progressive emasculation of state agency through limited readings of the human rights regime by the state's own agents and cannot be exclusively attributed to the perceived restrictions imposed under WTO law.

To no one will we sell, to no one will we refuse or delay right or justice.

Article 40, *Magna Carta*

No society can surely be flourishing and happy, of which the far greater part of the members are poor and miserable.

Adam Smith, *The Wealth of Nations*

I took one draught of life,
I'll tell you what I paid,
Precisely an existence-
The market price they said

Emily Dickinson, *Further Poems*

Chapter 1
Regulatory Diversity and the Patenting of Life

> Its peculiar character, too, is that no one possesses the less, because every other possesses the whole of it. He who receives an idea from me receives instruction himself without lessening mine; as he who lights his taper at mine, receives light without darkening me. That ideas should freely spread from one to another over the globe, for the moral and mutual instruction of man, and improvement of his condition, seems to have been peculiarly and benevolently designed by nature, when she made them, like fire, expansible over all space, without lessening their density in any point, and like the air in which we breathe, move, and have our physical being, incapable of confinement or exclusive appropriation. Inventions then cannot, in nature, be a subject of property. Society may give an exclusive right to the profits arising from them, as an encouragement to men to pursue ideas which may produce utility, but this *may or may not be done, according to the will and convenience of the society, without claim or complaint from anybody*.[1]
>
> Thomas Jefferson

> What's past is prologue.
>
> William Shakespeare

Context

My high school geography teacher kept a shelved row of labelled jars containing seeds. When asked about this curiosity, he spoke apocalyptically of an approaching dystopian future when seeds would no longer have the capacity to self-propagate due to human genetic intervention. Farmers, reduced to agri-serfs, he claimed, would become dependent on large corporations (agribusinesses) that, in pursuit of greater profit, would dole out single harvest seeds at monopolistic prices under strict licences. I dismissed these doomsday comments as typical of the political hyperbole characterizing the environmental movement of the late 1980s. But the comments were ever

1 Emphasis added. Letter from Thomas Jefferson to Isaac McPherson (13 August 1813) in *The Founders' Constitution*, Thomas Jefferson to Isaac McPherson', vol. 3, art. 1, s. 8, cl. 8, doc. 12, The Founders' Constitution <http://press-pubs.uchicago.edu/founders/documents/a1_8_8s12.html>.

present as the supporting evidence mounted in the following two decades. Now, over 20 years later, these statements reveal a prescient awareness of the then-nascent issues that have come to dominate the debate surrounding intellectual property protection and human rights, the patenting of life (biopatenting) from genes of all species, to plant and animal varieties, and the impact on human rights of growing, converging, sometimes overlapping, and generally excessive intellectual property rights (IPRs) protection. Peter Drahos and John Braithwaite in their review of the maladies of an overbearing intellectual property (IP) protection system observe,

> [we] experience these restrictions not as a mass of individuals living in a totalitarian society, but as members of smaller communities who find strands of intellectual property law settling on and changing the customary ways in which we have accessed and exchanged information. Farmers who follow ancient practices of saving, swapping, bartering or selling seeds to each other find that these practices have to take place in the shadow of patent claims over those seeds. Just what a farmer may or may not do with plants containing patented genes becomes a lawyer's game.[2]

The game is no longer national. Because of the number of countries affected and the scope of their application, the most significant of IP protections are entrenched by international agreement within the rules-based multilateral trade system of the World Trade Organization (WTO). Stronger protections have been negotiated under bilateral and regional agreements but are limited to negotiating parties rather than the entire WTO membership. Membership now depends on implementation and enforcement of universal minimum standards for IP along with other trade obligations that may effectively require these stronger protections to be extended to WTO members.

Meanwhile, developing countries protest these new legal and technological restrictions on their sovereign right to self-determination and domestic regulatory space. They complain of an apparent hypocrisy on the part of developed countries: historically, the same developed states that today push for stronger IPRs in multilateral and bilateral agreements did not observe strict IPRs until they had reached a high level of development; an economic maturity facilitated by the liberal exchange of ideas and transfer of technology. Some developing countries assert that they would simply like to be extended the privilege to determine domestic industrial policy in accordance with other regulatory priorities and social values without suffering potential economic costs for non-compliance with trade obligations – just as developed countries had before them. Sovereignty and the right to self-determination, they argue, require that governments be free to exercise appropriate discretion to

2 Peter Drahos and John Braithwaite, *Information Feudalism: Who Owns the Knowledge Economy?* (London: Earthscan Publications, 2002) at 4.

regulate the patenting of life domestically and without threat of trade-related interference.

Intellectual property protection is no longer simply about domestic industrial policy but in fact impacts on a broad scope of public policy arenas affecting numerous sectors of vital importance to a country's socio-economic and technological development. As intellectual property has expanded to absorb developments in the life sciences, the regulation of biotechnology portends to be the most controversial and central challenge facing governments in relation to diverse areas of public policy in the twenty-first century. These include governance of public and population health genomics; alternative energy technology (biofuels) and waste management; environmental protection; food production and food security; sustainable development; national security and bioterrorism; and foreign policy with respect to aid as well as fair and free trade. The regulatory trajectory of biopatenting began with the significant discovery of the structure of DNA over half a century ago and followed the attendant shift in study from genetics (genes) to genomics (entire genomes). Deoxyribonucleic acid (DNA) provides the genetic coding for all life. Its imperative role explains why it is referred to in such monumental terms as the 'building blocks of life', the 'blue print' or 'Manual' of Man, the 'Code of Codes' and the 'Book of Life' itself.[3] There are many arguments for and against patenting life, ranging from the economic to the ethical, any of which may appeal to culturally stratified citizens and their governments. At its core, however, the debate is about free access to genetic *information* and the propriety of its regulation and redistribution through market restrictions designed for *inventions*. Genetic information has a number of public good uses and is valued both as its own commodity as well as the basis of knowledge in the information economy.[4] How this information may be used, whether its possibilities will be realized, and who has access to this information and under what terms are all social and legal issues resolved by intellectual property regulations which are derived from statute and thus territorial.

Biopatenting: The Case for State Agency

Patents, a form of intellectual property had, until recently, appropriately remained within the exclusive jurisdiction of each nation state to regulate in accordance with its own phase of social, economic and industrial development. With some exception for United Nations based agreements under the World

3 Kevin Davies, *Cracking the Human Genome: Inside the Race to Unlock Human DNA* (New York: The Free Press, 2001).

4 J.H. Reichman and Jonathan A. Franklin, 'Privately Legislated Intellectual Property Rights: Reconciling Freedom of Contract with Public Good Uses of Information' (1999) 147 *U. Pa. L. Rev.* 875 at 884.

Intellectual Property Organization (WIPO), countries were essentially left to determine the existence and extent of their domestic IPRs schemes. To attract protection for intellectual property, the author or inventor would have to register or otherwise comply with the legal institutional requirements within each individual jurisdiction where rights were sought. This meant that patent law was essentially country-specific and largely variable; although an effective means for optimizing regulatory governance, it was not necessarily conductive to profitable industry. With the shift to a 'knowledge-based' economy and facilitated transnational capital flows, the IP generating industries successfully lobbied for a corresponding shift to substantive international IP protection negotiated within the multilateral trading regime. Established Western IP standards were used as the baseline through which protection, and therefore financial returns, were secured in foreign markets.

Not all states are at the same level of development, however; nor are they in an equal position to capitalize on the gains of the knowledge (or bio) economy. The institutionalization of IP in the world trading system under the *Agreement on Trade-Related Aspects of Intellectual Property Rights* (TRIPS)[5] rewards innovation while ignoring the comparative advantage enjoyed by poorer nations, because of their cheap labour, in imitation. TRIPS' universal minimum standards also ignore the historical territoriality and state derived scope of IPRs and are ostensibly at odds with the institutional history and normative, welfare-promoting justifications for the development of the human rights and trade regimes.

WIPO remains an important international institution specialized in and exclusively dedicated to IP administration despite the fact that, having lacked a formal dispute resolution mechanism, it seems to have been displaced by the WTO's jurisdiction under TRIPS. In fact, WIPO has tried to revitalize its position in the international framework for IP governance.[6] In 1995, WIPO concluded the WIPO-WTO Cooperation Agreement and since 1998 has overseen a technical assistance programme for developing countries in order to facilitate compliance with TRIPS obligations.[7] WIPO continues to strengthen its relationship with other institutions such as the WTO and the Convention on Biological Diversity Conference (CBD) of the Parties of the Secretariat. WIPO acknowledges the significant relationship between health

5 Agreement on Trade-Related Aspects of Intellectual Property Rights, 15 December 1993, Marrakesh Agreement Establishing the World Trade Organization, Annex 1C, Legal Instruments – Results of the Uruguay Round, <http://www.wto.org/english/docs_e/docs_e.htm>.

6 See Graham Dutfield, *Intellectual Property Rights, Trade and Biodiversity: Seeds and Plant Varieties* (London: Earthscan Publications, 2000) [Dutfield, *IPRS, Trade and Biodiversity*] 95–7.

7 Joint WTO-WIPO press release, 21 July 1998, Press/108, <http://www.wto.org/english/news_e/pres98_e/pr108_e.htm>.

and IPRs and 'supports the initiatives taken by the Secretary-General of the United Nations, and is cooperating with the World Health Organization, UNAIDS, and the World Trade Organization in areas where it can offer its experience and expertise regarding the question of patents and health care'.[8] In fact, WIPO's Global Intellectual Property Issues Division was developed in 1998 to address

> [t]he challenges facing the intellectual property system in a rapidly changing world, such as accelerating technological advancement, the integration of the world's economic, cultural and information systems, and the expanding relevance of intellectual property issues in trade, culture, investment, human rights, health and the environment spheres.[9]

WIPO's norm setting and capacity building initiatives keep it relevant as an organization in the post-TRIPS lawmaking environment. But institutionalizing IP in the WTO has created greater capacity for IP enforceability, making the WTO the forum of choice. The substantive minimum standards mandated for all WTO members are also perceived impediments to regulatory diversity. I argue that in relation to patenting life these restrictions are minimal and are rooted in misgivings of what trade obligations require or corresponding rights permit. In actuality, states continue to enjoy significant latitude in the ability of their domestic patent laws to respond to the question of whether life and its genetic building blocks are or ought to be patentable.

Some governments may view biopatenting as favourable to their (economic) policy objectives and development agendas. Canada's federal government, for example, has embraced the currency of industry rhetoric and openly endorsed the global proliferation of patents on the assumption that growth in the number of patents issued is a measure of increased innovative activity good for the economy. Given the strong push by developed countries for higher IP protection, former Canadian Prime Minister (then Finance Minister) Paul Martin was likely not alone in the view expressed in his Budget Speech:

> Today the strength of a nation is measured not by the weapons it wields, but by the patents it produces; not by the territory it controls, but by the ideas it advances; not only by the wealth of its resources, but by the resourcefulness of its people. It such a world, successful nations will only be those that foster a culture of innovation. They will be those that create new knowledge and

8 See WIPO Publications, 'Striking a Balance: Patents and Access to Drugs and Health Care', <http://www.wipo.int/about-ip/en/studies/publications/health_care.htm>.

9 Quote based on unpublished information provided to Graham Dutfield by Mr Wend Wendland and Mr Shakeel Bhatti of WIPO, cited in Dutfield, *IPRS, Trade and Biodiversity* supra note 6 at 96.

> bring the product of that knowledge quickly to market. Our goal … must be to lead the way.[10]

Governments of developing states may, conversely, find that restrictive patent protection hampers technology transfer and knowledge dissemination critical in the early stages of an economy's development and supplants the welfare interests of their constituents with the property interests of foreign nationals. They may find that internationally mandated strong protections also pervert national cultural policies. States may have very differing rationales for preserving independent views on whether patents can issue to new subject matter and the instruments required to provide optimal incentives for innovation. Different cultural values may also inform popular disagreement as to what constitutes 'life'. Such disputes would have to be legally mediated through national representative governments.

It is erroneously believed that TRIPS unduly restricts a sovereign nation's ability to place its citizens' rights over the private interests of patentees. Overcoming such misapprehension is the major hurdle confronting civil society in its effort to convince representative government to act equitably in the patenting life debate, balancing individual human rights and the public interest with an efficient, coordinated set of laws and policies. While states 'have considerable freedom to choose different social policy responses to a global era',[11] there seems to be significant reservation regarding the scope and restraints on this freedom due to increasingly fragmented international legal obligations. In some cases, governments look for legitimate means of overcoming perceived impediments to domestic policy formation while in others national governments may use their international obligations to justify or excuse a failure to act. Politically charged and controversial issues are instead abdicated to a judiciary often less concerned with achieving optimal patent law and policy than with mitigating the economic and political costs of divergence from dominant, primarily American-based, patent law.

Present jurisprudence surrounding biopatenting is more the result of patent-friendly judiciaries, and administrative exercise of discretion that favours extending patents to life and its genetic building blocks, than of legislative action stemming from robust public debate. Where a state opposes the issue of biopatents in order to fulfil its international human rights obligations, its government may adopt domestic TRIPS-compliant measures available under current law. Governments could engage in policy reform and legislative amendment to broaden statutory exclusions and available defences, and issue

---

10 Paul Martin, Finance Minister, 28 February 2000 Budget Speech, full text <http://www.fin.gc.ca/budget00/speech/speech1e.htm>.

11 Keith G. Banting, 'The Social Contract in a New Century: Choice and Social Cohesion in a Global Era', Donald Gow Lecture, School of Policy Studies, Queens University, Canada, 25 April 2003 at 2–3.

interpretative patent practice guidelines through national patent offices as a way of establishing more stringent application of the patentability requirements. Alternative regulatory models to patent law should be considered as should the increased use of competition law to curb escalating anti-competitive behaviour.

A current view of patent practice and a comparative review of the history of patenting life in two key but diverging common law jurisdictions, Canada and the United States, demonstrates that by doctrinal standards life and its genetic building blocks should not be patentable. Because of the significant ethical, religious, social, cultural and economic issues generated by this topic, however, the main purpose of this book is not to impose this substantive view of patentability as one to which all nations should adhere. Rather, it is to empower states to act in a self-determined manner to settle this debate within their own cultural and regulatory milieu instead of simply co-opting the international protectionist trends. The examples provided in this text demonstrate how states, through their own agents – namely, courts, administrative patent offices, law and policy makers, and trade technocrats – progressively emasculate their own agency through limited readings of human rights regimes. Accordingly, this text strives to answer a central principled question that cuts across state motivation and stratification on the biopatenting debate in an effort to preserve critical regulatory space:

How Should a State Respond to Competing International Obligations where the Patenting of Life is Concerned?

In an age of deep economic integration reaching beyond borders to eliminate non-tariff regulatory barriers to trade, how can a state, compelled to attain greatest compliance with its own human rights obligations or desirous of achieving a different balance in domestic public policy, adopt domestic measures that *prima facie* forebode non-compliance with international trade obligations? Can a state domestically prioritize human rights without finding itself defending this position as a respondent to a WTO complaint and subject, ultimately, to countermeasures by way of legitimately imposed trade sanctions?

The text prescribes a framework for appropriate state agency in light of competing international obligations, one that fosters democratic accountability, is responsive to the private interests of citizens over corporations, and displays fundamental respect for sovereignty. While many of the examples drawn upon are Canadian, the need to establish a framework for state agency is important for all countries and their citizens. International trade law provides skewed incentives for prioritizing industrial policy and trade obligations over health policy and human rights by creating cost consequences associated with defending a WTO complaint against the contested measure. Effectively, human-rights promoting measures may paradoxically be subjected to welfare-reducing trade sanctions. No similar external threat exists to influence state agency towards compliance with United Nations (UN) human rights instruments. There are no such cost consequences where sovereign discretion is exercised to give trade obligations priority over human rights. Moreover, developing countries (DCs) and least developed countries (LDCs) may appreciate the additional scope offered in terms of their own development agendas.

The state is integral to the domestic realization of international human rights commitments. Most states are also members of the WTO and obligated under its agreements to advance consumer welfare by helping to ensure reduced costs of consumption through commitments to eliminate trade barriers. I examine how states can exercise the right of self-determination to order domestic law and policy over biopatenting to better accord with objectives promoting welfare and competition under trade law. State agency is also central to the domestic national coordination between such cross-ministerial public policy objectives, such as industry and health and towards greater regulatory coherence and plays an imperative role in the realization of the social and economic rights of citizens located within national boundaries, in particular the right to health and food and to benefit from scientific progress. Protection of minority and cultural rights, such as indigenous claims to the protection of their traditional knowledge or of biodiversity through the creation of new plant genetic resources (PGRs), also depends on state action. While there may be moral obligations binding states, where governments have committed to the protection of human rights they are legally mandated to act on these despite any ambiguity as to how such rights are to be conceived. The state plays a central role, then, managing the intersection of international and national obligations as they pertain to biopatenting. Measures a state takes to improve industrial policy to achieve the progressive realization of human rights such as health, even if they are trade-related, should be immune from liability under international law.

The hegemony of trade and private property rights over domestic public policy and human rights has invited significant scholarly criticism and forceful arguments presenting reasons why human rights should always trump trade obligations. A commitment to a state's right to self-determination, however, would ask a broader, and yet more modest, question: How can the trade disincentives for making progress in human rights protection domestically, and thus internationally, be neutralized and associated costs and risks for the state minimized? There should be no *down-side risk*, at least, for states honouring their human rights obligations. The WTO will have an instrumental role in responding to human rights related claims and defences if it is to survive the challenges surrounding the extent of its governance over substantive areas that continue to threaten the conclusion of the current Doha Round of negotiations.

Theoretical Premise

The scholarly literature on trade and human rights has typically focused on how restrictive border measures on imports such as quotas or tariffs may be used as trade sanctions to coerce human rights compliance in *other* jurisdictions – to eliminate for example indentured child labour in an exporting country – rather than on how trade obligations may fetter the discretion for human rights compliance in a Member's home jurisdiction. Alternatively, the focus has been to assess how monetary aid and debt relief delivered in tandem with institutional reform of the multilateral trade regime may facilitate (sustainable) 'development' rather than the ways in which trade-related IP obligations may interfere with *homegrown* development that comes from imitation, transfer of knowledge and liberal access to technology. Little attention has been paid to the state's role in mediating the validity of plural perspectives infusing substantive normative claims for IPRs and the balance they are to achieve when culturally and politically located. This book departs from the existing scholarship and attempts to fill the gap by prescribing and defending a pragmatic two-step approach to state agency in response to these questions because, if law is intended to reflect community values, regulatory options are nevertheless increasingly shaped by global regulatory trends. A normative approach to state agency is more conceptual, procedural, and pragmatic than content driven; the content will necessarily vary with the socio-economic conditions of a state, the political will of its constituents, and desired public policy objectives democratically determined. The ensuing analyses make implicit a liberal rights-based approach.

The late legal philosopher John Rawls, in his *The Law of Peoples*, speaks of the attainment of a 'Realistic Utopia' and states that 'in the Society of Peoples, the parallel to reasonable pluralism is the diversity among reasonable peoples with their different cultures and traditions of thought …'.[12] Rawls's fundamental respect for regulatory and cultural diversity within certain limits – maintaining the priority of universal human rights – defers to state sovereignty.[13] A *realistic utopia* is an achievable goal with universal appeal in that it celebrates human difference as the common characteristic of our

12 John Rawls, *The Law of Peoples and 'The Idea of Public Reason Revisited'* (Cambridge, MA: Harvard University Press, 1999) at 12.

13 Rawls identifies a number of *familiar and traditional principles of justice among free and democratic peoples* as evidenced from the global welfare promoting institutions of the United Nations and the WTO which include the following: Peoples are 1. free and independent, and their freedom and independence are to be respected by other peoples; 2. to observe treaties and undertakings; 3. equal and are parties to the agreements that bind them; 4. to observe a duty of non-intervention; 5. to observe human rights; and have 6. a duty to assist other peoples living under unfavourable conditions that prevent their having a just or decent political and social regime. Ibid. at 37.

existence: '[a] (reasonable) Law of Peoples must be acceptable to reasonable peoples who are thus diverse; and it must be fair between them and effective in shaping the larger schemes of their co-operation.'[14] This text embraces a Rawlsian view of international law as not just the *the law of nations* but more importantly the *law of peoples.* A principled blend of historical, doctrinal, and interpretative textual analyses supports this commitment and seeks to shelter it.

On Audiences and Agents

Canada is a liberal democratic member state of the WTO and the United Nations (UN) and a ratified signatory of the twin Covenants protecting human rights. Canada has recently been embroiled in patent violation controversies and could have found itself a potential respondent to a trade-related IP complaint by the United States (US) over unauthorized government use of a patented cancer-related gene and method for its diagnostic testing in the course of public health delivery. The Canadian example is a good one for demonstrating how the issues of biopatenting do not belong only to poor economies. Governments of developed countries committed to ideals of the welfare state can no longer remain the passive audience of globalization forces but must assume a leadership role in authoring a global society not disconnected from its humanist aspirations.

Nations poorer than Canada and in the process of developing their patent systems have the most to gain from the proposals advanced for the preservation of regulatory diversity. Deference to internal policy priorities and the efficient mediation of competing if not conflicting international obligations at the intersection of biopatents is even more critical for governments with limited capital, institutional, and human resources and generally lacking in infrastructure. The framework prescribed allows them to address the impact of patent policy on their development agendas. But there is also something in this book for legal professionals in any jurisdiction.

The patenting of life was established by the agency of proactive judges apparently endorsing industry-friendly developments in patent offices. Biopatent laws in many common law jurisdictions arise from administrative fiat and judicial law-making rather than by legislative activity. This book serves to caution against the potential consequences of inappropriate judicial agency in expanding proprietary exclusivity in life and warns of the risks of pursuing legal harmonization based on developments in other jurisdictions once stripped of their relevant cultural and political contexts and attendant public-interest safeguards. The common law extension of patents to life is consistent with transjudicial trends in domestic courts where judges seek convergence of

14 Ibid. at 12.

their decisions with foreign judgments and sources of law.[15] Judicial agency has encouraged biopatenting and the globalization of higher protections in the biotechnology sector. Yet, whether life is patentable is a moral, political, and economic question of significant technical complexity and fundamental national importance; it merits legislative attention.

Non-governmental organizations (NGOs) and civil society may also be moved to strategic action by this text. In dispelling perceived legal obstacles to the realization of human rights, a renewed basis for increased advocacy by appropriate non-governmental agencies, working in tandem with governmental ones, is invited. Already, human rights advocates in the developed world have joined the social movements of developing countries, indigenous peoples and the global NGOs that represent them in challenging the neoliberal rhetoric that posits the primacy of private property rights now entrenched under the multilateral TRIPS and bilateral TRIPS Plus agreements. The patents-human rights relationship is often geopolitically manifested as a polarized and politicized debate between the *North* and *South* or the *developed* and *developing* countries respectively. This may be because liberal democracies of the North/developed countries provide more stable economic and legal environments for corporations and are, relative to the governments of South/developing countries, more likely to be subject to government capture by strong industry lobby as evinced by the history of TRIPS. Yet, '[i]t is difficult to suggest' writes William Quigley, 'that current economic arrangements, dependent as they are on large international corporations, are working for the common good'.[16] The human rights interests of citizens in rich, north, developed countries are often neglected even though these are equally at risk as those in the poor, south, or developing countries, albeit with important differences. Accordingly, this book may offer activists a new arsenal for promoting 'bottom-up' state accountability transcending the needs of the developing world by suggesting that the human rights of those in the developed world deserve no less protection by their governments from trade obligations that serve to assure private profit rather

---

15 See e.g. Anne-Marie Slaughter, 'A Typology of Transjudicial Communication' (1995) 29 *U. Rich L. R.* 8 at 99. Slaughter defines transjudicialism as 'the communication among courts – whether national or supranational – across borders'. See also Ann-Marie Slaughter 'Court to Court' (1998) 92 *Am. J. Int'l L* 708; 'Judicial Globalization' (2000) 30 *Va. J. Int'l L* 1103; and 'A Global Community of Courts' (2003) 44 *Harv. Int'l L. J.* 191. See also the Court's reasoning in the first major case to allow patents on life in Canada, *Re Application of Abitibi Co.* (1982), 62 C.P.R. (2d) 81 [*Abitibi*], discussed in Chapter 3.

16 William Quigley, 'Catholic Social Thought and the Amorality of Large Corporations: Time to Abolish Corporate Personhood' (2004) 5 *Loy. J. Pub. Interest* 46. For a discussion of the subversion of civil society by corporate interests through law and the profoundly undemocratic influence and existence of corporations, see e.g. Harry Glasbeek, *Wealth by Stealth: Corporate Crime, Corporate Law, and the Perversion of Democracy* (Toronto: Between the Lines, 2002).

than the welfare promoting functions for which these obligations were designed. Civil society in rich and poor countries may unite, mobilized by a common cause: to assert the primacy of human rights and the need to account for these as part of the costs and consequences of industrial and cultural policies.

Overview of Text

Six substantive chapters follow this introduction. Chapter 2, 'The Promise and Perfidy of Patents', examines the vast welfare and human rights potential of biotechnology and the disproportionately positive gains promised to the poorest and worst off in both developed and developing countries. This chapter reflects on the 'surge' in patenting in all fields and emphasizes that the proliferation of patents in the biological, biomedical, and life sciences (biopatenting) should be of concern since the process of discovery is often cumulative, the subject matter contentious, and the scope of impact broad. Significant moral, ethical, and human rights considerations are integral to discussions of owning and manipulating life and should be contrasted with the growing anti-competitive uses of patents now treated as assets.[17] Today, patents are used as security in attracting R&D, venture capital and skilled employees, or to secure promotion and tenure in public (academic) institutions. Companies are formed and merged based on IP assets, shifting the focus from incentives to innovate to incentives to accumulate patent instruments as legal commodity. The conceptual dissonance between the accretive nature of the scientific paradigm by which invention and discovery are made and the predatory uses of patents makes acute our discomfort in granting private exclusive rights to public collaborative research in the life sciences. Alternative regulatory approaches and more immediate reforms are considered with an overall emphasis on the thoughtful democratic deliberation required in optimizing the balance of trade-offs.

Chapter 3, 'Origins of the Patented Species: Of Mice and Men, and Sincerity of "Invention"', builds on the prior analyses of flawed patent regimes through a historical and comparative discussion of judicial attitudes towards biopatenting. The instrumental role of a predominantly patent-friendly judiciary in extending patents to life through the courts is considered with a case law review that demonstrates the incremental and surreptitious nature by which biopatenting has occurred and the difficulty in reconciling this outcome with the auto-activity of living matter. The jurisprudence in Canada leads to an incredibly odd legal conclusion from which other nations should take note. Two important Supreme Court of Canada (SCC) decisions demonstrate the unintended consequences of judicial activism and the valuable lessons to be learned by other jurisdictions. The Supreme Court decisions positively

17 See e.g. Curtis Cook, *Patents, Profits and Power* (London: Kogan Page, 2002) at 142.

signal a growing acceptance by the country's highest (though non-specialized) court to preside over very technical IP cases but also reinforce the criticisms directed towards the piecemeal evolution of patent law by an unelected and unaccountable judiciary with resulting *ad hoc* governance of patentable subject matter and related human rights. Agency in the courts continues to erode important regulatory and policy space reinforcing the need for legislation as misplaced political and legal pressures for judicial convergence continue to influence the common law expansion of patents to life.

Chapter 4, 'The Global Governance of Trade: Trading in Political Ethics?', establishes the contradiction of including a substantive monopoly-creating code in a system designed to reduce consumer costs and increase welfare gains through free competition and liberal access to markets. Whereas the last two chapters demonstrate how nationally biopatenting resulted from inappropriate administrative and judicial agency, this chapter contextualizes these results within broader globalization trends that facilitated the entrenchment of IPRs by technocrats in a multilateral trade regime starkly lacking in political ethics. With the proliferating reach of international agreements into sovereign jurisdiction, the lack of democratic global participatory processes, and the increasing means of enforceability of international trade obligations, individual human rights and public interests may be usurped by competing private economic proprietary claims of transnational corporations (TNCs), without an opportunity for meaningful political dialogue. The lack of a compelling global morality or political ethic makes strong national policies imperative. This chapter builds on the prior two and evidences the argument that national legal developments restrict important policy space more than they are credited for, challenge the ethos of the multilateral regime, and may in fact exceed international obligations of a state as required under the TRIPS universally mandated minimum standards for IP protection. But, as a mirror, they reflect more wholesale patterns of a rising Corporate Republic in which public participation in important policy decisions is increasingly diminished and political accountability for regulatory choices difficult to achieve. TRIPS, as a significant departure from trade policy, is explained by the strong impetus for its inclusion in the multilateral regime by a powerful private lobby, because

> [f]or the trade world, IPRs are one of the pillars on which the entire trade edifice is built. Without widespread recognition of property rights, without considerable degree of harmony in how these rights are recognized, and without an assurance that these rights will be respected and – where necessary – enforced, the confidence necessary for international commerce is difficult to muster.[18]

18 Mark Halle, Project Director, IUCN Project on the Convention on Biological Diversity and the International Trade Regime, foreword to Dutfield, *IPRS, Trade and Biodiversity*, supra note 6.

TRIPS should never have become part of the package of trade agreements in the Uruguay Round. But, a review of its obligations makes apparent that despite confusion to the contrary the text and its ambiguities support substantial discretion specific to biopatenting.

Chapter 5, 'TRIP'ing over Human Rights: A Legitimacy Crisis at the WTO' provides a parallel history of the international human rights protection regime and considers how these obligations of a state invariably translate the universal minimum trade-related IPRs protections of TRIPS as equally 'human-rights related'. This chapter serves three important functions. First, it provides examples of specific human rights engaged by biopatenting and thereby concretely demonstrates the need for state accountability for protecting these rights as part of coherent domestic industrial policy. Second, IPRs are now central to the multilateral trade regime and raise important issues of domestic public policy. For example, despite their institutional character, are IPRs also *human rights*? This question is explored in order to determine whether the discourse in debates over patents and human rights should be concerned with competing *human rights* such as a right to IP and a right to health or with competing obligations of a state which pit human rights (such as the right to health) against trade obligations (the protection and enforcement of patents). Third, this chapter anticipates the need to reconcile trade- and human-rights obligations in dispute settlement at the WTO. A discussion of the human rights engaged by biopatenting demonstrates potential sites of vertical conflict for governments (between domestic human-rights promoting laws and international trade rules) as well as the possibility of horizontal conflict internal to the human-rights regime (between competing human rights, to health and IP, for example). The chapter concludes by asserting that IPRs are *not* human 'property' rights of exclusion dominantly accepted. If IPRs exist as human rights, they are rights of *inclusion*, *participation* and *access* that reasonably complement, rather than undermine, the realization of other human rights. The right of exclusion is an institutional denomination requiring state justification and subject to state discretion of its limits. The experience with the first TRIPS amendment offers direction on potential remedial mechanisms towards the joint realization of human rights, welfare-promoting objectives of states, and WTO legitimacy.

Chapter 6, 'The Sovereign Right to Self-Determination: From Agency to Development and Dignity', considers the role and rights of the state and advances the argument that these are the central criteria framing both the human rights and the trade regimes. Since IPRs are cultural and commercial policy instruments designed to achieve specific desired outcomes within a geopolitical territory through institutional variability, the debate over the patenting of life fundamentally affects the basic exercise of sovereignty necessary for human development, dignity, and cultural rights. Self-determination ensures that the complementarity of the global governance regimes may be realized in a culturally contingent manner. It is also the point from which political accountability can be generated and vertical conflict minimized. The array of international instruments has increased the potential sites of horizontal conflict making the state's role central and imperative in defending diverse domestic regulatory measures as part of an overall commitment to a broader *system* for coherent global (self-) governance. Sovereign discretion to respond to international human rights obligations is, accordingly, both a *desirable* and *defendable* position against trade complaints.

Chapter 7, 'Reconciling Competing International Obligations: The Equitable Conduct Defence and the Stewardship of the WTO', examines how principles of treaty interpretation can be used to assist states in positive efforts to interpret competing obligations in a conflict minimizing manner. Requiring the grant of patents for all biotechnology creates potential conflict with human rights and other international instruments such as the Convention on Biological Diversity (CBD). Certain conflicts can be read out of international texts through proper interpretation of the rights and obligations conferred. A state prioritizing human-rights obligations under ratified Conventions should do so defending its conduct, I argue, by invoking its rights and obligations under human rights and trade instruments as first and foremost commitments to its citizens. Given the rules-based model of the WTO, it is only appropriate that attention be directed to available justifications and defences for state conduct such that even if the allegations made against a state are substantiated, a state may nevertheless be absolved of liability. The WTO Dispute Settlement Body should accept its stewardship over trade-related human rights and remain open to an *equitable conduct defence* (ECD), so long as the primary trade principle of non-discrimination is met. The ECD is designed to enable governments to protect their public's interest in relation to biopatenting, particularly since the public have no legal standing in the WTO complaint process. The proposed framework for reconciling human rights and trade obligations in disputes would serve the ends of social justice and allay criticism related to substantive and procedural deficiencies at the WTO.

Appropriate state response may be domestic and/or international; a dual strategy is essential, however, in a global society, as an anticipated international approach will complement national strategies and help protect against a successful challenge under WTO instruments. The first branch of my framework for state agency proposes states adopt TRIPS-compliant measures to embrace existing public policy space (and allow cross-policy coordination) for domestic regulation of biopatenting. Reform measures should be trade-compliant in order to avoid attracting patent litigation or a WTO complaint. Chapters 2, 3 and 4 articulate various recommendations to this end. The second branch engages in treaty interpretation to conceptualize a TRIPS defence (Chapter 7) for states in case domestic measures are challenged at the WTO and provides contextual, historical, and interpretive analyses (Chapters 5 and 6). The bifurcated framework safeguards the accountability of representative governments to their constituent 'publics' and should empower state agency through a new reading of international obligations that reduces the potential for external costs of public policy preferences and reflects the common commitment of states to dignity and development within a coherent system of *global self-governance*. Domestic policies that prioritize IPRs over human rights are due to regulatory failure and a state's progressive emasculation of its agency potential by its own agents; it cannot be exclusively attributed to the perceived TRIPS restrictions as states may adopt TRIPS-compliant domestic measures to reconcile competing obligations and should be able to defend these, so long as the measures are not discriminatory, at the WTO.

The paradox of TRIPS within the multilateral trade regime highlights a historical false dichotomy of national obligations: classifying them as either exclusively trade *or* human rights (HR) related based on the source from which they are derived. This has in turn created a distinct duality in our socio-legal and political identities, a likely vestige of our past, as either *merchants* (capitalists) in pursuit of trade related gains or *missionaries* (colonialists) engaged in cultural imperialism with 'charitable' forms of support aimed to convert societies, so-often by fostering dependency on foreign 'assistance' conditioned on assimilation and the obliteration of cultural difference. Some argue that institutional human rights are themselves a manifestation of cultural imperialism accentuating the tenuous relationship between culturally diverse value systems and universal claims asserting themselves as shared concerns and values. Others, such as anthropologist Kirsten Hastrup, observe that HR culture 'draws upon all kinds of ancestry' and offer:

> [w]hether we agree to the view of liberal justice which Rawls (1971) has identified simply as 'fairness', the international community agrees surprisingly well across differences in local culture, belief, language and metaphysics on the demands for justice and mutual benevolence, and universal standards of rights that are increasingly specific are put forward and subscribed to as part of the international legal instrumentalization …[19]

Universal declarations of HR, for Hastrup, are a pursuit for the 'common good' on a global scale. So too is trade liberalization, although the tools of choice differ. The analyses here strive to eliminate this contrived divisibility

19 See Kirsten Hastrup 'Accommodating Diversity in a Global Culture of Rights' in Hastrup, ed., *Legal Cultures and Human Rights: The Challenge of Diversity* (New York: Kluwer Law International: 2001) at 7, 9 and 25. At 1–2, she writes 'Culture and human rights are often seen as antagonistic entities, culture positing people according to particular semantic schemes and value systems, and human rights referring to shared concerns and a universal standard of evaluation … [T]here is no contradiction …anthropology itself presupposes both a shared humanity, without which the comparative study of societies and cultures would make no sense, and a sensitivity to cultural distinctions …[;] different ways of practicing universal capacities'.

of institutions (and the legal identities they foster) in order to correct any misgivings about the common purpose of trade and HR systems based on their significantly divergent approaches (the former, utilitarian in focus, the latter, individualized). States can no longer disregard the effects of this mutually exclusive taxonomy (and its insular expansion) on distributive justice. Biotechnology promises the means for addressing the paucity and disparity of natural endowments contributing to famine and disease but also underscores the important economic, cultural, religious, ethical, and human rights issues to be weighed by regulatory governments of nation states against industrial policy objectives for defining domestic private property rights, their scope and limits.

Chapter 2
The Promise and Perfidy of Patents

Promise: a reasonable ground for hope and expectation, especially of future excellence and satisfaction.

Perfidy: the act of violating faith, trust …, treachery.[1]

The Promise of Biotechnology

> Using new biological tools, researchers have developed a wide range of possibilities for using living organisms, or parts of living organisms, to produce new products or processes. Biotechnology has applications in many sectors, including healthcare, agriculture, environmental protection, and aquaculture. In healthcare, for example, research based on biotechnology has resulted in new diagnostic tools and treatments for cancer, atherosclerosis, osteoporosis, asthma and AIDS. In agriculture, the industry has created disease resistant plants that are helping developing economies respond to the food needs of their growing populations.[2]

Biotechnology has expanded what is possible through molecular engineering and emancipated us with promises for a better *geneticized* future.[3] Biotechnology can be defined as 'any technique that utilizes living organisms (or parts of organisms) to make or modify products, to improve plants and animals or to develop micro-organisms for specific purposes'.[4] It is the interaction of human

---

1 *Funk and Wagnalls Standard Dictionary*, *s.v.* 'promise'; 'perfidy'.

2 Ronald Hirshhorn and Jock Langford, 'Intellectual Property Rights in Biotechnology: The Economic Argument', prepared for The Canadian Biotechnology Advisory Committee [CBAC] Project Steering Committee on Intellectual Property and the Patenting of Higher Life Forms, March 2001 at 12.

3 Abby Lippman coined 'geneticized' to pejoratively reflect the reduction of all qualities to a function of genetic codes. See also Melanie Rock, 'Genetic norms, eugenic logic and UNESCO's International Bioethics Committee' (1997) 7 Eubios J. of Asian and Int'l Bioethics 108 and A .Robertson, 'Procreative Liberty in the Era of Genomics' (2003) 29:4 *Am. J. L. and Med.* 439.

4 US Office of Technology Assessment, 'Commercial Biotechnology, An International Analysis' quoted in Henk Hobbelink, *Biotechnology and the Future of World Agriculture* (London: Zed Books, 1991) at 25 cited in Victoria Tauli-Corpuz,

technology with biology for the purposes of exploiting biological systems, or parts thereof, for productive ends.[5] Traditional cross-breeding methods may be included in this definition which does not necessitate molecular genetic manipulation, although the growth of the biotech industry is certainly linked to that capacity and its corresponding private commercial enclosure.

Today, biotechnology promises to feed the world and cure the sick, extending the quality and quantity of our lives. Scientists have heralded genetic and biological manipulations that enable genetic diagnosis, gene therapies, plant derived edible vaccines and biopharmaceuticals that will tailor medical health treatment to an individual's biological needs and thereby improve metabolism and treatment efficacy. The allure of these promises is strengthened through the medical mantra of diagnosis and treatment to which bioethical centres contribute.[6] Biotech also promises nutrient enriched, more resilient and highly adaptable crops as 'improved' varieties. Biotech products being developed include caffeine-free coffee beans, vitamin C enriched oranges and cancer-fighting broccoli. As well,

> [a] potato that contains a vaccine that fights against a killer diarrheal disease has also been developed, with the idea it could prevent a disease that affects millions of people in developing countries [although] ... there are still questions of dosing, and ... the fact that the vaccine is lost if the spuds are cooked.[7]

Researchers at the Science University of Tokyo are using a new breed of genetically modified (GM) rice plants to develop the hepatitis B antibody to produce immunity to a virus that has over 300 million carriers worldwide. In the past, the inoculation for this epidemic had to be manufactured using blood from hepatitis B carriers. Jo Chiba, a university professor and the team's leader, believes this is a great discovery since it will reduce manufacturing

---

'Biotechnology and Indigenous Peoples' in Brian Tokar, ed., *Redesigning Life? The Worldwide Challenge to Genetic Engineering* (Montreal: McGill-Queen's University Press, 2001) at 253.

5 Engineer Karl Ereky was first to use this term in 1919. See Mark J. Fecenko, *Biotechnology Law: Corporate-Commercial Practice* (Markham, ON: Butterworths, 2002) at 4.

6 See e.g. A.S. Daar, H. Thorsteindóttir, D.K. Martin et al. 'Top 10 Biotechnologies for Improving Health in Developing Countries' (2002) 32 *Nature Genetics* 229 [Daar et al., 'Top 10 Biotechnologies'].

7 See Pippa Wysong, 'Breaking Ground', *Financial Post* (13 November 2000) E1at E-2. For a discussion of food insecurity, nutritional deficits, and biotechnology's response, see Ellen Messer, 'Food Systems and Dietary Perspective: Are Genetically Modified Organisms the Best Way to Ensure Nutritionally Adequate Food?' (Fall 2001) 9:1 *Ind. J. Global Legal Stud* (Symposium on Sustainable Development, Agriculture, and the Challenge of Genetically Modified Organisms) at 65–91.

costs, decrease the chance of other viruses entering the products, and increase access by developing countries that may lack the funds to purchase costly antibody products.[8] Since for many poor countries food means rice, Vitamin A enriched rice, known as golden rice, was an earlier genetically engineered crop designed to reduce massive annual child deaths and cases of blindness in the third world; Monsanto holds the patent.

The West African Rice Development Association (WARDA) with the support of partners including the United Nations Development Program (UNDP) has developed new rice varieties that combine the best qualities of African and Asian species (New Rice for Africa – NERICA) and are said to yield up to 50 per cent larger crops without fertilizer:

> In addition to the significant gains in production, the new varieties also mature 30-50 days earlier than the currently grown varieties…is also substantially richer in protein, more tolerant to disease, drought and acid soils, resists some of the most damaging insect pests in West Africa, and can out-compete weeds. Because the rice was designed for resource poor farmers, it can help farmers reduce poverty and save developing countries millions of dollars in rice imports.[9]

More efficient use of decreasing land through drought and pestilent-resistant agriculture is made possible and purports to allow for sustainable agriculture where traditional agricultural production methods are no longer tenable for sustaining population growth.

The idea of biotech-as-panacea extends beyond the scientists and others employed in its industry and serves to strengthen support for research and development (R&D) at the higher institutional and governmental level. United Nations Development official Peter Malton worked on the African rice project. He contends that 'NERICA is an excellent example of how science can be put to work for poverty reduction. It is already having a real impact on improving poor farmers' incomes and increasing their food security'. Using science as a tool to overcome food security challenges was reinforced at the June 2008 Food Summit convened by the United Nations Food and Agriculture organization (FAO).

One hundred and eighty countries participated in this Summit; 43 represented by their Head of State or Government and 100 by high-level Ministers, along with 60 non-governmental and civil society organizations. The Food Summit concluded with the adoption (by acclamation) of the Declaration of the High-Level Conference on World Food Security: The

8 *Daily Yomiuri*, 'Hepatitis Antibody made from GM Rice' (1 November 2000), Crop Biotechnology News, <http://www.agbios.com/static/news/NEWSID_1763.php>.

9 United Nations Development Program, Press Release, 4 April 2001, <http://www.undp.org/dpa/pressrelease/releases/2001/april/4apr01.html>.

Challenges of Climate Change and Bioenergy Declaration, and called on international assistance, finding '[t]here is … an urgent need to help developing countries and countries in transition expand agriculture and food production, and to increase investment in agriculture, agribusiness and rural development, from both public and private sources' and '[i]t is essential to address the fundamental question of how to increase the resilience of present food production systems to challenges posed by climate change. In this context, maintaining biodiversity is key to sustaining future production performance'.

Human health requires food and is inextricably linked with nutrition. Positive gains to human health will result from medical applications of this new technology but also, it is claimed, through the commitment to improving food quantity, quality, and adaptability despite increasingly limited land resources and growing environmental stresses such as drought, floods, and changing soil composition.[10] The world's population is an estimated 6 billion and, 'by 2050, the UN estimates, it will probably near 9 billion. Almost all that growth will occur in developing countries. At the same time, the world's available cultivable land per person … will decrease by half over the next 50 years …'.[11] Since the gains are at the population level, biotech promises to reduce reliance on foreign aid by promoting the necessary social conditions, healthy work force, knowledge diffusion, and technological capacity required for economic and sustainable growth and development.

Critics claim the problems biotech purports to solve are unrelated to production capacity. It is a failure not of the market but of politics, they contend, that sufficient food is produced to feed the world but is not appropriately distributed. Philanthropist Bill Gates writes,

> [t]aken as a whole, the world produces enough food to feed everyone – but much of it is simply in the wrong place. Especially in countries with undeveloped transport infrastructures, geography restricts food availability as dramatically as genetics promises to improve it [and] biotech has its own 'distribution' problems.[12]

10 See FAO, Press Release, 'Agriculture in the Near East likely to suffer from climate change', Rome/Cairo, 3 March 2008 <www.fao.org/newsroom/en/news/2008/1000800/index.html>.

11 The International Service for the Acquisition of Agri-Biotech Applications (ISAAA) cited in Bill Gates, 'Will Frankenfood Feed the World? Genetically modified food has met fierce opposition among well fed Europeans, but it's the poor and the hungry who need it most' (19 June 2000) v.155 Vision 21 No. 25, <http://www.time.com/time/magazine>.

12 See e.g. Mark Sagoff, 'Biotechnology and Agriculture: The Common Wisdom and Its Critics' (Fall 2001) 9:1, *Ind. J. Global Legal Stud.*, Symposium: Sustainable Development, Agriculture, and the Challenge of Genetically Modified Organisms, at 13–35.

Moreover, it is claimed that the biotechnological approach is short-sighted in that 'patented techno-fix seeds will not provide the adaptation strategies that small farmers need to cope with climate change' but will instead 'concentrate corporate power, drive up costs, inhibit independent research, and further undermine the rights of farmers to save and exchange seeds'.[13] Still, biotechnology's promise to solve resource disparities in the world is a compelling industry narrative and it is supported by the technical capacity to engineer transgenic organisms custom-designed to express adaptable traits of one or both species.

The long history of plant and animal husbandry confirms that genetic manipulation is not new in food and agriculture. Animal domestication reportedly had an important impact on decreasing biological diversity and raising the risk of species extinction.[14] There are now similar concerns over the impact of molecular manipulation on the erosion of biological and genetic diversity. Yet others claim that genetic manipulation may actually increase biodiversity by contributing new species or to their varieties. Whereas in the past such efforts and their consequences were invariably limited by Mother Nature's rules of propagation, today's routine isolation of genes and determination of sequence functions for all species is coupled with the ability to transplant genes from one organism into another organism of a completely *different* species with often unknown consequences. Transgenic organisms have at least one other species' DNA inserted into their own through genetic engineering. In addition, 'synthetic genes' coding for 'synthetic proteins' may also be thrown into the mix. Synthetic genes are increasingly used as surrogates for naturally occurring genes in crops. While 'the genes actively expressed in bacteria or humans are not very active in crop plants', Joe Cummins warns, the synthetics are not 'substantially equivalent' and may have unforeseen consumer effects, such as toxicity.[15] Preternatural and inter-species genetic manipulation is unique to modern biotechnology but the capacity for it developed some 50 years ago with the basic research of publicly funded scientists.

---

13 Action Group on Erosion, Technology, and Concentration Communiqué, Issue 99, 'Patenting the "Climate Genes"… and Capturing the Climate Agenda', <http://www.etcgroup.org/en/materials/publications.html?pub_id=687>.

14 'How Humans Invented Animals' June (2003) CBC Radio One Interview, archived, <http://www.cbc.ca/ideas/features/invented-animals/index.html>.

15 Joe Cummins, 'Synthetic Genes in Food Crops', Institute of Science in Society, 1 September 20/04, <http://www.i-sis.org.uk/sgigmc.php>. See also Roland Pease, 'Artificial Life Comes Step Closer' 18 December 2004, <http://news.bbc.co.uk/1/hi/sci/tech/4104483.stm>.

The Double Helix and the Scientific Process

> If you steal from one author, it's plagiarism; if you steal from many, it's research.[16]

Mapping technologies and the discovery of the structure of DNA was essential for advancements in molecular genetics and the intensification of the biotech industry. DNA is 'the fundamental hereditary material of all living organisms',[17] comprised of amino acids arranged in sequences that code for genes, the basic functional and physical units of heredity that in turn contain coded information necessary for understanding the functions of proteins (the study of proteomics). Genes affect everything from intelligence and physical appearance to the propensity to develop specific diseases, differentially metabolize drugs, or respond to environmental or social changes such as diet and exercise:

> Each new gene is a potential target for drug development – to fix it when broken, to shut it down, to attenuate or amplify its expression, or to change its product, usually a protein. Finding a gene gives investigators a molecular handle on problems that have proven intractable.[18]

In 1953, James Watson and Francis Crick discovered that the structure of DNA was a double-stranded helical molecule characterized by complementary base pairings of nucleotides fixed with hydrogen bonds and with the capacity to self-replicate; the two base pairs, Adenine to Thymine and Guanine to Cytosine, were 'identical in shape';[19] it earned them the Nobel Prize. 'Discovering the nature of the gene 'was the most important objective in biology'', writes Victor McElheny in his detailed history, but the actual discovery came, he claims, after a long and equally intertwined scientific process during which other necessary and cumulative advancements were made by leading scientists in the field.[20] Watson had attended a conference which proved to be life altering. He sat in on a lecture by Maurice Wilkins, a scientist who had left physics for biology, on the study of crystalline nucleoproteins in living cells. Wilkins presented an X-ray diffraction photograph of DNA fibres he had taken and believed to be of help in determining gene structure. On his way back to Copenhagen,

16 Wilson Mizner, as cited in Cook, *Patents, Profits and Power*, at 93.

17 Purves et al., *Life: the Science of Biology* v. 1, *The Cell and Heredity*, 5th edn (Sunderland, MA: Sinauer Associates Inc., 1998) Glossary.

18 Dr Robert Mullan Cook-Degan, 'Origins of the Human Genome Project', <http://www.piercelaw.edu/risk/vol5/spring/cookdeeg.htm>.

19 Victor McElheny, *Watson and DNA: Making a Scientific Revolution* (New York: Basic Books, 2004) at 56.

20 Ibid. at 34, endnotes omitted.

Watson stopped in Geneva and learned of 'Linus Pauling's great achievement of finding large stretches of the amino acid subunits of complex proteins arranged in what he called the alpha helix. Helices were in the air'.[21]

Watson and Crick were not motivated by letters patent although by today's standards their discovery would have been a patentable invention. They were driven rather by what Thomas S. Kuhn in his seminal work *The Structure of Scientific Revolutions* has observed motivates all great scientists to persevere against the odds:

> To scientists ... the results gained in normal research are significant because they add to the scope and precision with which the paradigm can be applied. That answer, however, cannot account for the enthusiasm and devotion that scientists display for the problems of normal research ... Bringing a normal research problem to a conclusion is achieving the anticipated in a new way, and it requires the solution of all sorts of complex instrumental, conceptual, and mathematical puzzles. The man who succeeds proves himself an expert puzzle-solver, and the challenge of the puzzle is an important part of what usually drives him on.[22]

Watson seems to have shared this view; he once said: 'I quickly lose interest in a scientist if I discover that he lacks the virtual monomaniacal interest in his work.'[23] It has been said of Watson that he 'thought of himself as the only one in Cambridge who "lived solely to understand how DNA functioned as the gene, and who had first-hand practical experience in using bacterial viruses to get close to the self-replication of the gene"'.[24] Such characterizations significantly differ from legal understandings positing the essential and necessary service patent monopolies provide as incentives for stimulating creativity and invention. Kuhn captures cogently the scientific ambition, noting that one

> may be attracted to science for all sorts of reasons. Among them are the desire to be useful, the excitement of exploring new territory, the hope of finding order, and the drive to test established knowledge ... What then challenges him is the conviction that, if only he is skillful enough, he will succeed in solving a puzzle that no one before has solved or solved so well ...

---

21 Ibid. at 29.

22 Thomas S. Kuhn, *The Structure of Scientific Revolutions*, International Encyclopedia of Unified Science, v. 2 No. 2 (London: The University of Chicago Press Ltd., 1962) at 36–7. Kuhn adds, '[t]hough intrinsic value is no criterion for a puzzle, the assured existence of a solution is' (ibid.).

23 James D. Watson, 26 May 1982, quoted in McElheny, *supra* note 19, at 179.

24 Ibid. at 33.

> On most occasions any particular field of specialization offers nothing else to do, a fact that makes it no less fascinating to the proper sort of addict.[25]

Global activist Vandana Shiva, Director of the Research Foundation for Science, Technology, and Ecology and one of the leaders of the International Forum on Globalization, contends that the 'patents as incentive' myth is based on a false construction of knowledge disconnected from any temporal, spatial, or social specificity, even though the nature of knowledge, as the Watson and Crick example demonstrates, is a collective and cumulative enterprise based on the dialectic between scientists and their community.[26] Watson and Crick's decision to emulate Pauling's cardboard models for their structure benefited from 'know how' – skills, capacity, and methodology for executing a task – as well as 'know-who' in working with each other and other experts who may have assisted, intentionally or not. Such tacit knowledge is as beneficial as the substantive knowledge written up in a science journal but is often limited by proximity.

The Organization for Economic Cooperation and Development (OECD) has found that a 'fundamental aspect of learning is the transformation of tacit into codified knowledge and the movement back to practice to develop new kinds of tacit knowledge'.[27] This is consistent with the trend in science towards regional technology and knowledge based clusters, such as the Biotech Belt built around Duke University in North Carolina, USA or Canada's Technology Triangle (a system of networked municipalities) whereby innovative gains can be realized by the synergy of tacit and codified knowledge that develops through collaboration and team work[28] and reinforces the *spirit* of science is 'communitarian' and open, requiring access to the knowledge and techniques it creates to further its objectives.[29]

'Science', according to Shiva, 'cannot be used to refer only to modern Western science. It should include the knowledge systems of diverse cultures in different periods of history'.[30] Kuhn too concluded from his observations on the scientific paradigm that 'normal science is ... a highly *cumulative*

---

25 Kuhn, *supra* note 22 at 38.

26 By popular accounts, the 'Paulingesque model-building approach of Crick and Watson was virtually the opposite of the very *slow*, methodical way that Maurice Wilkins and Rosalind Franklin were working at the Medical Research Council-supported biophysics lab ...'. McElheny, *supra* note 19, at 37.

27 OECD 1996 Knowledge Based Economy Report at 229–56. See generally, Michael Gibbons et al., *The New Production of Knowledge* (London: Sage, 1994).

28 Jeffrey Roy, 'Canada's Technology Triangle' in *Local and Regional Systems of Innovation* (The Hague: Kluwer Academic Publishers, 1998) at 239–55.

29 See the classic work of R. Merton, *The Sociology of Science: Theoretical and Empirical Investigations* (Chicago: University Press, 1973).

30 Vandana Shiva, *Protect or Plunder: Understanding Intellectual Property Rights* (Halifax: Fernwood Publishing Ltd, 2001) at 20.

enterprise, eminently successful in its aim, the steady extension of the scope and precision of scientific knowledge'.[31] Yet, for incentivizing new innovations, the instrument of choice amongst dominant regulatory systems is letters patent that confer exclusive rights based on private 'intellectual' contributions 'built on the fiction of totally individualistic scientific innovation'.[32] Richard Gold concludes that the patent system, which focuses more on commercialization than innovation, produces skewed results, rewarding the final codification according to specific institutional prescription of prior existing or newly emerging knowledge.[33] One cannot test the validity of the normative incentive claims that patents are premised on because 'it is difficult, if not impossible, to isolate the effect of patents on innovation'.[34] What can be evinced is that science operationalized is ostensibly at odds with the patent system. If the patent system was truly focused on spurring inventions through the recognition of the individual's right to the products of his genius or his labour as the case may be, then independent inventors would share the benefits and, as with the law of copyright, enjoy exclusive rights that are good against everyone save other independent authors/inventors of the same 'work' – especially since the paradigmatic structure of the scientific process means that simultaneous independent invention is not uncommon:

> To see how closely factual and theoretical novelty are intertwined in scientific discovery examine a particularly famous example, the discovery of oxygen. At least three different men have a legitimate claim to it, and several other chemists must, in the early 1770's, have had enriched air in a laboratory vessel without knowing it ....[35] This pattern of discovery raises a question that can be asked about every novel phenomenon that has ever entered the consciousness of scientists. Was it Priestley or Lavoisier, if either, who first discovered oxygen? In any case, when was oxygen discovered?[36]

31 Kuhn, *supra* note 22 at 52.

32 Shiva, *supra* note 30 at 22.

33 E.R. Gold, 'Biomedical Patents and Ethics: A Canadian Solution' (2000), 45 *McGill L. J.* 413. See also Ikechi Mgbeogi, *Global Biopiracy: Patents, Plants, and Indigenous Knowledge* (Ithaca, NY: UBC Press, 2006).

34 E.R. Gold et al., 'The Unexamined Assumptions of Intellectual Property: Adopting an Evaluative Approach to Patenting Biotechnological Innovation' (2004) 18 *Public Affairs Quarterly* 299. See also Matthew Herder and E.R. Gold, 'Intellectual Property Issues in Biotechnology: Health and Industry' Report for the OECD International Futures Project on 'The Bioeconomy to 2030: Designing a Policy Agenda', Third Meeting of the Steering Group, 7–8 February 2008.

35 Referring to the classic discussion of oxygen's discovery, A.N. Meldrum, *The Eighteenth-Century Revolution in Science – the First Phase* (Calcutta: Longmans, Green and Co., 1930), ch. V. in Kuhn, *supra* note 22 at 53.

36 Kuhn, *supra* note 22 at 55. Kuhn adds '... As a ruling about priority and date, an answer does not at all concern us. Nevertheless, an attempt to produce one

Science historians may find it difficult to identify exact dates of discovery or to ascribe these to specific scientists; they may agree with Kuhn that '[p]erhaps science does *not* develop by the accumulation of *individual* discoveries and inventions'.[37]

In an economy where knowledge is capital, privatizing the ownership of knowledge and 'discovery', whether of naturally existing oxygen or DNA, through patents translates into a number of distributive inequities creating frustration for many academics, scholars and public interest groups. Canada has a first-to-file requirement like most developed countries[38] whereas the United States stands virtually alone on the 'first to invent' system but both deny to different degrees the cumulative and often communal process of science by creating a 'winner-take-all affect' from the patent grant. David Vaver dispels the romantic mythology surrounding incentive theories for granting patents and demonstrates the anomaly created by patent rules:

> The decision of who gets the monopoly right where two or more persons invent something independently, without knowing of the other's work, is often more a matter of *luck* than anything else: the history of science and invention suggests that the phenomena of simultaneous discovery is the rule, not the exception. *The sower who first turns up at a patent office will reap; the other sower will rue.*[39]

Some proposals for reform aim to remedy resulting inequities. For example, it has been suggested that an interim patent system to overlay national first-to-file and first-to-invent systems with a first-to-post (on the internet) system into a globally distributed database searchable for prior patent files publicly laid open and subject to a global novelty determination would help

---

will illuminate the nature of discovery, because there is no answer of the kind that is sought ... The fact that it is asked – the priority for oxygen has repeatedly been contested since the 1780s – is a symptom of something askew in the image of science that gives discovery so fundamental a role ... Clearly we need a new vocabulary and concepts for analyzing events like the discovery of oxygen ... [T]he sentence, "Oxygen was discovered", misleads by suggesting that discovering something is a single simple act assimilable to our usual (and also questionable) concept of seeing. That is why we so readily assume that discovering, like seeing or touching, should be unequivocally attributable to an individual and to a moment in time. But the latter attribution is always impossible, and the former often is as well'.

37 Kuhn, *supra* note 22 at 2, emphasis added.

38 See s. 28.1 of the *Patent Act*, R.S.C. 1985, c. P-4, [CPA].

39 David Vaver, *Intellectual Property Law: Copyright, Patents, Trade-Marks* (Concord, ON: Irwin Law, 1997) [Vaver, *IPL*] at 7, footnotes omitted. This criticism can extend to prizes as well; Rosalind Franklin's contributions to the DNA structure discovery was acknowledged only after the Nobel Prize was awarded.

to overcome 'winner-take-all effects'.[40] Though helpful, these suggestions do little to address the broader conceptual disharmony between the nature of the scientific paradigm and the normative assumptions institutionalized within patent law.

The failings of our current regime warn against taking a primarily Western conception of science and IPRs, with internal flaws best resolved nationally with contextualized safeguards, and transplanting them to international fora as standards to which *all* countries must adhere. Rosemary Coombe cautions:

> [t]he maps imagined and inscribed by the global harmonization of intellectual property rights – level playing fields upon which equal partners reciprocally exchange rights recognized in homogeneous jurisdictions – have obvious distributional consequence (which … adversely affect the creative works of indigenous peoples, women, and the vast majority of peoples in the Third World generally) … Not only has the author/work relation been expanded in scope, but just as significantly, authorial rights have been exported *without* the defenses, the exemptions, or the constitutive role of the public interest that provided their historical justification.[41]

We should be wary of racing to provide more, stronger, and increasingly restrictive protection paradigmatically incongruent and cognitively dissonant with the scientific process. Market stimulated and guided invention frequently depends on the strength of educational systems and the science base that feeds into applied R&D. It is the generous availability of publicly funded basic research that provides the necessary inputs for the development of product applications; it is fair to argue that the results of such research should remain in the public domain. Richard Nelson offers some suggestions to help prevent the further erosion of the scientific commons and to create a system of incentives and control for generating productive science: keep scientific findings in the public domain, tie the reward of scientists to the acclaim of fellows, and provide public funding of research based on peer review of the scientific promise of both the proposal and the scientist.[42] And, Frederick Abbott outlines how 'societies, for centuries, evolved on the basis of informal

40 See Paul Edward Geller 'An International Patent Utopia' (2003) 85 *J. Pat. and Trademark Off. Soc'y* 582

41 Rosemary J. Coombe, 'Authorial Cartographies: Mapping Proprietary Borders in a Less-than-Brave New World' (1996) 48.5 *Stan. L. Rev.* 1357 [Coombe, 'Authorial Cartographies'] at 1365.

42 Richard R. Nelson, 'The Market Economy, and the Scientific Commons' LEM Working Paper Series 2003/24, November 2003, <http://www.lem.sssup.it/WPLem/files/2003-24.pdf>.

transfers of knowledge and technological advances in know-how, from masters to students, from fathers to sons, from mothers to daughters'.[43]

Coordination of regulatory policies to maximize welfare gains would have to entail a review of public education institutions and existing national and international 'reward' systems for scientists because knowledge production, dissemination, and transfer has historically been the essence of universities. But it will also entail a review in tandem of our patent laws that ignore the very nature of the scientific process, shaping it while purporting to promote it. The DNA structure was discovered in the course of doctoral research during a time when property and profit were not yet fully subsumed in the hegemonic imperatives that define the current *zeitgeist.* Scientists did not yet know very much about how to decode the messages in the double helix or their meaning but the discovery of the structure was critical for future understanding of its function in conveying genetic information, mutation, and self-replication and effectively spurred the genetic revolution. Amongst the scientific community, it was widely anticipated that biotech would enable the improvement of human, social, and health conditions. Global public multisectoral collaborative projects were organized to see this happen.

The Human Genome Project, HGDP, HAPMAP and GAIN

The next stage of genomic R&D was characterized by private/public competition and commercialization issues relating to the identification of genes coded by DNA sequences. Genes, the specific sequences of base pairs that encode the instructions for making proteins, are contained in chromosomes but only a fraction are protein-coding.[44] The remainder of the genome consists of non-coding regions which may function to provide chromosomal structural integrity, regulate the activation and de-activation of other genes, control where and what proteins are made, or serve some other unknown function.[45] The Human Genome Project (HGP) initiated in 1988 created a conglomerate of coordinated efforts by international scientists to map the entire human genome that formally commenced two years later. Eighteen countries with national human genome programmes participated in the HGP with the common goal to complete the sequence for the three billion DNA subunits (bases) and share

43 Frederick Abbott, Thomas Cottier and Francis Gurry, eds, *The Making of the International Intellectual Property System* (The Hague: Kluwer Law International, 1999) at 130.

44 'Genes are responsible for directing … [the] subsequent activity of the human body's 100 trillion cells … [but] account for only about ten percent of each cell's DNA'. Fecenko, *supra* note 5 at 10.

45 http://www.ornl.gov/sci/techresources/Human_Genome/project/info.shtml>. See also 'Cracking the Code' *TIME Magazine* 156:1 (3 July 2000).

the results for further study for the benefit of all. With Watson directing the US project, the government approved a 15-year $3 billion dollar plan for the effort with funding from the US National Institutes of Health (NIH) and the Department of Energy (DOE). The link with the Department of Energy seemed tenuous then but is clear now with the discussion of engineered 'food' such as GM corn for biofuels. The American government has initiated new policies for encouraging biotech genetic research to 'produce enzymes that will convert agricultural biomass or waste into fuels, plastics, and other bio-based products'. The US Department of Agriculture and the Department of Energy have allocated $250 million for such projects and one possibility is to use 'genetically engineered plants that contain higher than normal sugar and starch contents or make plastic precursors in their stalks'.[46] Others associate the birth of the HGP with creating a funding opportunity.[47] After only a few short years, Watson resigned from the HGP over disagreement with Bernadine Healy, the Director of NIH on gene patenting. Kevin Davies provides a detailed chronicle of events and explains:

> Healy strongly supported a controversial NIH decision to seek patents for hundreds of gene fragments identified by NIH scientist Craig Venter, if for no other reason than to obtain clarification from the Patent Office on the legitimacy of patenting genes of no known function … To add insult to injury, Healy asked Venter to consult on the future of human genome research at the NIH while instructing Watson not to go public with any further criticisms.[48]

In 1992, Craig Venter established his own non-profit research institute, The Institute for Genomics Research (TIGR)[49] as the NIH withdrew his funding.[50] By 1998, near the halfway point of the HGP with only 3 per cent of the targeted sequencing in the HGP complete, Venter abandoned the public project and formed Celera Genomics to pursue a rival private for-profit effort to map the human genome with his 'shotgun' technique and a smaller budget of $300 million for three years. Celera Genomics and other private enterprises followed the NIH lead. As early as 1991, NIH had undertaken mass patent applications for approximately 2000 genes, some of which were for anonymous gene fragments or raw genomic DNA sequences, despite the fact that the

46 Fecenko, *supra* note 5 at 34. See also the official US government website <http://www.genomics.energy.gov/>.

47 See e.g. Daniel S. Greenberg, *Science, Money, and Politics: Political Triumph and Ethical Erosion* (Chicago: University of Chicago Press, 2001).

48 Davies, *Cracking the Human Genome*, at 31.

49 The Edge Foundation, <http://www.edge.org/3rd_culture/bios/venter.html>.

50 'The Top 25 Managers of the Year' (2001) *Business Week*, <http://www.businessweek.com>.

functions of these were often unknown. The NIH established a dangerous precedent of DNA patentability from which Venter seized a ripe and *big* business opportunity. Watson was succeeded at the NIH by medical geneticist Francis Collins, a devout believer in the public-good quality of genomics, who continues to champion open access to genetic data. The NIH ultimately changed its position on DNA patenting in the interests of promoting science, medical progress and new scientific collaborations but it was too late from a legal perspective; the dynamics were set:

> Venter's intent was for his company, Celera Genomics, to sequence the human genome years before expectation (leaving thousands of gaps if need be), to be able to patent hundreds of genes and sell precious information about the genome sequence for a gene king's ransom to the pharmaceutical industry. Collins's task was to kick-start an unwieldy federal program to keep pace with Venter's private effort and deliver the complete, gold-standard sequence years earlier than projected, all the while releasing its DNA data every night to make the human genome unpatentable.[51]

Kevin Davies reports that by the summer of 2000, for political reasons related to maintaining public perception on the integrity of science and in order to 'restore a measure of dignity to the quest for the human genome', the race 'was declared over' and Collins stood together with Venter in a ceremony at the White House as President Clinton announced '[t]oday we are learning the language in which *God created life*'.[52] But despite the religious tribute, by popular account more than 4,000 of the approximately 24,000 human genes have been claimed in US patents – that is 20 per cent of the human genome patented, 63 per cent of which are assigned to private firms as compared to 28 per cent to universities[53] even though 'the functions are unknown for over 50% of *discovered* genes'.[54] Apparently, God was not alone and kept company with so-called 'genetic engineers'– an ill-fitting metaphor according to Drahos and Braithwaite:

> One could liken the synthesis of new compounds to invention in mechanical engineering. The use of the metaphor becomes more problematic in the case of organic chemistry where the chemist finds molecules that exist in nature and have useful properties. In the case of patent claims over DNA

51 Davies, *supra* note 48 at 6.

52 Ibid., emphasis added.

53 Stefan Lovgren, 'One-Fifth of Human Genes Have Been Patented, Study Reveals' *National Geographic News*, 15 October 2005, <http://news.nationalgeographic.com>.

54 Emphasis added. Human Genome Project Information, <http://www.ornl.gov/sci/techresources/Human_Genome/project/info.shtml>.

> instructions and their corresponding proteins … [i]t is hard to claim an entitlement to the DNA code on the basis that it had been engineered. It, after all, had been in existence for thousands of years before the genetic engineer and corporate laboratories. It had been uncovered or found rather than designed and built.[55]

If God created life, it was Venter who commercialized it. In so doing, he took for granted the public community based efforts that preceded and enabled his success. Why not? The patent system allowed it; it invited it. In a 1999 interview with CNN, Venter claimed his was an 'information company' and added,

> [w]e consider that we are making discoveries, we're not making inventions, with the primary sequencing. In the US, patent law makes no distinction between inventions and discoveries. European law does make that decision – it has to be an invention, not a discovery. So it is an important difference in different parts of the world and that's why there are different cultures with it.[56]

By 2001 Venter was named one of the top 25 managers of the year in the United States. Celera Genomics generated a reported $43 million not through drug related revenue, but by selling subscriptions to its online database of sequenced DNA; clients included drug companies, universities, medical centres and the Australian government.[57] His 'success' story reminds us of the discordant policy of a legal proprietary regime that rewards the 'first to sow' without considering who cultivated the land. Venter could not have done it without the prior communal efforts or the complicit operation of an administrative system that generously nurtured his efforts while denying those of others. The administrative expansion of private rights of patent holders rewarded labour expended, not innovation, to discover naturally existing genetic sequences even where their utility was unknown or, if known, was linked only to a potential propensity. The public HGP was completed in 2003, two years ahead of schedule, and the 'full, finished sequence of the genome was deposited in public databases and made available to all'[58] without impact on the existing gene patents.

55 Drahos and Braithwaite, *Information Feudalism*, at 157.

56 Stephen Frazier, 'Blueprint of the Body', 3 June 1999, CNN, <http://www.cnn.com>.

57 Ibid.

58 Francis Collins, 'Genomic Medicine: Opportunities and Responsibilities', Genomics and Public Health: 4th International DNA Conference, Montreal, Canada, 4–7 June 2006, abstract of speech.

The Human Genome Diversity Project (HGDP), a subset of the HGP, was approved by the Council of the Human Genome Organization (HUGO) in 1994. Anthropologists, geneticists, doctors and linguists undertook the HGDP in order to expand the focus of the HGP to determine the genomic variation amongst a greater diversity of the world's peoples by broadening the scope of populations studied. The HGDP was to provide a database for the further study of human biological and migratory histories but failed for funding reasons tied to the significant controversy it garnered. It was poor politics to sanction mapping efforts targeted primarily at 'isolated *indigenous* communities' especially without proper protocols for informed consent, proposals for benefit sharing, and means to manage patent controversies. One controversy related to patent for a cell line developed from blood taken from a member of the Hagahai, an isolated indigenous tribe in Papua New Guinea. The tribe enjoyed immunity from leukaemia, even though they carried a gene that predisposed them to it, because of the presence of a T-lymphotrophic virus which their peoples also commonly carried. The patent was later abandoned.[59] In the end, the HGDP drew attention to the need for regulatory instruments to sort out the social, ethical and legal issues relating to the scope and implication of this new knowledge; particularly given its capacity to transcend time by predicting probable futures while also carrying 'the secrets of our past' in documenting migratory patterns of populations and evolutionary journeys of species. Genetic information is special in providing what has been called a 'unique molecular fingerprint of the living and the dead'.[60] As a social experiment, the HGDP proved useful in creating significant controversies that highlighted the need for common principles of understanding for use and access to genetic information and led to the international emergence of model ethical protocols to ensure that the benefits of such information will be harnessed for *all* world peoples. However, as Amani and Coombe contend,

> until these are enshrined in legislation, they have little legal weight, except perhaps as 'customary norms' that might be appealed to in domestic courts as relevant international law. Principles of informed consent, maximizing benefit, minimizing harm, confidentiality, and the avoidance of conflict

59 See e.g. Bita Amani and Rosemary Coombe, 'The Human Genome Diversity Project: The Politics of Patents at the Intersection of Race, Religion, and Research Ethics' (2005) 27.1 *Law and Policy* 152 at 179–80; and WIPO Magazine, 'Bioethics and Patent Law: The Cases of Moore and the Hagahai People' Issue 5, September 2006.

60 Davies, *supra* note 48 at 9. See also 'Of Jaws and Man: Initial decoding of elephant shark genome helps uncover ancient DNA in human genome', Press Release, 21 December 2006, J. Craig Venter Institute, <http://www.venterinstitute.org/press/news/news_2006_12_21.php>.

of interest became accepted as guiding principles amongst the regional committees of the Project.[61]

The HAPMAP project (haplotype map) launched in 2002 grew out of the HGP and effectively, though quietly, replaced it. The HAPMAP began with $100 million in funding and involved a new consortium of nine research groups from private and public labs in five countries using blood samples from people in Nigeria, Japan, China, United States and Europe with the aim to understand how DNA in the genome is organized into the 10 million sequence variations common to all persons called SNPs:

> single nucleotide polymorphisms or SNPs (pronounced 'snips') are DNA sequence variations that occur when a single nucleotide (ATC or G) in the genome sequence is altered … to be considered a SNP, it must occur in at least 1% of the population. SNPs, which make up about 90% of all human genetic variation, occur every 100 to 300 bases along the 3-billion-base human genome[,] … can occur in both coding and noncoding regions of the genome[,] … [and] [m]any SNPS have no effect on cell function, but scientists believe others could predispose people to disease or influence their response to a drug … This makes SNPs of great value for biomedical research and for developing pharmaceutical products or medical diagnostics. SNPs are also evolutionarily stable – not changing much from generation to generation – making them easier to follow in population studies.[62]

SNPs cluster in neighbourhoods described as haplotype blocks; these genes are closely grouped together and are usually inherited as a group such that variation within the group might explain diseases inherited from that specific genetic source. Collins characterizes this 'flagship' international effort as the latest tool in genetics to:

> begin translating the wealth of data generated through the [HGP] into knowledge and tools applicable to understanding human health. The fundamental goal … was to provide scientists around the globe with the tools necessary to fast-track analysis of the 0.1% of human DNA that varies between individuals by making the data produced through the Project *freely available.* It is this small percentage of the total DNA complement that conveys the genetic basis for observed differences in disease susceptibility, as well as differential responses to therapeutic drugs, toxic substances, and environmental factors. With the application of HAPMAP to carefully phenotyped case-control studies, we are poised to identify the genetic

61 Amani and Coombe, *supra* note 59 at 166.

62 Human Genome Project Information, SNP Fact Sheet, <http://www.ornl.gov/sci/techresources/Human_Genome/faq/snps.shtml#snps>.

> variants that contribute to common disease such as diabetes, mental illness, and heart disease.[63]

Single letter insertions or deletions in the 3.1 billion base pairs can eliminate an entire gene or act as a disease-causing mutation. The HGP and HAPMAP projects have facilitated the discovery of disease-related genes, which has in turn led to the development of predictive tests to identify genetic predisposition, diagnostic tests for cancer detection in its earliest stage, and therapies targeting cellular abnormalities. Some 3,000–4,000 hereditary diseases are linked to errors in our genetic code including cystic fibrosis and muscular dystrophy, and specific genes have been correlated with the propensity of developing a number of common or chronic illnesses like diabetes (G6PD) and a variety of cancers including breast and ovarian cancer (BRCA1 and BRCA2), stomach cancer (CDH1), and colorectal cancer (HMPCC). Given that most diseases and all cancers arise from genetic alteration, many hereditary,[64] the potential for improving health outcomes is enormous but may be limited by the fact that such discoveries and methods for their testing are almost always patented.

In 2006, the Genetic Association Information Network (GAIN) was launched and is an international research collaboration based on a private-public partnership between the Foundation for the National Institutes for Health (FNIH), the NIH, and Pfizer Global Research and Development with a goal 'to unravel the genetic causes of common diseases over the next three years … The information derived from GAIN will be *publicly available* to researchers world-wide'.[65] GAIN signals overdue recognition that genetic data is a public good that should be *pre-competitive* and not subject to IPRs claims, even when research efforts involve private (corporate) actors. In addition,

> [b]ecause benefiting regions may not coincide with political boundaries[,] … [t]he benefits of public goods provided by one state may spill over into others. As a result provision becomes an international and, with unlimited benefit space, global issue … The more global is the region over which benefits extend, the greater the need for global policy instruments.[66]

63 Emphasis added. *Supra* note 58.

64 See 'Research at Stanford Links Stomach Cancer Deaths to a Single Gene', 11 December 2002, <http://mednews.stanford.edu/releases/2002/december/stomachcancer.html>; 'Genetic Cause of Stomach Cancer', 25 April 2002, at <http://news.bbc.co.uk/1/hi/health/1949530.stm>.

65 National Human Genome Institute, 'Novel Public-Private Partnership Created to Unravel the Genetics Of Common Disease Through Whole Genome Association Studies', February 2006, <http://www.genome.gov/17516722>.

66 Inge Kaul et al., eds, *Providing Global Public Goods: Managing Globalization* (Toronto: Oxford University Press, 2003) at xi.

Global institutions like the United Nation's World Health Organization (WHO) have emphasized the importance of free access to genetic information for allowing the full potential of genomics to be realized for all. In a 2002 WHO Report, it was observed that

> [t]here are profound concerns that current practices in intellectual property, particularly regarding the granting of patents on fundamental genetic information will place many Member States at a considerable disadvantage in realizing equitably the health care potential of genomics. *WHO should adopt a proactive role as an advocate for health equity in international debates on intellectual property issues.*[67]

WHO has reaffirmed this position in its recent draft Global Strategy on Public Health, Innovation and Intellectual Property which the 61st World Health Assembly adopted in May 2008 in tandem with the Plan of Action on Public Health, Innovation and Intellectual Property after two years of negotiation in the Intergovernmental Working Group. The growing relationship between IP and public health requires more active mediation; this time what was both strengthened and made explicit was WHO's role in shaping national IP policy at its intersection with public health.[68] In the meantime, public scientists, the HGP, HAPMAP and Genome Canada continue to publish new and emerging genetic data on public websites and databases as soon as possible to ensure that 'data are being made swiftly and freely available to public databases',[69] in this manner attempting to foreclose gene patenting on the basis that 'public domain' information lacks novelty. Researchers are encouraged to access this data through the HapMap Data Coordination Center, the NIH-funded National Center for Biotechnology Information's dbSNP and the JSNP Database in Japan.[70] These efforts seek to subvert the *de facto* policy of patent offices in expanding IPRs through the privatization of genetic information. Conversely, other private-public research and commercialization efforts portend growing IP and 'IP-related coordination issues' that relate to access, control, ownership, licensing and transfer agreements, as well as multi-jurisdictional legal compliance requirements.

---

67 WHO 'Genomics and World Health Report 2002', <http://whqlibdoc.who.int/hq/2002/a74580.pdf>.

68 Sangeeta Shashikant, 'WHO: WHA Strengthens WHOs Mandate on IP and Health', Third World Network, 4 June 2008, <http://www.twnside.org>.

69 See Genome Canada, <http://www.genomecanada.com>; Canadian Program on Genomics and Global Health, University of Toronto, <http://www.utoronto.ca/jcb/genomics/index.html>; http://www.utoronto.ca/jcb/genomics/documents/Convergent_Biotechnology.pdf

70 <http://www.hapmap.org>; <http://www.ncbi.nlm.nih.gov/SNP/index.html> and <http://snp.ims.u-tokyo.ac.jp/>.

The proposed Canada-California 'Cancer Stem Cell Consortium' (CSCC) for example, is a cross-border research project initiated in 2007 and jointly funded by the Canadian government funding agencies and California Institute for Regenerative Medicine (CIRM) in the amount of US$ 250 million each and is endorsed by governments in both jurisdictions. CSCC is emblematic of why we should be concerned about mixing private/public research according to a new OECD Report:

> Consider, in particular the provisions intended to provide a return of financial benefits to the State of California. Any research grouping under the CSCC that accepts funding from CIRM would be required to return 25% of revenues earned from licensing agreements in respect of any patented inventions developed with CIRM funds ... [T]he optics of Canadian tax dollars (in the form of government grants) flowing directly into the coffers of the State of California is likely to attract criticism. Should Canadian funding agencies therefore institutionalize a similar payback obligation into their respective funding agreements? How would doing so impact the commercialization process?[71]

Practical business realities contribute to the perfidies of patent practice and are inconsistent with the historical account of the role and function of patents in society.

Patent Privileges: An Institutionalist's Perspective

Despite the misnomer of 'intellectual property rights' and the compelling force of various rights-based theories justifying the grant of patents on the basis of moral claims to individual genius or alternatively the results of one's labour, it is quite well established that there is no *natural* legal right to intellectual property. Edith Penrose long ago rejected the natural rights theory of patents, opting to characterize them instead as a matter of national 'social policy' that takes into account the costs and benefits to a given society.[72] This is fundamentally a utilitarian approach to rationalizing government grants of limited-term exclusive rights and focuses less on the interests staked by the inventor and more on the broader interests of society. It is consequentialist in asserting that laws should be measured by the effects they produce and utilitarian (Benthamite) in the mandate to produce the greatest good for the greatest number; from this perspective patents are instrumentally essential as a policy tool for maximizing social welfare and addressing distributional and

71 Herder and Gold, *supra* note 34 at 31.

72 See Edith Penrose, *The Economics of the International Patent System* (Baltimore, MD: Johns Hopkins University Press, 1951).

equity issues. When not bogged down by the moral imperative of rights claims, this is the more compelling and alternative rationale justifying IPRs cited by the courts. The utilitarian perspective rationalizes the granting of limited-term monopolies for 'inventions' on the basis that these ultimately serve the public interest; rights, if any are deduced from welfare-maximizing principles. In Canada, the Supreme Court in *Cadbury-Schweppes v. FBI Foods* looked not to traditional *rights* rhetoric but to the *bargain* theory with the Crown:

> A patent is a statutory *monopoly* which is given in exchange for a full and complete disclosure by the patentee of his or her invention. Accordingly, at least one of the policy objectives underlying the statutory remedies available to a patent owner is to make disclosure more attractive, and thus hasten the availability of useful knowledge in the public sphere *in the public interest.*[73]

Scientific 'progress' would not stand still in the absence of a patent system but a limited 'trade-off' is considered to be worthwhile for society in expediting research and because the inventor may otherwise be inclined to keep her discovery a trade secret. The early disclosure patents invite help to promote further innovative activity during the patent period and thereafter, upon expiry of the patented invention; both advance the public's interest because they improve social welfare in the long term. The majority of the Federal Court of Canada in the *Harvard OncoMouse* decision made a similar finding, concluding that 'the object of the *Patent Act* is to promote the development of inventions in a manner that benefits *both* the *inventor* and the *public*'.[74] Such utility encourages adaptive reform and conforms to 'our sense of social responsibility; that is, the idea that the well-being of humans matters, and moral rules must be subjected to tests for their consequences on human well-being'.[75]

---

73 [1999] 1 S.C.R. 142 at 171–2. Rothstein J.A. notes that while this was not a patent case, the Court discussed patent protection in order to distinguish it from the law of breach of confidence (at para. 26).

74 *President and Fellows of Harvard College v. Canada (Commissioner of Patents)*, [2000] 4 F.C. 528, 189 D.L.R. (4th) 385 [*Harvard,* FCA], <http://decisions.fct-cf.gc.ca/fct/2000/a-334-98.html> at para. 26. See also Lisa Austin and Bita Amani, 'Patents on Genes: Identifying Issues and Responses', discussion paper prepared for the Provincial Advisory Committee on New Genetic Technologies. Toronto, Ontario October 2001 [Austin and Amani] reproduced in Trudo Lemmens et al., *Reading the Future?: Legal and Ethical Challenges of New Predictive Genetic Testing* (Montreal: Les Éditions Thémis, 2007), pp. 105–39.

75 See Canadian philosopher Will Kymlicka, *Contemporary Political Philosophy: An Introduction* (Oxford: Clarendon Press, 1990) at 11, cited in Robert L. Ostergard Jr, 'Intellectual Property: A Universal Human Right?' (1999) 21.1 *Human Rights Quarterly* 156–78, at 163.

Sir Neil MacCormick's institutionalist theory for understanding the idea of intellectual property is consistent with the history of patents as limited *privileges*. McCormick contends that the positive law for IP protection requires three kinds of provisions to explain the extent and nature of the rights of IPRs holders. IP protection exhibits a triadic rule-structure that MacCormick asserts is characteristic of legal institutions according to the Institutionalist Theory of Law. To summarize: first, the law must prescribe the circumstances in which and the means by which the named IPRs comes to exist and vests in a person as property; second, the law must define the scope of the right by providing what privileges and other rights belong to the holder of the IPR. This would include licensing powers and the conditions precedent for authorized uses and the nature of the trespassory rules that define infringement of the right and restrict the conduct of all others; finally, the law must prescribe how and when the right is extinguished, transferred, or expired. The nature and scope of IPRs is therefore entirely dependent on the particular legal regime which constructs it, the interpretation of the conditions from which it will arise, the extensiveness of the trespassory rules recognizing infringement, and the number and scope of exceptions that qualify the right. Not only is the content of intellectual properties culturally articulated as an artifact of legality but so are the very laws by which IPRs are juridically defined, limited, constituted, and mediated. For example, *what* counts as IP addresses the first point and *why* IPRs (the normative claim to IPRs) addresses the latter. Western IPRs are based on a theory of exclusion not only in the property sense as a means of ensuring market exchange value in the underlying invention or expression of an idea, but also in the politics of excluding specific forms of works culturally incongruent with Western normative legal prescriptions, such as traditional knowledge (TK) of plant genetic resources (PGRs). This has led some commentators to conclude that Western IP systems need to accommodate the protection of TK as an *exception* to the general recommendation against the enclosure of the 'public domain' or commons through unqualified expansion of IPRs.[76] The claim to legitimacy alienates the contributions of indigenous communities and renders their stories, traditional (medicinal and ecological) knowledge and even biological contributions of (economic) value free for the taking as 'public' inputs for the 'private' outputs of those culturally conditioned for compliance.

When genetic information is located within this *public domain*, 'through an act of isolation and purification, [DNA could] transform a naturally existing product into an invention ... even though the purified DNA coded for the same

76 Graham Dutfield, 'The Public and Private Domains: Intellectual Property Rights in Traditional Knowledge' (2000) 21 *Science Communication* 274–95. See generally, James Boyle, 'Second Enclosure Movement' (2003) 66 *Law and Contemp. Probs.* 33 [Boyle, 'Second Enclosure'].

protein as the naturally occurring sequence'[77] While consent may be required for the purpose of extracting one's DNA, that same DNA is considered nothing more than biochemicals in which property rights may vest to those who invest labour. Only certain kinds of labour are legally recognized, however; to purify and isolate, for example, rather than to procreate or nurture. Such a legal system ignores as material a form of labour, creativity, and contribution nonexistent in the mappings of (or yet to be 'discovered' by) Western IP law. The dynamic has been played out repeatedly on various platforms but most notably with the privatization of the human genome and demonstrably with the now defunct Human Genome Diversity Project (HGDP) that was colloquially dubbed the 'vampire project' for the capitalist drive by Western scientists to patent genetic information derived from the samples taken from indigenous communities with little regard to the social conditions that indeed rendered them 'vanishing isolates' worth archiving for their comparative research value.[78] The genetic samples extracted were not conceived of as property although any valuable cell-line so derived would be property on the basis of the labour vested in it, once again rendering the labour of the individual in maintaining the physical self *invalid*; effectively, '*immaterial*' if property, and obviously not 'intellectual' in kind.[79]

A historical review by Adam Mossoff is consistent with the institutionalists' account of a 'bargain theory' as the origins for patent grants and asks a pivotal question:

> [t]he history of patents does not begin with inventions, but rather with royal grants by Queen Elizabeth (1558–1603) … that advanced her economic and industrial policies. Approximately 200 years after the end of Elizabeth's reign, however, a patent represents a legal right obtained by an inventor providing for exclusive control … What accounts for this radical shift from a grant by royal prerogative to common-law property right?[80]

Crown granted *privileges* were a strategic instrument of the Crown that served to encourage the importation of foreign or unknown (new) technology and promote its dissemination. These privileges were socially useful as a means to educate a domestic labour force in the trades through tutelage with a foreigner who was willing to come to England[81] but were limited in

77 Drahos and Braithwaite, *supra* note 2 at 157.

78 Amani and Coombe, HGDP, *supra* note 59 at 155.

79 See *Moore v. Regents of the University of California*, 51 Cal. 3d 120, 125; 793 P.2d 479 (Supreme Court Cal., 1990) creating the legal precedent for this and is discussed also in Amani and Coombe, ibid. at 153.

80 Adam Mossoff, Rethinking the Development of Patents, 52 *Hastings L.J.* (2001) 1255 at 1257 [Mossoff, 'Rethinking'] at 1255.

81 Ibid. at 1258.

term within a limited territory and simply created first-mover advantage by providing lead time to industrialists. Fewer than 55 monopolies[82] were granted between 1561–1603; 21 of these went to foreigners and some had conditions: the patent had to be practiced within two months or else it would be void or that the French patentee train Englishmen in his trade, for example.[83] The first patent granted was for a *manufacturing monopoly* in England as a means of stimulating domestic production of raw materials and manufactured goods; its elements were characteristic of the Queen's prerogative:

> Under the early grant of letters patent, a patentee was supposed to: (i) work the patent, i.e., bring a foreign industry into the realm (ii) not be inconvenient to other subjects, i.e., not interfere with established industries, and (iii) train apprentices, i.e., create a self-sufficient industry within the realm through which English subjects can make a living. Insofar as a patentee met all three conditions, the crown would exercise its prerogative and issue a letter patent for a monopoly in the respective trade.[84]

Anti-competitive behaviour was to be restricted as *inconvenient* to subjects:

> … a monopoly cannot displace an existing trade within the realm … [T]wo rules that derived from these arguments – no monopoly patents either for trades existing at the time of the grant or for mere improvements upon such existing trades constituted the only real limitation imposed upon the early patent monopolies.[85]

What the inventor shared in common with the importer was the introduction of a new industry and thus patents for inventions earned royal favour with the rights of the inventor being derivative of the importer's right. Crown discretion was to be exercised narrowly, 'subject to Magna Carta and the common law, both of which were aimed in terms of ideals, at the protection of the negative

82 Since substituted goods may reduce the price of the IPR below classic monopoly pricing, these are not true *monopolies* 'still, no one may sell, copy or manufacture … without the authorization of the owner … [B]ut for those rights there would be competition in delivery of that good … [E]conomic conditions under which these non-rival and largely non-excludable goods are distributed look considerably more similar to a monopoly …'. James Boyle, 'Forward: The Opposite of Property?' (2003) 66 *L. and Contemp. Probs.* 1 at 2.

83 Mossoff, 'Rethinking', *supra* note 80 at 1230.

84 Ibid.

85 Ibid. at 1231.

liberties of subjects'.[86] Therefore, the Crown grants were not to unduly restrain trade or the customary right to work, writes Mossoff, and:

> [a]s long as the patentee was not infringing upon the customary right to work in one's trade, i.e. as long as a patentee was not attempting to 'put a new button on an old coat' or to 'taketh away a mans trade', then he was accorded the status of 'inventor' insofar as he was the *first* to establish (or re-establish) his respective trade within the realm. [87]

Even if discovery of DNA sequences are considered 'inventive', the for-profit enterprises cannot claim to be *first* to establish them within the 'realm'. Moreover, as Vaver suggests,

> the modern agnostic stance of Canadian patent law is not a necessary feature of it ... Elizabethans, for example, feared the social disorder high unemployment might bring and so deemed it 'inconvenient' to patent machines that would throw workers out of jobs'. Commonwealth states, such as Australia, that 'retain the *Statute of Monopolies'* criteria test the patentability of medical or surgical treatments for public 'inconvenience'.[88]

The *Statute of Monopolies* of 1624 judiciously limited patents to a term of less than 14 years and excluded from patentability inventions that were 'mischievous to the state by raising prices or commodities at home, or hurt of trade, or [were] generally inconvenient'.[89] It was a way of containing the royal prerogative which, over time, was nevertheless abused and many patents issued much like today 'regardless of whether an industry was new to the realm or not',[90] were extended to manufactures and sales. The original justifications for patent grants surrendered to monopolies based on political capture and patronage, impinging on individual liberties. To overcome patronage and abuse, these privileges had to be juridically circumscribed[91] and ultimately led to transforming the discretionary royal grants to statutorily created contractual arrangements between the inventor and the crown as agent for society that gave

86 Peter Drahos, *A Philosophy of Intellectual Property* (Brookfield, VT: Dartmouth, 1996) [Drahos, *A Philosophy*] at 30.

87 Mossoff, 'Rethinking', *supra* note 99 at 7.

88 Vaver *IPL*, *supra* note 39 at 120.

89 *An Act concerning Monopolies and Dispensations with Penal Laws, and the Forfeitures thereof* 1623 (UK), 21 Jac. I, c. 3. s. 6.

90 Mossoff, 'Rethinking', *supra* note 80 at 1263.

91 See *Darcy v. Allen (The Case of Monopolies)* in Coke's Reports discussed in Mossoff, 'Rethinking', *supra* note 80 at 1264–6, denouncing the 'common-law rule for adjudicating the *legitimacy* of a grant of monopoly privileges' and the limitation on patents as odious monopoly.

rise to *rights* void of crown discretion. The question of whether the law should consider qualitative aspects of the invention, according to Brad Sherman and Lionel Bentley, was converted over time into a 'fear of judgment' and a 'fear of making qualitative judgments'. It was believed that the law should resist making determinations of aesthetics and value and, 'intellectual property law thus came to echo modernism's fear of being tainted by politics, morality and judgment'.[92] This choice, however, was itself political in establishing the template for our current patent system that all but abandons any ideal of *quality* patents and has almost entirely surrendered the *public's* interest in obsessive pursuit to increase the number of patents. But

> during much of the nineteenth century, it was the quality of what was registered that mattered most. Indeed, the multiplication of patents was seen as an evil that needed to be avoided. To this end, attention was given to increasing the cost of registration and to the introduction of an examination system as ways of ensuring that inventions of a trivial or undeserving nature were not patented.[93]

Patent term was one way of containing the 'evil' of these artificial monopolies and while Thomas Jefferson's writings have typically been interpreted to support the greater expansion of IPRs based on the view that 'ingenuity should receive a liberal encouragement', by other accounts it is said of Jefferson that he

> was never enamored of the patent system and throughout his life displayed a marked ambivalence toward it … A highly relevant point the Court has chosen to ignore totally is that [Jefferson] … expressed considerable skepticism concerning both its usefulness and its effectiveness. He clearly did not believe that patents promoted the progress of the useful arts … [he] privately believed that the patent system more often served to permit patentees to obstruct rather than promote the progress of useful arts.[94]

Jefferson advocated stridently for a constitutionally limited term of only 12 years to be included in the Progress clause, Article I, Sec. 8, cl. 8, of the American Constitution which gives Congress the power 'to promote the Progress of Science and the useful Arts, by securing for limited Times to

92 Brad Sherman and Lionel Bentley, *The Making of Modern Intellectual Property Law: The British Experience, 1760–1911* (Cambridge: Cambridge University Press, 1999) at 178 [footnotes omitted].

93 Ibid. at 177–8 [footnotes omitted].

94 See e.g. Edward C. Walterscheid, 'The Use and Abuse of History: The Supreme Court's Interpretation of Thomas Jefferson's Influence on the Patent Law' (1999) *IDEA* 195 at 196.

Authors and Inventors the exclusive Right to their respective writings and discoveries'.[95] Jefferson failed; patent term is determined by legislation.[96]

Malla Pollack has looked at this clause to determine the definition of 'progress', empirically investigating the public usage of the word in the United States of 1789. She finds the term formally was used to mean 'dissemination', 'distribution', and 'spread' as opposed to qualitative or quantitative improvements in the knowledge base. Pollack contends that 'judicial trimming of Congressional over-protection' is authorized:

> [t]o the extent that Congress chooses not to act under this clause, the default position is that each person in the United States has a *property right not to be excluded from publicly accessible knowledge and technology*. Congress has only a very limited power to create private quasi-property, i.e. rights to exclude the rest of the commoners. Congress may only create temporary individual rights for 'authors' or 'inventors' to exclude from use of 'their respective writings and discoveries' when such individual rights 'promote' the spread of knowledge (Science) and technology (useful arts).[97]

The public interest arguments favouring limited patent monopolies are consistent with Pollack's examination of Congressional power in the USA and Mossof's historical European account; it can also be used to argue for greater stringency in modern patent law.

Canadian Patent Law: A Primer

Most jurisdictions share common features of patentability with some variation over specific subject matter and related legal doctrine. This section examines Canadian patent law by way of example. A patent is an exclusive right under statute to use, manufacture, sell, or import an invention in Canada for a

95 US Constitution, <http://www.house.gov/Constitution/Constitution.html>.

96 See Malla Pollack, 'The Democratic Public Domain: Reconnecting the Modern First Amendment and the Original Progress Clause (A.K.A. Copyright and Patent Clause)' (2004) 45 Jurimetrics J. 23.

97 Malla Pollack, 'What is Congress Supposed to Promote? Defining 'Progress' in Article I, Section 8, Clause 8 of the United States Constitution or Introducing the Progress Clause, <http://cyber.law.harvard.edu/openlaw/eldredvashcroft/progress.html #Footref12>. See also Margaret Chon, 'Postmodern "Progress": Reconsidering the Copyright and Patent Power' (1993) 43 *DePaul L. Rev.* 97, 102–3. Chon takes the argument further by asserting the existence of a constitutional right of access to information. In Canada, parliament's power under s. 91 of the *Constitution Act 1867* (UK), 30 and 31 Vict., c. 3 to make laws in relation to patents for inventions and discovery is without regard to any purpose.

period of 20 years from the date of filing (s. 44 of the CPA).[98] The exclusive nature of this right, its transfer, sale, inheritance, and that the patent owner can exclude others from using, making or selling the invention, attracts the 'property' label despite the poor conceptual fit given that knowledge goods are both non-rivalrous and non-excludable. Unlike traditional personal property, intellectual property objects can be used by many individuals at the same time and without diminishing the use of the property holder in the traditional sense as with tangible property. Since patents allow the owner to exclude others who have independently invented a similar product or process, or even one who may have invented sooner but filed later, the nature of these rights are stronger than those found in other forms of intellectual property and are harder to justify on the basis of *fairness* alone. Fairness, according to romantic ideas of entitlement, would require inventors be compensated for the investment of their time, labour, and skill, with a return on their investment but the current system makes evident that legal protection can only be based on the public *benefits generated* by the systemic incentives for disclosure.

Disclosure and Claims: Two Parts to the Patent

A patent has two parts – the 'specification' (also called the 'description') or the 'disclosure', and the 'claims'. The disclosure essentially functions as a 'how-to' guide providing information so that others are able to make and use the invention. Section 27 of Canada's Patent Act (CPA) establishes the technical requirements for a patent application. Section 27(3) provides that the specification of an invention must (emphasis added):

> (a) correctly and *fully* describe the invention and its *operation* or *use* as contemplated by the inventor; (b) set out *clearly* the various steps in a process, or the method of constructing, *making*, compounding or using a machine, manufacture or composition of matter, in such a full, clear, concise and *exact* terms as to *enable any person skilled in the art or science in which it pertains or with which it is most closely connected, to make construct, compound or use it;* (c) in the case of a machine, explain the principle of the machine and *the best mode in which the inventor has contemplated the application of that principle; and*
> (d) in the case of a process, explain the *necessary sequence,* if any, of the various steps, *so as to distinguish the invention from other inventions.*

The statute mandates the inclusion of sufficient detail to make the invention disclosed enabling to one skilled in the art to make and use it for its stated

98 CPA, *supra* note 38.

purpose.[99] That means that skilled readers should have enough information to allow experimentation with and improvements on the invention during the patent term and for the invention to be practiced with little difficulty under licence or upon its expiry. This is an integral part of the crown 'bargain' that allows the invention to immediately enter the common stock of knowledge although exclusive control remains with the patentee for the patent term. If the disclosure is insufficient or otherwise flawed in what it reveals, the whole patent should be invalid. In Canada, '[a] patent is void if any material allegation in the petition of the applicant in respect of the patent is untrue, or if the specification and drawings contain more or less than is necessary for obtaining the end for which they purport to be made …' (s. 53(1), CPA). In other jurisdictions, as the United States, misrepresentations made to the patent office may constitute inequitable conduct and compromise the validity or enforceability of patents.

The claims describe the invention to which the exclusive legal rights attach, providing a legal boundary enclosing the patentee's rights. Section 27(4) of the CPA provides that '[t]he specification must end with a claim or claims defining distinctly and in explicit terms the subject-matter of the invention for which an exclusive privilege or property is claimed'. The two parts of the patent create a tension ripe for technical and legal debate as well as abuse. On the one hand, the description must be sufficiently detailed to allow the inventor to distinguish the invention from others existing in the prior art and specific enough to allow persons skilled in the art to work the patent; it must be narrowly drafted to ensure that the inventor is not himself a patent infringer. 'Prior art' is a legal term that describes inventions, literature, or any other existing documentation and material that proves that the idea being protected by a patent is not new but has been previously publicly disclosed. The problem is that prior art may be difficult to find at the time of the patent application and this has led to many recommendations to create national and international databases of prior art. Claims will be found to be invalid upon challenge if they are broader than the disclosure made.[100] On the other hand, in order to ensure maximum protection of legal interests and to reduce the likelihood and degree of their competition, patentees have good reason to try and disclose the minimum necessary; a certain craftiness is required according to David Vaver:

> The game for patentees, especially in highly competitive industries, is to reveal as little and to claim as much as possible. The less disclosed, the more that can be retained as competitive edge. The wider one claims, the tougher it is for imitators. But the specification must stay clear of the known and

99 CPA, *supra* note 38, s. 28.3.

100 *Monsanto Canada Inc. v. Schmeiser,* [2004] 1 S.C.R. 902, 2004 SCC 34 [*Schmeiser* SCC] at para. 124, citing *Amfac Foods Inc v. Irving Pulp and Paper Ltd.* (1984), 80 C.P.R. (2d) 59 (F.C.T.D.) and examples at 80.

> the obvious. It must demonstrate and claim only something over and above existing technology. Much patent drafting involves trying simultaneously to achieve these aims.[101]

Optimal disclosure would be enough to secure a valid patent but not so much as to forfeit market dominance. Consequently, claims are typically very abstractly drafted, narrowing further in scope with each additional claim to help ensure the full spectrum of protection. Upon court challenge, any of the claims may be held invalid without compromising the validity of the remainder. And, confirms Vaver, 'what is not claimed is disclaimed'.[102]

The *person having ordinary skill in the art* (PHOSITA) in patent law is an ever-elusive fictional person creating a nebulous legal standard. The Supreme Court of Canada's Justice Binnie describes PHOSITA as

> a hypothetical person possessing the ordinary skill and knowledge of the particular art to which the invention relates, and a mind willing to understand a specification that is addressed to him. This hypothetical person has sometimes been equated with the 'reasonable man' standard in negligence law. He is assumed to be a man who is not going to try to achieve success and not one who is looking for difficulties or seeking failure.[103]

Much of the variance in patent standards across technologies (and sometimes jurisdictions) is attributed to PHOSITA. As a legal construct, it is used to determine obviousness and enablement by the courts and renders decision makers the gatekeepers of privatized new technologies based on how PHOSITA is personified. Dan Burk and Mark Lemley find:

> [t]he more skill those in the art have, the less information a patentee has to disclose, but the harder it is to find an invention nonobvious. One reading of the biotechnology and computer software cases is that the Federal Circuit believes computer programmers are extremely skilled, while biotechnology experts know very little about their art. We do not challenge the idea that the *standards in each industry should vary with the level of the skill in that industry* … [T]he use of the PHOSITA provides needed flexibility for patent law, permitting it to adapt to new technologies without losing its essential

101 Vaver, *IPL*, *supra* note 39 at 139.

102 Ibid.

103 *Free World Trust v. Electro Santé Inc.*, [2000] 2 S.C.R. 1024, 2000 SCC 66 at para 44. quoting from H.G. Fox, *The Canadian Law and Practice Relating to Letters Patent for Inventions*, 4th edn (Toronto: Carswell, 1969) at 184.

> character ... The level of skill in the art affects not just patent validity, but also patent scope.[104]

While PHOSITA may have some utility in the flexibility it provides patent law, the flip side is that the meaning of the patent is ultimately a question of law to be, as David Vaver lucidly states, 'decided by a judge who usually is not skilled in any art or science, let alone the relevant one'.[105] Judges have complained, on more than one occasion, of the linguistic ambiguity in claim drafting[106] and this is not helped by the scientific complexity in understanding the boundaries to technologically specific inventions. Yet the complaints seem to have little impact on judicial findings of claim validity and therefore on the drafting practices they critique.

Normally, a challenge to patent validity follows defensively as a counterclaim to a patent infringement suit. To a certain extent, the ambiguity in claims drafting is deliberate in order to benefit from the doctrine of *purposive construction* used by the courts in order to save litigated claims. This doctrine directs judges when reading the patent not to construct a claim too narrowly or literally such as to defeat it.[107] Patentees hardly need encouragement with their fence drawing exercise of claims drafting. 'Kitchen sink' patents,[108] also known as 'reach-through claims' reflect the tendency of patentees to claim too broadly, to reach through the invention to privately enclose more than that which is invented.[109] Reach-through claims are anti-competitive and reflect the

---

104 See Dan L. Burk and Mark A. Lemley, 'Is Patent Law Technology-Specific?' (2002) 17 *Berkeley Technology and Law Journal*, 1155 cited to PDF, <http://www.innovationlaw.org/pages/DanBurk.doc>, 2–3

105 Vaver *IPL, supra* note 39 at 140.

106 Vaver, Ibid. at 141 refers to *Xerox of Canada Ltd. v. I.B.M. Canada Ltd.* (1977), 33 C.P.R. 24 at 88, n. 14, where the judge complained that claims such as the one before the court passed from 'riddle to enigma' – the claim to collect used toner from photocopiers was a sentence made of 178 words with little punctuation. A 281-word claim was held valid after a nine day trial despite some earlier proclaimed doubts by the judge during interlocutory proceedings in *Risi Stone Ltd. V. Group Permacon In.* (1990), 29 C.P.R. (3d) 243 at 247–48 (Fed. T.D) (interlocutory); (1995), 65 C.P.R. (3d) 2 at 9 (Fed. T.D.) (trial).

107 See *Catnic Components Ltd. V. Hill and Smith Ltd. (1980),* [1981] F.S.R. 60 at 65-66.

108 See Kevin G. Rivette and David Kline, *Rembrandts in the Attic: Unlocking the Hidden Value of Patents* (Boston: Harvard Business School Press, 2000) at 21 which get their name 'because they [the patents] sometimes appear to be asserting ownership of everything under the sun ....

109 See Keynote Address, Rebecca S. Eisenberg, Dual Controversies of the Double Helix: Challenges of Regulating the Information and Property Aspects of Genetic Technology, *Reaching Through the Human Genome*. Webcast, Centre for Innovation Law and Policy, <http://www.innovationlaw.org/tip/pages/genetic_technology.htm>.

patentee's desire to demarcate as much of the market in a particular field of technology as falling within his or her legal rights. Patents in this field are directed at the DNA molecule and not necessarily a *use* of the molecule such that any subsequent utility discovered may also be captured within the patent claim. The patentee's spot is 'reserved' for later and broader applications of the disclosed technology (i.e. DNA sequences). The broad claims, if granted, are disingenuous and a breach of the *bargain with the crown* for granting monopoly rights. They are inefficient and upset the proverbial trade-off of short-term monopoly costs for long-term welfare gains because they attract more rights than are merited from the disclosed 'invention'. They are anticompetitive if granted but due to the presumption of validity and high cost of litigation, will nevertheless define the legal rights used for extracting royalties until successfully challenged. Not all inventions are legally recognized, however.

Common Patentability Requirements

A patent for Canada must be applied for at the Canadian Patent Office (CPO) and is granted by a patent examiner if it meets the requirements for 'invention'. Section 2 of the CPA defines 'invention' as 'any *new and useful* art, process, machine, manufacture or composition of matter' or any new and useful improvements to these. 'Invention' could cover *product* and *process* claims so long as it meets the statutory definition of invention and falls into the scope of patentable subject matter (it is not expressly excluded in the statute), is new, useful, non-obvious, and fully disclosed. The novelty requirement of the s. 2 definition is elaborated in s. 28.2(1) of the CPA which requires that the subject matter of the invention has not previously been disclosed in public, subject to limited exceptions. Therefore before a patent is granted, there is a search of the prior art to ensure that the application is in fact for a new invention and, as far as is known in good faith, it does not infringe any other patented inventions in that jurisdiction. The utility requirement means that the invention must have a known and described utility, often understood as 'industrial value' or in some contexts 'industrial application'. In either case, it 'must be apparent from the description to one skilled in the art'. No prototypes are required to display the claimed utility except in relation to perpetual motion machines. Even if new and useful, the invention claimed cannot be a development that would be obvious on the claim date to the PHOSITA or science to which it pertains or is information publicly available 'in Canada or elsewhere' (CPA, s. 28.3). This requirement was added by amendment in 1993 to codify the requirement of inventiveness imported through case law. Once the application has met these requirements and technical matters such as filing fees have been complied with, the patent *must* be granted without discretion of the Patent Commissioner (CPA, s. 27(1)). Of course, the lower the threshold for novelty, utility and non-obviousness, the more inventions will be patentable. Patent applications are published 18 months after they are filed on the website of the relevant patent office before the patent is granted. Since the proceedings and the publications are not *publicized*, however, the ability to challenge a patent generally does not arise until after the patent has been issued and public attention has been drawn to a potential conflict with public interest uses or human rights. Often, only the threat of a lengthy and costly infringement suit will bring the patent under expert scrutiny.

The Perfidy of Patents

The BRCA1 and BRCA2 genes are linked with the propensity to develop breast and ovarian cancers based on common mutations. The patent for genetic diagnostic testing of these genes are owned by Myriad Genetic Laboratories Technologies Inc. (MGL), an American company based in Utah, USA.[110] Controversy surrounding alleged unauthorized use of these 'inventions' forced gene patents into the centre of medico-legal and public health debates in Canada but with little regard for the foreign patentee. Ontario's then Premier, Mike Harris, urged the Province to continue to provide women with genetic cancer testing despite MGL's claims that this constituted infringement of the company's exclusive rights. Ontario had not obtained a licence from MGL or their exclusively authorized licenced Canadian partner, MDS Laboratory Services, for the publicly funded test, which cost the government $800 a person – a fraction of the cost ($3,850) MGL demanded.[111]

While the patents/public health relationship finally elicited the public, media, and regulatory attention it deserved,[112] the legal risks were too high and

110 The University of Utah and the National Institute for Environmental Health Sciences in the USA were co-applicants on the first patent filed on BRCA1 in that jurisdiction in 1994 on the isolated DNA coding for a BRCA-1 polypeptide and on a screening method. For BRCA2, the first patent was filed in 1998 and included the University of Toronto and the Hospital for Sick Children as co-applicants. It was agreed in the various private/public research teams that MGL would be the commercial licensee from the team and therefore often the patents are entirely attributed to them alone. MGL claims to support liberal research testing but not commercial testing. See The National Academies Fifth Meeting of the Committee on Intellectual Property Rights in Genomic and Protein-related Inventions, 11 February 2005, unedited verbatim transcript, <http://www7.nationalacademies.org/step/Genomics_Committee_Meeting_6_transcript.pdf>.

111 See Robert Benzie, 'Ontario to defy U.S. patents on Cancer Genes: Province Will pay for $800 test, not $3850 version by Myriad Genetic Laboratories Inc.', *National Post*, 20 September 2001: A15, discussed in Bita Amani, 'Patents and Public Health' (2002) 22:3 *Health L. Can.* 76 [Amani, 'Patents and Public Health'].

112 See e.g. Karen Palmer, Public Health Reporter, 'Battle over cancer gene test U.S. company's patenting claim called 'abhorrent'' Organic Consumers Association, 01/09/2003 online www.organicconsumers.org/Patent/010903_patent.cfm; Canadian

some provincial governments complied with the 'cease and desist' letters sent. British Columbia (BC), for example, referred cancer patients to Ontario for two years. BC has resumed testing but now has a waiting list of approximately two years. The event clearly demonstrates how private proprietary interests can dictate provincial health policy. In Ontario, Mike Harris refused to back down and in his speech to the Ontario Advisory Committee on Predictive Genetic Technologies implored that Canada needs to amend its laws to prevent privatization of human genes and that '[u]nlike new drugs, genes aren't invented – they are discovered. They have always existed'. Moreover, Harris claimed that

> [t]he benefits of a world-wide effort such as the human genome project should not be the property of a handful of people or of companies. Our genetic heritage belongs to everyone. We must share its benefits fairly. We must do what we can to make genetic tests and therapies affordable and accessible … if we have the ability to save a life, we have the responsibility to do so.[113]

Other international governments took a similar position as they grappled with the cost of MGL's patent on public health delivery. In France and later Europe, MGL's patents were successfully attacked on the basis that they failed to meet statutory requirements. The original dispute between the Institut Curie and MGL eventually spilled over into all of Europe and involved the French government, Switzerland's Social Democratic Party, Greenpeace Germany, the Netherlands as represented by their Ministry of Health, the Austrian Federal Ministry of Social Security, and the European Parliament amongst others. MGL's claims were broadly drafted and reached through to limit alternative and possibly more accurate genetic testing by assuming a monopoly on *all* BRCA analyses. Rather than continue as a potential infringer or subvert health policy, the Institut Curie, the Assistance Publique-Hopitaux de Paris and the Institut Gustave-Roussy filed an opposition notice with the European Patent Office (EPO) which, following public hearings on 17 May and 18, 2004, resulted in the revocation of MGL's patent for BRACAnalysis® (European Patent 699754). The attack asserted that the claimed invention lacked novelty, inventive step and industrial application and that the disclosed invention failed to enable a PHOSITA to carry it out. There had been errors in the original patent application that had not been corrected until the gene

Cancer Society (CCS) and National Cancer Institute of Canada (NCIC) Position Paper on the Patenting of BRCA1 and 2 genes, 8 March 2002; James Meek, 'US firm may double cost of UK cancer checks', *The Guardian*, 17 January 2000, <http://www.biotech-info.net/US_firm.html>; Council for Civil Liberties, 'Patenting Breast Cancer'.

113 Benzie, *supra* note 111.

sequences were in the public domain; Myriad's own publications thereafter constituted prior art and served to defeat their patent on the basis that it lacked novelty by the time full disclosure of the invention was made in an amended application.[114] Their BRCA2 patent was maintained in amended form and MGL continues to hold the patent over HMPCC linked in carriers with the greater likelihood of colorectal and endometrial cancer. MGL's appeal of the decision to amend two other of its BRCA-1 related patents to exclude diagnostic methods was unsuccessful at the Technical Board of Appeal of the European Patent Office which in September 2007 confirmed the outcome of the first instance opposition procedure. The original patent granted in 2001 had covered the isolated human BRCA1 gene, its mutant forms, and use in diagnosis of predisposition to breast and ovarian cancer. Herder and Gold offer that

> in hindsight, Myriad Genetics' main mistake appears to have been in failing to effectively communicate its business plans; it claims never to have intended to exercise its patent rights against health care providers. But because of how its actions were perceived, this instance is rightly characterized as a patent problem in our view. [115]

The Canadian Cancer society reports that since the EPO 'has not yet issued a written decision that describes the specific reasons it revoked this patent … it is difficult to know what impact this will have, if any, in Canada'.[116] But revocation is neither the most desirable or efficient regulatory response to health policy needs. As an after-the-fact remedy, revocation may actually undermine the integrity of the patent system in the long term as well as the securities value of patents by increasing the level of uncertainty amongst patentees and the public as to the validity of patents. There are numerous examples of absurd patents granted by the USPTO that evince how patents are being rubber-stamped for 'inventions' that are not new and are obvious even to the unskilled layperson. Resource deficits, discussed next, are partly to blame. However, 'bad patents' can also be attributed to lax application of the substantive standards for patentability. A system that is porous up front unfortunately relies on private litigants as a means of maintaining quality assurance. In short, we rely upon industry competitors to subsidize a poor administrative system and to ensure that judges patch up its leakages. But, according to Arti Rai,[117] in the United States, when patents do receive judicial

---

114 'Bioethics and Patent Law', *supra* note 110.

115 Herder and Gold, *supra* note 34 at 8.

116 See CCS, 'Background on the Patenting of BRCA1 and 2 Genes', <http://www.cancer.ca>.

117 Arti Rai, 'Addressing the Patent Gold Rush: The Role of Deference to PTO Denials' (2000) 2 *Wash. U. J. L. and Pol'y* 199.

scrutiny, judges are more amenable to intervene (in *favour* of the patentee) where the patent application has initially been rejected. Rai observes that:

> [f]irst, the Court of Appeals for the Federal Circuit's (CAFC) reversal of PTO decisions denying patent protection to certain biotechnology and computer program inventions has been a major reason for the proliferation of patents. Second, given the CAFC's frequent exercise of its ability to *reverse* PTO patent denials, PTO reform will, on its own, likely be insufficient. Third, at least for the time being, the most prudent course for addressing patent proliferation may not be a significant change in substantive patent law.[118]

Rai attributes the proliferation of patents in 'technologically complex' and 'rapidly expanding industries' not to mere changes in substantive law, but also to the immediate impact of institutional actors and the roles that patent offices and courts play. She argues that a re-examination of the relationship between courts and patent offices is needed with greater attention to be directed at the application of administrative law principles:

> With respect to this relationship, considerations of institutional competence – particularly institutional resources and expertise – suggest that the CAFC should be wary of reviewing independently PTO's decision denying patentability. The CAFC's review should be particularly deferential when the denial is based on a determination that the invention is 'obvious' – that is, not truly new. Deference to patent denials is warranted moreover, even if the PTO continues to have skewed incentives as well as limitations on its own resources and expertise: these limitations will tend systematically to produce errors in patent grants, not patent denials. Indeed, as a consequence, PTO reform will be much more important for ensuring valid PTO patent grants than for ensuring valid PTO patent denials.[119]

Lax application of patent standards and institutional deficits are interrelated phenomena contributing to the creation of secondary incentives towards predatory market behaviour by patent holders. Evidence of so-called 'bad' patents (*bad* because they should never have issued if proper attention was given to patentability requirements) in sectors of little comparative importance should serve as a cautionary tale mandating extreme restraint and due diligence in biopatenting. They draw attention to the urgent need for regulatory and institutional reform, especially given the sensitivity of the subject matter, its importance for a country's socio-economic and technical development, and the regulatory discretion needed to meet the cultural and moral values infusing the public's interest.

118 Ibid. at 201.
119 Ibid. at 202.

Quality v. Quantity of Patents: The Rising Price of Presumptive Validity

Until only two decades ago the discussion of patents was obscure and highly technical. It was left for the specialized few who were more interested in the underlying invention than the broader social, ethical, and welfare issues concomitant with any emerging technology. As a result, stronger protections and their extension to new subject matter occurred with stealth through patentee correspondence with the patent office rather than with public debate about the merits or impact of expanding protection into other areas important for public policy. The shift in emphasis in the issuing of patents, away from the quality of the invention and towards the sheer number of pertinent patents, has been subtle.

The World Bank reports a 'surge' in patenting which, between 1985 and 1995, saw the number of applications for US patents by US inventors rise in absolute and percentage terms faster than in any other decade (effectively doubling); by 1995, more than 120,000 patent applications were made in the US.[120] The European Patent Office estimates that 700,000 applications for all fields of inventions are filed annually, with patent protection being sought on average in four countries per invention. The current number of patents in force worldwide today is more than 4 million. Licensing revenues from patented products reached an estimated US$100 billion worldwide by 2000, increasing tenfold since 1990.[121] The 2003 Canadian Biotech Advisory Committee's Annual Report found that Canada hosts 417 core biotechnology companies, with combined revenue of $2.7 billion.[122] Canada is ranked second only to the United States for biotechnology[123] According to OECD data, Canada accounted for 4 per cent of all biotech patents granted by the USPTO in 2000 and 3 per cent of all patent applications filed with the European Patent Office (EPO) in 1999 and the third highest rate of growth in the number of biotech patents filed with the EP between 1990–1999.[124] The OECD has found that 'biotech contributed more than average to the overall surge in patenting. Between 1991 and 2001, biotechnology … patents filed at the EPO increased 8.3%' compared to the 5.7% for all EPO applications'.[125]

120 The World Bank (Discussion Paper No. 412) *Intellectual Property Rights and Economic Development* (2000).

121 Cook, *Patents, Profits and Power*, at 6.

122 <http://cbac-cccb.ca/epic/site/cbac-cccb.nsf/vwapj/annua_report_2003.pdf/$FILE/annua_report_2003.pdf>.

123 <http://www.wd.gc.ca/8208_ENG_ASP.asp>.

124 Canadian Trends in Biotechnology, <http://www.biostrategy.gc.ca/StatusreportE/sum_e.html>.

125 OECD Compendium of Patent Statistics (2005), <http://www.oecd.org/dataoecd/60/24/8208325.pdf>.

WIPO recently celebrated its one millionth patent under the Patent Cooperation Treaty (PCT), applauding this 'success' with plans for electronic filing in the near future to facilitate a greater number of applications.[126] The PCT provides a unified procedure for applying for a patent in various jurisdictions by allowing a single search for legal opinion even though the actual patent would still have to be granted by each jurisdiction separately. This facilitates the process of obtaining a patent within contracting states but does not grant a single internationally viable patent, as such an instrument has yet to exist. WIPO reports:

> PCT applications soared as businesses woke up to the strategic importance of intellectual property assets. It took 18 years to reach 250,000 applications but only 4 years to double that figure, and another 4 to double again.[127]

WIPO has assumed, as have the EPO, United States Patent and Trademark Office (USPTO), Canadian Patent Office (CPO), OECD, and World Bank, that the world will benefit simply from *more* patents on the dominant rationale that patents incentivize innovation without considering the implication of *overprotection* or the need to ensure that the patents granted are in fact merited. The assumption is also deeply shared in domestic governmental policy making and judicial decision making.[128] There is consistent growth in the number of patent applications and 'patent grants have also risen during this period, but their level is affected by the degree to which Offices can cope with the rising tide of applications'.[129]

A paucity of institutional and human resources at most patent offices in the developed world means that the current regime is incapable of providing full and thorough consideration of the growing number of applications before it. Patent offices in North America are so backlogged with applications that they can neither appropriately perform their delegated administrative duties nor function as effective gatekeepers of validly claimed monopolies. Patent

126 See 'PCT 2005 One Million and Counting' *WIPO* Magazine (February 2005) at 2 ['One Million'], <http://www.wipo.int/freepublications/en/general/121/2005/wipo_pub_121_2005_01-02.pdf>.

127 Ibid. at 4.

128 See e.g. Gold et al., 'Unexamined Assumptions' *supra* note 34 and discussion in Chapter 3.

129 William Cornish, *Intellectual Property: Omnipresent, Distracting, Irrelevant* (Oxford: Oxford University Press, 2004) at 5. Cornish explains that the adjectives in his title highlight the different dimensions of these problems: '*omnipresent* – to capture the cases where IPRs appear to be spreading like a rash, particularly across new technologies, and threatening to leave few patches of unblemished, open skin; *distracting* – to indicate cases where the rights achieve little of their essential purpose, but cause persistent itching; *irrelevant* – to refer to technology which in practice seems to be rendering IP nugatory' at 1.

fees are a major source of revenue for financing patent office administration in all jurisdictions. Yet in Canada filing fees remained at their 1989 levels and did not change for well over a decade until 1 January 2004 and then only in response to the Canadian Intellectual Property Office's (CIPO) new role as one of ten national offices with status as an International Search and Preliminary Examination Authority under the PCT.[130] The updated fee schedule applies to patents pending and granted patents regardless of when they were filed and marks a 33 per cent increase from existing fees. PCT transmittal fees have also increased by 50 per cent prompted by the growing complexity of applications, the demands for services, and the corresponding inability to meet examination and execution needs. While Canadian data is comparatively scant, the American statistics are telling. The Director of the USPTO advised in a 2005 report that while theirs is still the fastest and least expensive of the major patent offices of the world,

> the volume and complexity of patent applications continues to outpace current capacity to examine them. The result is a pending – and growing – application backlog of historic proportions. Patent pendency – the amount of time a patent application is waiting before a patent is issued – now averages more than two years. In more complex art areas ... average pendency stands at more than three years.[131]

On average, only 18 to 36 hours are expended in the USPTO in reviewing an application and conducting the entire examination including searching prior art, interviewing applicants and engaging in correspondence.[132] Weak prior art searches are known to generate low quality patents but may account for high quantity. The USPTO grants approximately 75 per cent of patent applications. Courts invalidate some 46 per cent of litigated patent claims but only a negligible 2 per cent of patents, however, ever get litigated and 'less than two-tenths of one percent of all issued patents actually go to court'.[133] The costs of a full trial through appeal might approximate $1.5 million for each party according to a study by Mark Lemley.[134] The reported growth in the number of patent applications and grants in the world's largest market reflects broader social trends that demonstrate a "sharp increase' in the level

130 CIPO, <http://cipo.gc.ca>.

131 USPTO, 'Fiscal and Accountability Report 2005', <http://www.uspto.gov.>

132 Mark Lemley, 'Rational Ignorance at the Patent Office' (2001) 95.4 *Nw. U. L. Rev.* 1495 [Lemley, 'Rational Ignorance'] at 1508.

133 John R. Allison and Mark A. Lemley, 'Empirical Evidence on the Validity of Litigated Patents' (1998) at 208, SSRN, <http://papers.ssrn.com/sol3/papers.cfm?abstract_id=118149>.

134 Lemley, 'Rational Ignorance', *supra* note 132 at 1501–2.

of patent activity across the world'.[135] Presumably the resource problems would be compounded by a shortage of skilled labour and capital resources in similar offices in developing countries. What can we deduce from these trends regarding patent quality?[136]

Presumptively valid patents have been granted in the United States for peanut butter and jelly sandwiches (No. 6004596), a method of using the washroom (No. 6329919), a method of swinging, side to side (No. 6368277), and even a method for drafting patents (No. 6574645).[137] IBM boasts it is the only company to be issued more than 3,000 patents in a given year, which it has done for three consecutive years. For 11 consecutive years, it has ranked as the top recipient of the number of patents issued;[138] its portfolio includes the obvious method of using a washroom (on aeroplanes) on a first come first serve basis – is IBM better skilled at playing the patent game or actually this much more innovative?

The proclivity to patent has been coupled by a trend in stockpiling patents. Microsoft's Bill Gates publicly reported in 2004 that the company planned to file 3,000 applications that year – up from the approximately 2,000 filed the previous year and 1,000 filed in years prior – as part of their plan to boost licensing of their IP to other companies.[139] Microsoft's patent portfolio includes a recent patent issued to Robert Burchinal and assigned to Microsoft over a rare apple tree[140] – a case of patenting plant life that wryly comments on the margin for mistake at the patent office. Yet growing concentration of corporate power renders Microsoft's horticultural interests in 'apple' plausible as part of the continuous private enclosure of *life* and makes possible the ability to simply *decide* to increase a patent portfolio. Bad patents where the technology is simple allow the inference of deteriorating *quality* for all fields and disproportionately where subject matter is complex and technical,

135 See the OECD's Compendium of Patent Statistics 2005, <http://www.oecd.org>. See also 'PCT 2005 One Million and Counting' WIPO Magazine (February 2005).

136 Criticism of the patent system often corresponds with specific industries (i.e. biotech or software). See e.g. Shubha Ghosh and Jay P. Kesan, 'What do Patents Purchase? In Search of Optimal Ignorance in the Patent Office' (2004) 40 *Hous. L. Rev.* 1219.

137 See Bita Amani, 'Has Patenting Gone too Far?', *Innovate* (Spring 2004) 15 [Amani, 'Too Far'].

138 See Martin LaMonica, 'IBM retains patent crown' (12 January 2004), CNET News.com

139 See Ina Fried, 'Gates wants patent power' (29 July 2004), CNET News.com.

140 There are only approximately 1000 samples of the Burchinal Red Delicious growing in a rural community near Microsoft's home base in Redmond (Patent No. 14757). The assignment appears to be an error for which a certificate of correction is being sought. See David Becker, 'Apple patented by Microsoft: Apparently, intellectual property does grow on trees' (4 May 2004), CNET News.com.

as with biotechnology. Specific examples are less likely to draw immediate public scrutiny, explaining the 1980s patent 'gold rush' Arti Rai refers to during which the USPTO issued patents on genes that closely corresponded to foreseeable commercial products such as therapeutic proteins or diagnostic tests for recognized genetic diseases. Sometimes the only utility of the claimed sequence was its asserted use as a genetic probe or an expressed sequence tag (EST). While most opposition to gene patenting is based on the fact that genes exist as a product of nature and are not new, many believe 'the hundreds of thousands of patent applications that have been filed on gene sequences of unknown function should be denied on obviousness grounds'.[141] Instead of incentives to invent and disclose, the patent regime is said to threaten a *tragedy of the anti-commons.*

Literature on the commons grew out of the seminal classic of Garret Hardin, 'The Tragedy of the Commons',[142] in which Bentham's utilitarian thesis was criticized for its failure to recognize that 'a finite world can support only a finite population …'. In the absence of a private property regime, Hardin warned, the privilege of use would lead to rational overuse and underinvestment in the commons resulting in the tragedy of depleted environmental resources. The absence of the right to exclude would ultimately detrimentally affect everyone's use:

> [T]he inherent logic of the commons remorselessly generates tragedy … Adding together the component partial utilities, the rational herdsman concludes that the only sensible course for him to pursue is to add another animal to his herd. And another … But this is the conclusion reached by each and every rational herdsman sharing a commons … Each man is locked into a system that compels him to increase his herd without limit – in a world that is limited. Ruin is the destination toward which all men rush, each pursuing his own best interest in a society that believes in the freedom of the commons. Freedom in the commons brings ruin to all.[143]

Michael Heller and Rebecca Eisenberg were among the first to draw attention to the growing transaction costs of doing biomedical research due to its private enclosure *vis-à-vis* letters patent. Unlike the tragedy of the commons, the anti-commons are characterized by too many rights to exclude without corresponding rights to use.[144] In their compelling 1998 article, 'Can Patents

141 Rai *supra* note 117 at 216.

142 Garret Hardin, 'The Tragedy of the Commons', 162 *Science* 1243 (13 December 1968).

143 Ibid. at 1244.

144 Michael A. Heller and Rebecca S. Eisenberg, 'Can Patents Deter Innovation? The Anticommons in Biomedical Research' *Science* 280:1364 (1 May 1998) 698–701 [Heller and Eisenberg], *Science*, <http://www.sciencemag.org/cgi/reprint/280/5364/698.

Deter Innovation: The Anticommons in Biomedical Research'[145] the authors argue that biomedical information may effectively become underutilized by rational individuals acting separately under the exclusive rights that patents confer. This in turn may effectively cause a collective *waste* of scarce resources. The dangers of over-patenting, they argue – particularly in basic biomedical research – arise from the need to bundle multiple licences, raising the barriers to entry, and increasing transaction costs, with long term deleterious effects on downstream (applied/developmental) innovation:

> Under the commons model, the federal government sponsored premarket or 'upstream' research and encouraged broad dissemination of results in the public domain. Unpatented biomedical discoveries were freely incorporated in 'downstream' products for diagnosing and treating disease. In 1980, in an effort to promote commercial development of new technologies, Congress began encouraging universities and other institutions to patent discoveries arising from federally supported research and development and transfer their technology to the private sector ... Privatization can go astray when too many owners hold rights in previous discoveries that constitute obstacles for future research ... The result has been a spiral of overlapping patent claims in the hands of different owners, reaching ever further upstream in the course of biomedical research.[146]

What results is 'patent thickets', defined by Carl Shapiro as 'a dense web of overlapping intellectual property rights that a company must hack its way through in order to actually commercialize new technology'.[147] Patent pools and cross-licences are two common and effective business arrangements for cutting through the ticket, although they have their own transaction costs and can slow down the commercialization of new technologies.

---

pdf>. See also Michael Heller, 'The Tragedy of the Anticommons: Property in the Transition from Marx to Markets', <http://www.bus.umich.edu/KresgeLibrary/Collections/Working papers/wdi/wp40.pdf>; Andrea D. Brashear, 'Evolving Biotechnology Patent Laws in the United States and Europe: Are they Inhibiting Disease Research?' (2001) 12 *Ind. Int'*; and *Comp. L. Rev.* 183 at 186–87. Herder and Gold, *supra* note 34 at 36 contend that while biotechnology began this trend, stem cell technology, synthetic genomics and nanotechnology push it even further.

145 Heller and Eisenberg, op. cit.

146 Ibid. at 698. They argue that privatization of biomedical research must be more carefully deployed to sustain both upstream research and downstream product development; 'otherwise, more intellectual property rights may lead paradoxically to fewer useful products for improving human health' (ibid. at 701).

147 Carl Shapiro, 'Navigating the Patent Thicket: Cross Licences, Patent Pools, and Standard-Setting' (March 2001), <http://faculty.haas.berkeley.edu/shapiro/thicket.pdf> at 2.

Ironically, more IPRs would translate into fewer useful products for improving human health. Heller and Eisenberg contend that excessive fragmentation of patent rights in the technological base for commercially oriented biomedical innovation may deter smaller private competitors from investing in follow up innovation, especially when the patents are held by large corporate actors in merging agrochemical and biotechnical industries. More importantly, universities, the very institutions whose mandate has been to do property-less research, may have to pay disproportionately greater transaction costs. As public institutions, universities may find the new costs of conducting research too prohibitive to pursue, slowly transforming into dissemination sources of private labs under licence:

> some universities and other nonprofit research institutions have balked at terms DuPont Corporation has offered for licenses to use patented oncomouse and cre-lox technologies, although others have acquiesced to the license terms. These patents cover genetically engineered mice useful in research that could result in products falling outside the scope of the patent claims. DuPont has offered noncommercial research licenses and sublicenses on terms that seem to require licensees to return to DuPont for further approval before any new discoveries or materials resulting from the use of licensed mice are passed along to others or used for commercial purposes. DuPont thereby gains the right to participate in future negotiations to develop commercial products that fall outside the scope of their patent claims. In effect, the license terms permit DuPont to leverage its proprietary position in upstream research tools into a broad veto right over downstream research and product development.[148]

A major factor aggravating the under use problem of excessive biopatenting is the shift in judicial attitudes causing the law in almost all developed country jurisdictions to shift significantly in favour of biotech patenting. In England, for example, the court in *Genentech Inc. v. Wellcome Foundation*[149] held that an isolated DNA sequence encoding human t-PA could not be patented in 1989 but less than a decade later, in *Biogen Inc. v. Medeva plc.* it was found that an isolated DNA sequence coding for the hepatitis B virus could.[150] While there are continued doubts regarding the ability of biopatent applications to satisfy strict doctrinal standards, there is no containing this maelstrom as it pulls in other critical sectors for public policy, such as biofuels, where there has been an increase of 150 per cent in patent applications over the past two years.[151]

---

148 Ibid., footnotes omitted.

149 1989 R.P.C. 147 (Eng. C.A. 1988).

150 36 I.P.R. 438 (H.L. 1996). See generally Justine Pila, 'Bound Futures: Patent Law and Modern Biotechnology' (2003) 9 *B. U. J. Sci. and Tech. L.* 326, fn 7.

151 Ronald Kamis and Mandar Joshi, 'Biofuel Patents are Booming', January 2008, <http://media.cleantech.com/2329/biofuel-patents-are-booming>, report that a

Since knowledge is both consumable as a good and a necessary input in the creation of more knowledge, patent portfolios serve to block new competition in the particular R&D field of the patent holder; they enable coercive rent extraction by creating the need to secure technology licensing; and are a means of leveraging more favourable terms in cross-licensing negotiations (defensive patenting).[152] Today, there are 'patent development firms, intellectual property licensing enterprises, even trading exchanges for patent licenses – 'there's plenty of money to be made when the 'product' itself is intellectual property'.[153] But, expanding patents rights to life without the public participatory mechanism attendant on positive legislative action merits particular scrutiny not only because of the collaborative and cumulative character of the scientific paradigm, but also because neither the subject matter nor the consequences of biopatenting are commensurate with private monopolies traditionally held in 'art, process, machine, manufacture or composition of matter'.[154] Unless we are speaking of synthetic life, life is neither built nor invented. Moreover, life offers valuable genetic *information* for future life. Geneticist and science writer Kevin Davies asserts:

> [w]ithin a decade or two, we may be carrying this [genetic] information on our own personal DNA DVD, replete with information on our genetic susceptibility to disease and our tolerance to drugs. Clinics will be able to select genetic traits in human embryos by screening DNA before implantation and employ novel gene-based therapies to replace or repair faulty genes to cure inborn illnesses and cancer. And by the end of the game, we may know even enough about the secrets of our own genome to associate genes with elements of human character.[155]

Genetic information offers a vast scope for application. This feature means upstream patenting can create a dangerous *chill* exercise for the production of new knowledge according to Graeme Dinwoodie and Rochelle Cooper Dreyfuss:

'patent boom' is occurring in biofuels. In the past six years almost 3,000 biofuel-related patents have been published in the United States. Of those published in the 2006–2007 period, 57 per cent were owned by corporate entities and 11 per cent by universities or other academic institutions, discussed in Herder and Gold, *supra* note 34.

152 See generally Stuart Macdonald, 'When Means Become Ends: Considering the Impact of Patent Strategy on Innovation' (2004) *Information Economics and Policy* 16:1, <http://else.hebis.de/cgi-bin/sciserv.pl?collection=journals&journal=01676245&issue=v16i0001>; and Rivette and Kline, *supra* note 108 at 4.

153 Ibid. at 10.

154 Section 2 of the CPA, *supra* note 38.

155 Davies, *supra* note 48 at 9.

> Patents may now confer power not only in *product* markets, but also in *innovation* markets … Because second comers can often invent around end-use inventions, patents rarely monopolized product markets. In contrast, a patent on, say, the structural information of a protein, or a metabolic pathway, or a computer operating system, could give the patentee control over all work involving that protein or pathway, or all opportunities to create applications of that system.[156]

Informally, 'working solutions' that attempt to bypass the rights of the patentee or somehow work with them have been used in order to overcome access issues.[157] More formal attempts to address such concerns relate to greater administrative stringency with the application of tighter patentability requirements. In 2001, for example, the USPTO tried to regain control over the number and quality of genomic patent applications by issuing new Utility Guidelines that imposed a stricter threshold for applicants. Consequently, applicants now have to disclose a 'specific, *substantial*, and credible utility'. However, while the new guidelines allegedly raised the bar, they also implicitly confirmed that applications for patents on gene sequences meeting the higher utility requirements *will* be granted and not defeated *a priori* on the basis of novelty or obviousness. Whether isolated genes should be considered inventions or mere discoveries is not contemplated by the guidelines but has important consequence for the intersection of public policy and biopatenting. Utility standards in Canada also favour patentees in allowing some speculative utility with the doctrine of *sound prediction*:

> Utility is an essential part of the definition of an 'invention' (*Patent Act*, s. 2). A policy of patent first and litigate later unfairly puts the onus of proof on the attackers to prove *invalidity*, without the patent owner's ever being put in a position to establish validity. Unless the inventor is in a position to establish utility as of the time the patent is applied for, on the basis of either *demonstration or sound prediction*, the Commissioner 'by law' is required to refuse the patent (*Patent Act*, s. 40).[158]

---

156 Graeme B. Dinwoodie and Rochelle Cooper Dreyfuss, 'International Intellectual Property Law and the Public Domain of Science' (2004) 7(2) *JIEL* 431 at 433, footnotes omitted. The authors consider three examples of domestic legal reform, whether they would remain compliant with TRIPS, and the impact on protecting the public domain of science pursuant to TRIPS: '(1) excluding certain discoveries from the subject matter of eligible patent protection; (2) creating a statutory exemption that gave courts discretion to permit unauthorized uses of sufficient significance; and (3) varying the right to relief' at 432.

157 See e.g. J.P. Walsh, A. Arora and W.M. Cohen, 'Sceince and the Law: Working Through the Patent Problem' (2003) 299 *Science* 1021

158 *Apotex Inc. v. Wellcome Foundation Ltd.* [2002] 4 S.C.R. 153, 2002 SCC 77, para. 46. Paras. 41–44 provide for administrative deference that is to be owed to the

Binnie J. for the Supreme Court expounds:

> The doctrine of sound prediction has three components. Firstly, as here, there must be a factual basis for the prediction … Secondly, the inventor must have at the date of the patent application an articulable and 'sound' line of reasoning from which the desired result can be inferred from the factual basis … Thirdly, there must be proper disclosure. Normally, it is sufficient if the specification provides a full, clear and exact description of the nature of the invention and the manner in which it can be practiced … It is generally not necessary for an inventor to provide a theory of *why* the invention works. Practical readers merely want to know that it does work and how to work it. In this sort of case, however, the sound prediction is to some extent the *quid pro quo* the applicant offers in exchange for the patent monopoly.[159]

TRIPS is essentially silent on the content of patentability requirements although requiring certain minimum criteria of WTO members. Restricting or expanding the legal definitions of preconditions for patentability or raising the bar for the degree of skill of the PHOSITA in biotech applications are TRIPS-compliant means of managing desired outcomes for patent policy. Tightening up the requirements to raise the standards for patentability to ensure that only quality patents be issued is a positive means of directing domestic patent policy towards technological goals and is an easy and effective means of quelling the excessive proliferation of biotechnology patent applications based on basic research that may, ironically, stifle applied research.[160]

The Nuffield Council on Bioethics Discussion Paper, 'The Ethics of Patenting Life' (1997) made a similar recommendation critical of current patent systems.[161] The report found that changes in technology now made DNA sequencing routine and suggested that DNA claims are no longer 'inventive', if they ever were, and therefore should be denied a patent. The Council suggested a re-examination of the requirements for patentability, concluding that gene patents ought to be the exception to a general rule against their issue. And, DNA sequences as research tools (i.e. ESTs, SNPs) should be completely excluded based on proper application of patentability requirements. Some patents, such as for general methods of genetic diagnostics, they argue, should

---

Patent Commissioner.

159 Ibid. at para. 70.

160 Jerome Reichman, 'From Free Riders to Fair Followers: Global Competition Under the TRIPS Agreement' (1997) 29, Nos 1–2 *J. Int'l L. and Pol.* 11 at 37 [Reichman, 'Fair Followers'].

161 Nuffield Council on Bioethics Report, *The Ethics of Patenting DNA: A Discussion Paper*, 20 July 2002, Nuffield Bioethics Council, <http://www.nuffieldbioethics.org> [Nuffield Bioethics Report].

also not be issued at all because what is being claimed is an obvious association between a gene and a disease (some allowance is made for particular means of testing). The report also contends that the use of genes in gene therapy is considered an obvious use once the relevant gene is identified. As statutory schemes, most patent regimes provide not only exceptions to patentability but also exemptions from infringement.

In Canada, section 55.2(6) of the Patent Act deals with exemptions to infringement and states that the

> provisions do not affect any exception to the exclusive property or privilege granted by a patent that exists at law in respect of acts done privately and on a non-commercial scale or for a non-commercial purpose or in respect of any use, manufacture, construction or sale of the patented invention solely for the purpose of experiments that relate to the subject-matter of the patent.

Effectively, the section preserves common law rights that exempt private, non-commercial acts as well as experiments attempting to improve the patented invention.[162] In the field of genomics, this is a particularly useful provision given that the patent over a gene sequence is often useful as a research tool for further biomedical scientific research. In the absence of substantive legislated changes in patent law, however, the most obvious way for patent offices and the courts to contain patents for less than sincere innovations is by tightening up the main requirements for patentability to make the system less porous.[163] As Michael Meurer writes, low patent quality reflects current patent practice and can be interpreted as an indication of a possible host of additional problems such as '... overlapping claims, inappropriately broad claims, slow patent prosecution, and patents on obvious inventions'.[164] 'The problem of overlapping claims' he adds, 'is inherent to patent floods because of the likelihood of near simultaneous invention and multiple patent applications covering the same invention'.[165] A patent flood is defined as

> a dramatic jump in the number of patents filed covering a specific class of inventions ... A flood of related patents in a new market creates special problems for competition in addition to the usual problems that arise from

162 *Micro Chemicals Ltd.* v. *Smith Kline and French Inter-American Corp.* (1971), [1972] S.C.R. 506.

163 See e.g. J.H. Barton (2001), 'Non-Obviousness' (2003) 43 *IDEA: The Journal of Law and Technology* 475; J.H. Barton, 'Reforming the Patent System' (2000) 287 *Science* 1933.

164 Michael J. Meurer, 'Business Method Patents and Patent Floods' (2002) 8 *J. L. and Pol'y* 309 at 310, 323.

165 Ibid.

> market power associated with individual patents [by] … strain[ing] the resources of the US Patent and Trademark Office (PTO) and adversely affect[ing] the quality of issued patents.[166]

Patent floods exacerbate the existing difficulty – due to lack of resources, time and examiner expertise – faced by the PTO as it searches the prior art to assess patent applications. Patent floods compound institutional resource deficiencies by placing additional demands on an already strained system and thereby lead to continuously lower quality patents which in turn contributes to patent thickets, uncertainty about the scope and validity of patents, and increased frequency in litigation. According to Meurer, '[t]he fragility of the many start-ups in new markets makes them vulnerable to strategic patent litigation. The threat of patent litigation may deter entry or induce exit from the market'.[167]

Patent stacking is particularly a problem in genetics since a single genomic sequence may be patented in several different ways i.e. as cDNA, EST, a gene, a SNP etc., discouraging further R&D in product development because of the royalty costs owed to all the various patent owners of that sequence and the sheer number of licences to be sought from these patent owners.[168] These costs are invariably passed on to consumers individually and as tax payers for publicly funded education and health care systems. With improved sequencing technology, genomic patents will over-compensate the patentee by rewarding routine discoveries considered to be 'the easiest step in the process' rather than the true innovations arising from downstream application. Current patent practice is far worse than the anti-commons predicted by Heller and Eisenberg. The patents that issue today are of such poor or dubious quality that we question their validity but actually refrain from doing so legally because of the cost of litigation. Bad patents constitute a fraud on the public and usurp the public trust inherent in the 'bargain theory' of crown. As odious monopolies, they challenge our faith in the patent system *in toto*.

A System of Secondary Incentives: The IP/Competition Interface

A patent does not grant the inventor the right to make, use, or sell the invention – only to *exclude* others from doing so. The focus on exclusion rather than commercialization is a character of the privilege conferred which institutionally contributes to anti-competitive rather than competitive behaviour. Patents, even the bad ones, could be used to extract licensing fees from users and smaller competitors who would want to avoid the cost of

166 Ibid. at 309–10.

167 Ibid.

168 Ibid.

litigation, if not liability.[169] The IP 'paradox' or the 'trade off' is between the social value of providing incentives to innovate by restraining the free rider and the public costs of the monopoly right, which may in fact excessively limit 'fair followers'.[170] Even where the patents are *good,* the regime's current level of incentive for inducing sufficient biomedical research is questionable at best, contends Richard Gold:

> *The argument for greater patent protection should be understood for what it is: an attempt to maximize profit, not to maximize levels of innovation.* Clearly, a company would prefer to have as large a monopoly as possible ... But patent law is not about individual profit maximization as it is about maximizing the overall level of innovation in society. *The two do not necessarily go together.*[171]

Modern patent systems create unintended incentives for patentees to claim works of lesser and lesser innovativeness as the basis for garnering maximum protection under statutes.

When the state, through its agents, uses the patent system to protect investment rather than true non-obvious innovation, it reneges on the implicit promise it made to its citizens to improve welfare in exchange for public acceptance of the new technology. As institutional deficits almost ensure that patent quality remains unchecked, investors have come to realize that the utility of a patent is not exclusively as a *shield* to commercial competition, protecting related R&D costs from would-be free riders or appropriators of the invention. Patent regimes typically offer a legislated presumption of validity attached to the patent. Section 43(2) of the CPA provides: '[a]fter a patent is issued, it shall, in the absence of any evidence to the contrary, be valid and avail the patentee and the legal representatives of the patentee for the term mentioned ...'. Presumptively valid patents therefore, coupled with the exorbitantly high cost of litigation, offer strategic potential for patents as a *sword* for unfair competition based on the exclusive property rights conferred. These rights may then be used to extract regular royalty income under the threat of litigation with attendant costs.

Predatory practices seem legal under patent law but may unduly hamper competition, as the example of Jerome Lemelson demonstrates. Lemelson was an American inventor and philanthropist known for his portfolio of 600

169 Rai, *supra* note 117.

170 Reichman, 'Fair Followers' *supra* note 179. See also 'Database Protection at the Crossroads: Recent Developments and Their Impact on Science and Technology' (1999) 14.2 *Berkeley Tech. L. J.* Spring 793–838; and Suzanne Scotchmer, 'Standing on the Shoulders of Giants: Cumulative Research and Patent Law' 5 *Journal of Economic Perspectives* No. 1 Winter (1991) 29–41.

171 Gold, 'Biomedical Patents', *supra* note 33 at 423, emphasis added.

patents. He is said to have made a fortune from 'submarine patents' based on the filing of a vague claim which would remain secret (underwater) until the technology became available. By some accounts, the claims in the pending patent application would be amended to add new claims reflecting emerging technology, and once the patent issued (surfaced), he would assert his rights against others in the industry, seeking royalties and extolling settlements as a way for would-be defendants to avoid the significant cost of litigation.[172] A new phenomena is also emerging from modern patent practice. The growing anti-competitive behaviour by *patent trolls* who purchase weak patents in the open market for the exclusive purpose of litigation is intensified and has led to the stockpiling of low quality and easy to get patents on which a false edifice of proprietary claims to exclusivity is built. *Patent trolls* is a pejorative American reference reserved for predatory practices at their extreme as trolls, often firms, direct all their efforts to purchasing patents, sometimes even from bankrupt businesses, for the sole purpose of 'licencing out these patents and enforcing them against those who refuse to take a licence'.[173] In addition to the fact that trolls are great saboteurs due to the difficulty in detecting them, they create a special problem according to Morag Macdonald:

> The novel difficulty with these patent trolls is that, because they do not have a manufacturing or research base of their own, the normal strategy for countering demands to take licences does not work. They will have no interest in entering into any cross-licences with those they wish to licence. They will not be concerned by a threat to any of their product lines due to the patent portfolio of their potential licensees.[174]

The result is patent practice that is both predatory and pre-emptively anticompetitive, effectively barring market participation by smaller resourced firms. Hope Shand writes:

> The power of exclusive monopoly patents is giving these companies the legal right to determine who gets access to proprietary science and at what price. Participation in industry isn't possible unless a company holds patents or has the money to license them … Pioneer Hi-Bred, the world's largest seed company (now a wholly owned subsidiary of DuPont), claims that one of its new, genetically engineered, insect-resistant corn hybrids requires access

172 See Teresa Riordem, 'Patents argued in the Court Creating and Invention Can Be Murky Business', 2 February 2004, <http://query.nytimes.com>.

173 Morag Macdonald, 'Beware of the Troll', *The Lawyer.com*, 26 September 2005.

174 Ibid.

> to thirty-eight different patents controlled by sixteen separate patent holders … [S]maller enterprises will find it increasingly difficult to compete.[175]

For smaller enterprises, it may be simply cheaper to pay the licence fee than be taken to court over infringement. Even if convinced that a patent is invalid, it costs too much money to be proven right.[176] And, not all firms patenting genes attract the kind of negative media publicity around their patent practices that MGL did with the BRCA1 and BRCA2 patents, attributed by some to their aggressive licensing. In fact, MGL has indicated that whenever a potential competitor wants to do work that MGL considers promising but might impinge on their blanket patents in the United States, they are invited instead to join the MGL team and accept MGL funding of that research.[177]

Acquiescence by smaller firms to the demand to either pay rent or merge only, encourages IP holders to further stockpile patents and bully their competitors. In addition, larger companies expend gross funds defending the validity of patent claims in order to discourage resistance to their licensing demands or similar future attacks by other opponents. While taxpayers subsidize such litigation, private firms may seek other ancillary forms of IP protection, like trademarks or copyright, in attempt to extend property rights in perpetuity ('evergreen'):

> The most recent attempt to use copyright to garner protection has been to encode DNA sequences as MP3s or other such music files and then copyright these 'genetic songs' as creative works of authorship. This method of attracting DNA copyright protection through the encoding of the genetic information in technological hardware is itself the subject of a patent application, which if granted, would deny anyone other than the original inventor of such a method, from storing or retrieving genetic data in this way.[178]

---

175 See Hope Shand, 'Gene Giants: Understanding the "Life Industry"', in Brian Tokar (ed.), *Redesigning Life? The Worldwide Challenge to Genetic Engineering* (Montreal: McGill-Queen's University Press, 2001) [Shand, 'Gene Giants'] at 226.

176 Strategic uses of patents include rent extraction through licensing of technology or franchising that enable income earnings directly from the IPRs. IBM, for example, generates $1 billion in patent royalties per year. See Christopher M. Kalanje, WIPO-WASME Special Program on Practical Intellectual Property Issues, Geneva, 6–9 October 2003.

177 Fifth Meeting of the Committee on Intellectual Property Rights in Genomic and Protein-related Inventions, The National Academies of the Science, Technology and Law Program, 11 February 2005, transcript, <http://www7.nationalacademies.org/step/Genomics_Committee_Meeting_6_transcript.pdf>.

178 Amani and Coombe, *supra* note 59, endnote 2. See also Viva R. Moffat, 'Mutant Copyrights and Backdoor Patents: The Problem of Overlapping Intellectual Property Protection' (2004) 19 *Berkeley Tech. L. J.* 1473.

What these practices highlight is the growing utility of patents in industry to quash competition and coerce smaller rivals to capitulate to demands for rent under threat of infringement. Patents are also used as barter chips in negotiations to secure cross-firm or reach-through licences; particularly in the genomics and biotech fields where upstream patenting has made licence bundling imperative for scientific research and the commercialization of new technology. The alternative would be to risk not only the cost of litigation but the threat of punitive damages (treble damages in the United States for patent infringement if the defending party becomes aware of a specific patent but fails to take affirmative action to avoid risking infringement). While transaction costs may be increased, these tend to be lower than the costs of litigation and the risks of liability. Such insights provide a different explanation for empirical findings of the relevant infrequency of biopatent litigation[179] despite the 'surge' in patenting and also help to explain another less than desirable trend known as 'defensive patenting'. Rosemarie Ziendonis explains:

> defensive patenting offers firms an alternative to getting 'fenced in' by owners of the technologies they use'.[180] This is because licencing agreements are 'facilitated by having a large patent portfolio of your own, so several firms, large and small, were engaged in defensive drives to increase their patenting rate. This had little to do with encouraging innovation, and in fact looked like a tax on innovative activity.[181]

The current legal climate raises the costs of patent policy and reinforces, by reward, bad behaviour from all industry participants:

> [f]rom a policy perspective, 'mutually assured destruction' games among firms that develop and produce complex technologies exacerbate the problem: to avoid being held-up by others, firms file more patents and, in turn, contribute to the proliferation of patents that others face.[182]

179 For an empirical review see Christopher M. Holman, 'The Impact of Human Gene Patents on Innovation and Access: A Survey of Gene Patent Litigation', <http://papers.ssrn.com/sol3/cf_dev/AbsByAuth.cfm?per_id=537838>.

180 Rosemarie Ziedonis, 'Don't Fence Me In: Fragmented Markets for Technology and the Patent Acquisition Strategies of Firms' (October 2003), <http://www.researchoninnovation.org/tiip/archive/2004_3c.html>.

181 See Bronywn H. Hall, 'Business Method Patents, Innovation, and Policy'. NBER Working Paper Series, Working Paper 9717, National Bureau of Economic Research, <http://www.nber.org/papers/w9717> at 10, referring to the findings of B.H. Hall and R.H. Ziedonis.

182 Ziedonis, *supra* note 180.

The result is a *secondary system* of *institutional incentives* (secondary to those espoused in normative, industrial and legal claims that patents are necessary incentives for R&D and thus a priority for regulatory governance). Patent trolls, floods, proliferation of thickets, patent stacking, defensive patenting and other current patent practices add to the perfidy of patenting life and all of its sub-components; these market practices undermine the veracity of biotechnology's narrative and threaten to compromise its very promises with overprotection and enforcement that invariably restricts access and use. Not only are potential public benefits reduced, but the public must also absorb additional costs in administering such an inefficient system. The glut of patents and their stratified anti-competitive uses are focused on accumulation of exclusion-based *commodities* and the ability to weed out competitors with monopolistic instruments that chill and drag innovation rather than invite it. They exemplify a mounting predatory ethos spurred on by the system itself and offer a rationale for the proliferation of patents, one that is based on a *property fundamentalism*.

Gary Rebak, a prominent American IP lawyer attributes much of growing anti-competitive behaviour to institutional design failure.[183] Other lead counsel, such as former IP counsel Charles Cella, have found an entrepreneurial market in correcting such failure. Cella started BountyQuest as a business dedicated to profiting from busting patents of questionable legitimacy. A patent, once issued, can be overturned only when validity is attacked by a challenger. The idea behind BountyQuest is that anyone who questions a patent's validity can offer a bounty (starting at $10,000) in order to attract relevant prior art that would help prove that the patent lacked novelty and thus failed to meet a critical requirement for patentability. This is a private-market solution for patent reform. When Cella was asked why he switched sides, he replied: 'Because there are a lot of bad patents out there, and there are a lot of valid patents too … What needs fixing is the system of "prior art".'[184] Still, opposition and review is expected to be more difficult to undertake and poor quality patents harder to detect the more technical the field. In such instances, PUBPAT (the Public Patent Foundation), a non-profit legal organization, steps in to help protect the public's interest from harms arising from 'undeserved patents and unsound patent policy' which, amongst other harms, 'causes prices for

183 Gary L. Rebak, 'Patently Absurd: Corporations are increasingly converting the Shield of Patent Protection into the Sword of Unfair Competition', *FORBES ASAP* 169:14 (24 June 2002) 44: '[t]here are those who view the patent system as the seedbed of capitalism – the place where ideas and new technologies are nurtured. This is a romantic myth. In reality, patents are enormously powerful competitive weapons that are proliferating dangerously, and the U.S. Patent and Trademark Office (USPTO) has all the trappings of a revenue-driven, institutionalized arms merchant.'

184 See Damien Cave 'Who ya Gonna Call? Patent Busters', <http://dir.salon.com/tech/view/2000/10/23/cella/index.html>.

goods to be artificially high, the advancement of science to be thwarted, and civil liberties to be inappropriately restrained. Most people still do not realize how significantly undeserved patents and unsound patent policy are assailing their health, their freedoms, and their wallets'. [185] PUBPAT is governed by an expert Board of Directors and is active in locating invalidating prior art, filing requests for reexamination with the PTO, criticizing patent policy, educating the public on the complex issues and their consequences, representing the public interest in court proceedings and promoting legislative patent reform. The organization helps fill a tremendous vacuum of represented stakeholder interest within patent law and policy and was formed in 2003 in the United States by another patent attorney, Dan Ravicher, at about the same time that the US Federal Trade Commission (FTC) released its report on the need for balancing patents with competition policy.

The FTC report resulted from a series of hearings conducted by the US FTC and the Department of Justice (DOJ) to decide whether there were too many patents that should never have issued. The objective of these meetings was to better align IP policies for the knowledge based economy given that this economy *relies* on free market competition as much as innovative enterprise.[186] The FTC report identified a number of aspects of US patent law and practice that 'exacerbate what already seems to be a zero-sum game'[187] and recommended reform to a) reduce the costs of challenging 'questionable' patents through the adoption of a post-grant opposition system, b) improve the rigour of the examination process while relaxing the evidentiary standards required to invalidate a patent, and c) raise the overall quality of patents by raising the bar of non-obviousness. An enhanced post-examination review system (re-examination) and the removal of the statutory presumption of validity are two other solutions that have been offered as a check against low quality or bad patents.[188] While functional, these would need to be coupled with efforts to address the embedded or institutionally fostered secondary incentives for anti-competitive behaviour that patent regimes can create.

The FTC recommends significant changes in the examination, prosecution and enforcement of patents. In response to this report and faced with mounting criticism of existing inefficiencies and systemic flaws, the USPTO proposed a number of institutional reforms, including increasing human resources, professional training, and other measures geared at producing quality patent

185 PUBPAT, < http://www.pubpat.org/About.htm>.

186 US Federal Trade Commission (2003), To Promote Innovation: The Proper Balance of Competition and Patent Law and Policy, <http://www.ftc.gov/reports/index.htm>. See also William Blumenthal, 'Aligning Competition Policy and Patent Policy: A Perspective From the Federal Trade Commission Staff', 30 June 2006, <http://www.ftc.gov/speeches/20060630FCBABackgroundPaper.pdf>.

187 Ziedonis, *supra* note 180.

188 Barton, 'Reforming the Patent System', *supra* note 163.

products. The resulting recognition that the cost of litigation is high and not an optimal way of testing validity or protecting the public interest is long overdue but recent changes at the patent office may also be an effort to insulate against liability. In 2005, the USPTO hired 978 new examiners, a hundred more than their target goal and intends on hiring 1,200 new patent examiners a year over the next several years to help it overcome current deficiencies that have embroiled it in two lawsuits and have resulted in a backlog of 800,000 unexamined patent applications on their shelves as of November 2007, with another 400,000 applications received each year.[189] It takes on average 32 months from the time of application to the grant or denial. For every two patent examiners hired, there is one leaving the USPTO. As a result, the USPTO is also pioneering an alternative business strategy with its telework policy programme directed towards overcoming difficulties recruiting and retaining skilled examiners in the Washington, DC area. The programme enables examiners to work in US cities and commute back to Washington for meeting and training, with the projection that by 2011 in excess of 3000 patent examiners will be teleworking. In fact, the USPTO reports that by the end of the 2007 fiscal year, it already had 3,609 of its employees engaged in some form of telework (59 per cent increase over the year before) which amounts to almost half (40.7 per cent) of its total workforce.[190] The USPTO has also proposed a number of training initiatives, including a one year academy approach, patent managers, and greater networks (through their Patent Hoteling Program) to address the growing need for highly skilled human resources and to restore some faith in the quality of patents that issue.[191] A second report has also been issued by the US Federal Department of Trade and Commerce and the Department of Justice examining patent licensing practices at their interface with competition policy.[192]

Post-grant opposition proceedings and improved human resources are important but should not substitute for a more thorough and thoughtful approach to policy development and legislative reform involving public debate. The Canadian patent office (CPO) has not pursued such full-scale measures as the USPTO but one can infer that it and many patent offices around the world generally experience the same or worse types of institutional deficits albeit on a different scale and volume. However, given the global patent proliferation trends reported and the ability to use one patent to secure priority rights in

189 Eileen McDermott, 'Examining the USPTO', November 2007, <http://www.managingip.com/Article/1629870/Examining-the-USPTO.html>.

190 'USPTO Embraces Virtual Workforce', 7 May 2008, <http://www.impactlab.com/2008/05/07/uspto-embraces-virtual-workforce/>.

191 See USPTO Performance and Accountability Report, Fiscal Year 2005, <http://www.uspto.gov/web/offices/com/annual/2005/2005annualreport.pdf>.

192 The FTC and DOJ, 'Antitrust Enforcement and Intellectual Property Rights: Promoting Innovation and Competion', <http://www.ftc.gov/reports/index.shtm>.

other jurisdictions, this neglect is not benign or inconsequential. Likely many patent offices worldwide are equally in need of reform given the stronger legal protection for IPRs; even a well run patent system may suffer from institutional deficits invariably based on limits to its financial resources.[193] In the interim, patent offices in developed countries for the most part do issue gene patents based on the ability to isolate and purify genes and in so doing they reverse, according to Linda Demaine and Aaron Fellmeth, 'the traditional principle in patent law of non-patentability of purified natural products and, in some cases, the principle of nonpatentability of natural products in their natural state'.[194] Gene patents and related therapies continue to be sought with increasing pressure for parity in the law of DCs.

Canadian patent legislation does have a remedial provision where there has been patent abuse. Three years after the grant of any patent, the Attorney General of Canada or any person interested can apply under s. 65 CPA to the Commissioner and request relief including a compulsory licence or patent revocation where any of the three statutory conditions deeming abuse have occurred. Historically, domestic (local) manufacturing and 'working' of patented inventions was required either by the patentee or its licence holder and the failure to do so constituted an abuse. However, because of Canadian obligations under the *North American Free Trade Agreement* (NAFTA), the local manufacturing requirement was repealed; today, local demand of a patented invention may be met entirely through imports or foreign providers.[195] The remaining provision governing abuse continues to be useful in Canada.

Section 65(2)(c)–(f) of the CPA may be used to prevent anti-competitive behaviour by patent holders and could be amended to deal with situations where patent holders prohibit domestic labs from providing genetic tests or to address anti-competitive practices in licensing, defensive patenting, and the array of other current practices using patents as a *sword* for unfair competition. TRIPS expressly allows for domestic regulation to curtail anti-competitive behaviour and abuse of patents. TRIPS Article 40.1 acknowledges that 'some licensing practices or conditions pertaining to intellectual property rights which restrain competition may have adverse effects on trade and may impede dissemination of technology', while Article 40.2 provides:

> Nothing in this Agreement shall prevent Members from specifying in their legislation licensing practices or conditions that may in particular cases constitute an abuse of intellectual property rights having an adverse effect on competition in the relevant market … [A] Member may adopt …

193 Lemley, 'Rational Ignorance', *supra* note 132.

194 Linda J. Demaine and Aaron Xavier Fellmeth, 'Reinventing the Double Helix: Novel and Nonobvious Reconceptualization of the Biotechnology Patent'. (2002) 55 *Stanford Law Review* 304 at 331.

195 Vaver, *IPL*, *supra* note 39 at 169–70.

> appropriate measures to prevent or control such practices ... in the light of the relevant laws and regulations of that Member.[196]

Such provisions allow important TRIPS-compliant safeguards to be built into domestic patent legislation but governments should also use competition law and other instruments of regulatory governance in crafting optimal biotech regulation for the creation and commercialization of new technologies. Canadian competition law provides against abuse of dominance position generally (s. 79) and specifically against abuse of IPRs (s. 32).[197]

Warren Grover, however, author of a report on this subject prepared for the Project Steering Committee on Intellectual Property and the Patenting of Higher Life of the Canadian Biotech Advisory Committee (CBAC), outlines the consistent ineffectiveness of Canadian competition authorities to appropriately address clear abuses of patent rights, and attributes this in part to a lack of understanding of the fundamentals of IP law by competition lawyers. The Intellectual Property Enforcement Guidelines, published in late 2000, clearly indicate that the competition authorities will not be vigorous investigators of alleged patent abuses. Instead, Grover recommends the Commissioner of Patents should be granted greater discretionary authority under the legislation to deal with patent misuse.[198] The remedial possibilities of competition law as a means to mediate industrial policy is being further considered by the FTC and DOJ following their two reports in the USA even while property's play continues with a full cast stockpiling, pooling and navigating against surfacing submarines and nasty trolls: '"Polyani's drama of innovation" goes on.'[199]

Alternative Models for Regulating Innovation

> [W]hile economists have long been able to show that *some* incentive to encourage innovation is needed, it remains contentious whether IP rights versus other rewards such as prizes and government grants offer the best alternative. Innovation will, moreover, not cease to occur without IP rights,

196 See also Articles 40.3–40.4.

197 Competition Act, Revised Statutes, 1985, c. C-34.

198 Warren Grover, 'The Interface of Biotechnology Patents and Competition Law' (2001), prepared for the CBAC Project Steering Committee on Intellectual Property and the Patenting of Higher Life, online under publications: Strategis Canada <http://strategis.ic.gc.ca>.

199 Ziedonis, *supra* note 180, referring to the work of Michael Polyani finding that 'invention, especially modern invention is a drama ... enacted on a crowded stage'.

> and not simply because of non-economic motives to invent or government subsidies.[200]

Patent regimes are designed to create incentives for innovation and their early dissemination. But, as Herder and Gold have found, the 'lack of patent protection of embryonic stem cells in Europe has not ... seemingly affected levels of research in Europe while the existence of patents in the United States has not increased innovation'.[201] Moreover, Wendy Gordon appropriately asks:

> patent[s] take the form of ordinary property ... [D]oes this similarity of form mask inconsistency of function? Justifications for tangible property typically refer to the internalization of both positive and negative effects, but justification for intellectual property tends to be more one-sided.[202]

If the social policy underlying this institution is to provide adequate incentives for knowledge production by ensuring fair compensation, then as international law and trade scholars Michael Trebilcock and Robert Howse have correctly acknowledged,

> [p]roprietary entitlements over one's creative product are, however, not the only form in which such compensation may be provided. For instance, in many countries, a large percentage of inventors or creators work in government laboratories and universities, and even literary and artistic activity is subsidized by the state. Invention and creation may be regarded as a salaried occupation like any other.[203]

Reform solutions may lie *outside* of the IPRs regime entirely and require us to ask: to what extent should instruments other than patent policy be used as forms of incentives to promote innovation? These might include subsidies, R&D tax incentives (credits) while minimizing the social and economic costs of the policies adopted for the use of inexhaustible goods.[204] The World Health

---

200 Herder and Gold, *supra* note 34 at 8, footnotes omitted.

201 Ibid. at 34.

202 Wendy Gordon, 'Of Harms and Benefits: Torts, Restitution, and Intellectual Property' (June 1992) XXI *J. Legal Stud.* 449.

203 Michael J. Trebilcock and Robert Howse, *International Trade Regulation*, 3rd edn (London and New York: Routledge, 2005) at 398. Strong university systems are integral to property-less innovative activity. A country may experience a scientific boom in the *absence* of so-called *necessary* incentives that patents provide, see e.g. cover story, Richard Stone, 'An Islamic Science Revolution?' (2005) 309 *Science* 1802–1804, <http://www.sciencemag.org>. See generally Drahos and Braithwaite, *supra* note 2.

204 For a discussion on alternatives to IP as incentives and their appropriate contexts, see Nancy Gallini and Suzanne Scotchmer, 'Intellectual Property: When

Organization (WHO) and World Bank (WB), for example, have recommended prizes to encourage the development of vaccines for developing countries where every year more than 5 million people die from malaria, tuberculosis and strains of HIV prevalent in Africa.[205] The initiative is to provide incentives for the development and distribution of such medicines beyond that offered under a proprietary system of patents, through prizes and procurement.[206]

Alternative policies may work equally well in tandem or independent of the patent system, providing incentives for basic research while reserving the patent system for applied research and development. Reforming incentives for genomics research may require governments to think 'outside of the box' by considering alternative legal models and entirely *different* legal mechanisms. In addition, it may be appropriate to ask whether the idea of property rights underlying popular conceptions of patent rights is entirely the wrong model conceptually for ensuring against harms that could better be addressed by the liability rules underlying compensatory systems.

Jerome Reichman,[207] Wendy Gordon[208] and other legal academics have considered alternatives to the proprietary or quasi-proprietary protections that IPRs afford by turning to other Western legal models such as liability where the owner, be it an individual or collective, has a right to receive a revenue stream from those who want access, but has no right to refuse it by exercising a right to exclude. A restitutionary model based on unjust enrichment or other legal protections against unfair competition provide additional options to consider. These conceptual frameworks, although in need of further exploration than this context will allow, are helpful in overcoming the threat

---

Is It the Best Incentive System?' in Adam B. Jaffe, Josh Lerner and Scott Stern, eds, *Innovation Policy and the Economy*, vol. 2 (National Bureau of Economic Research: MIT Press, 2002), c. 2, cited to PDF file <http://www.dklevine.com/archive/scotchmer-when-is-ip-best.pdf>; for optimal fixed duration, see W. Norhaus, *Invention, Growth and Welfare: A Theoretical Treatment of Technological Change* (Cambridge, MA: MIT Press, 1969).

205 Ed Crooks, 'Prizes for Saving Lives: Drug Companies will not develop new vaccines unless they are awarded for it', *Financial Times* (25 April 2001).

206 Michael Kremer, 'Creating New Markets for New Vaccines' (13 April 2000) makes the recommendation of turning a patent into a prize in order to avoid deadweight loss. In his model, a patent authority would repossess the patent and auction it to the highest bidder. Gallini and Scotchmer caution that this model of providing a 'prize' equal to the social value of an invention paid out of general revenue may be inefficient if there is more than a single researcher because it might attract other firms to race in which the firms over invest. Gallini and Scotchmer, *supra* note 204.

207 See J.H. Reichman, 'Of Green Tulips and Legal Kudzu: Repackaging Rights in Subpatenable Innovation' (2000) 53 *Vand. L. Rev.* 1743.

208 Gordon, *supra* note 202 at 449–50. See also W.J. Gordon, 'On Owning Information: Intellectual Property and the Restitutionary Impulse' (1992) 78 *Virginia L. Rev.* 149.

of the exclusion-thesis tied to the property paradigm of patents and their detrimental impact within the life sciences, particularly against patent holders acting unreasonably.[209] Other regulatory models and modalities should not be neglected as viable alternatives or complements by governments in reform efforts over patent law and policy.

Diverse models for innovation regulation expand the instrumental tools available for governments to create nuanced and culturally sensitive industrial policies. They may also facilitate the efficient coordination of cross-policy objectives to optimally advance economic growth without the current attendant costs of a strong proprietary and exclusion based system.[210] However, these alternative models, if imposed top down by some international order, may be subject to the same criticism of cultural imperialism that developing countries make against the importation and normative presumptive power of any developed world's legal regime (property, liability, or otherwise) to dissonant foreign contexts. Still, contemplating alternative models helps to distance the economic claim for regulatory oversight from the normative force of exclusion defining property as a concept integral to patent rights, and to patent rights in turn as a regime essential to ensuring innovation and creative production. Whether a better balance of interests may be achieved by focusing on compensation that is commensurate with the innovation for which it is rewarded needs to be fully considered. Because, '[o]nce we have established that the issue is actually the *level* of compensation to which a creator is entitled', write Trebilcock and Howse, 'then it is clear that at least implicitly the creator's claims are being balanced against other social interests'.[211] More immediate reforms for now lie *within* the existing patent as property paradigm and relate to practical considerations for improving the institutional system by which patents are issued, administered, enforced, and thus regulated.

Conclusion

The foregoing discussion helps us see biopatenting as a pandemic in need of an innovative cure. Advances in new technology offer a partial explanation of apparent 'innovative' activity when measured by patents as an indicator. Institutional deficits, anti-competitive behaviour, strategic business practices,

209 Guido Calabresi and A. Douglas Malamed, 'Property Rules, Liability Rules, and Inalienability: One view of the Cathedral' (1972) 85 *Harv. L. Rev.* 1089 stress the right of exclusion over that of alienation by emphasizing the essence of a property rule regime as the holder's right to name the price of access.

210 See also Jerome Reichman and Tracy Lewis, 'Using Liability Rules to Stimulate Local Innovation in Developing Countries: Application to Traditional Knowledge', <http://www0.gsb.columbia.edu/ipd/pub/Reichman_Lewis.pdf>.

211 Trebilcock and Howse, *supra* note 203 at 398.

a patent friendly judiciary (the subject of the next chapter) all offer additional, if not alternative, explanations for the proliferation of biotech patents. William Cornish warns that '[t]he number of patents being granted for inventions ... are threatening to become omnipresent in some [industries]'.[212] Institutional celebration of the exponential growth in patents is based on a false correlation of these instruments with true inventiveness and economic productivity and all but ignores the costs to competition and health policy. Meanwhile, the incentives created from this regulatory system have shifted over time away from imports and innovations to predatory anti-competitive behaviour, often used to *stop* imports, based on after-the-fact rewards. Patent offices must examine applications 'to test their assertions of inventiveness' before any grant is made; otherwise, all too often 'buccaneers will be armed with dubious rights with which to overpower the vessels of decent traders. Experience shows that they will use them to frighten others into taking expensive but unnecessary licences'.[213] Similarly, while post-grant opposition proceedings and improved human resources are important steps in improving the patent system, they do not substitute for coherent policy development ideally through a legislative process involving public debate *before* biopatenting becomes a crisis; more so if pursued globally. The human rights-promoting promises of biotechnology are often equivocated with patents; the perfidies of patents demonstrate too clearly how they will be betrayed if current practices continue unabated. In the absence of legislative reform, institutional reform of patent offices alone will not achieve the balance necessary for optimal patent protection. The courts have an important role in improving the system and should reinforce stricter standards for patentability, resisting any pressure for convergence.

212 See Cornish, *supra* note 129 at 5.

213 Ibid. at 6.

Chapter 3

Origins of the Patented Species: Of Mice, Men and the Sincerity of 'Invention'

> It is interesting to contemplate a tangled bank, clothed with many plants of many kinds, with birds singing on the bushes, with various insects flitting about, and with worms crawling through the damp earth, and to reflect that these elaborately constructed forms, so different from each other, and dependant upon each other in so complex a manner, have all been produced by laws acting upon us. These laws taken in the largest sense, being Growth with Reproduction; Inheritance ... Variability ... Ratio of Increase so high as to lead to a Struggle for Life ... Natural Selection ... Divergence of Character and the Extinction of less-improved-forms.
>
> Charles Darwin, *The Origin of Species*

> DNA, you know, is Midas' gold. Everybody who touches it goes mad.
>
> Maurice Wilkins

Introduction

The Canadian extension of patents to life has very much shadowed developments in American law. In addition to warning against too many IPRs, evidence of patent trends warns against emulating the substantive standards of other jurisdictions and harmonizing our laws with other more protective national regimes. Our social, economic and political context warrants differential regulation, but also because 'inventions' claimed in jurisdictions such as the United States of America have on more than one occasion proven less than sincere. There are growing inconsistencies between patent practice, doctrine and its rationales. Gene patents may confer inefficient overprotection of private interests providing a 'windfall' due to the initial and often significant prior investment of public sector resources. Genomic patenting is antithetical to the culture of collaboration and accretion model of science and surrenders control and access to private proprietors while emasculating the role of regulatory governments. These monopolies result from lax administrative interpretation, pin themselves to promises framed in human rights vocabularies, and represent the growing encroachment of the local by the global.[1] The promises of patenting must be weighed by law makers and policy analysts against their costs; if property attaches to life as new subject matter then the process of its objectification, its substance, conditions, and limitations need to be *institutionally* defined by our legislators and parliamentarians to maximize the utility of the monopolies for the promotion of welfare. Necessary review by policy analysts and lawmakers should occur regularly to ensure continued optimal operability. Instead, parliament has all but abdicated responsibility to safeguard the public's interest to a judiciary (unelected in Canada) that is crippled by the fact that the court's jurisdiction does not extend beyond the facts of a given case to address broader normative issues or public policy debates. The leap from the administrative decision to allow DNA patents to *higher life* was a judicial one void of the necessary participatory debate expected from liberal democracies on a vital bioethical issue.

Review of the Common Law on Life: From Sanctity to Scarcity

Patent offices in the western world have been granting patents on genes, cell lines, proteins, antibodies, hormones, micro-organisms and in some countries even plants and animals (although not in Canada) – this is now well established law.[2] The extension of patents to 'life' is based on the common law's recognition of biological and organic matter as patentable subject matter.[3] What seems to be most troubling from media accounts of public protest against biopatenting is the idea that that law has developed by stealth to prioritize property rights without public consultation. A 2002 news article read:

> Surprise! You may not hold full legal title to some of the genes in your body. The Canadian Intellectual Property Office, for some time, has been issuing patents on human genes, without any of the controversy the issue sparked in the United States and elsewhere … [T]here are hundreds of gene patents on the book, including a significant number covering human genes.[4]

---

1 See Coombe, 'Authorial Cartographies', op. cit. at 1364: '[t]hose borders which once defined national sovereignty and the spaces of legislative self-determination are now compromised in order to more fully protect and police the new sovereignty of intellectual property owners, who are once again clothed in the more appealing guise of the Romantic author struggling for unalienated survival'. Footnotes omitted.

2 See Nuffield Council on Bioethics Report, op. cit.

3 See generally Daniel Kevles, 'A History of Patenting Life in the United States with Comparative Attention to Europe and Canada', A Report for the European Group on Ethics in Science and New Technologies to the European Commission (2002), <http://ec.europa.eu/european_group_ethics/publications/docs/study_kevles_en.pdf>.

4 Dennis Bueckert, 'Patents Quietly Issued on Human Genes' *Toronto Star* (23, January 2002).

Negative, often intuitive, reactions relate to a sense that there has been a breach of public trust in granting gene patents using an antiquated system – designed to introduce new imports and new manufacture – for something so fundamental to our identities and existence. Opponents often state that 'patent holders are being allowed to patent a part of nature – a basic constituent of life; this allows one organism to own all or part of another organism'.[5] This claim is not technically accurate since the patent only relates to the intangible (the 'idea' of the invention) and not its tangible manifestation. The general public's difficulty in accepting gene patents is precisely because the search for an 'invention' seems futile for such products. Still, once patents on genes, comparatively mundane and obvious, were issued by the patent office without significant public or political debate, extension of IPRs to (higher) life was easier to make doctrinally since transgenic life at least contain some modicum, possibly a significant degree, of innovation. The valid ethical and social reasons why states may want to exclude life from patentability remain, however. Canadian courts struggled with these for more than a decade and the Seed Sector Review revitalized the issue.

The Seed Sector Review is an industry-led assessment of the seed sector by over 4500 pedigree Canadian Seed Growers and over 200 companies across Canada that belong to the Canadian Seed Trade Association; it was undertaken with the objective of facilitating consultation and consensus building in the Canadian seed and crop sectors and with the goal of creating a 'modern regulatory environment for Canada's seed sector'. The review extends to the nature of seed protection provided under patent law and plant variety protection and has alarmed NGOs and farmers' groups of potential erosion of farmer's rights to save seeds that may result from emerging recommendations.[6] The Review followed close on the heels of the MGL gene patent controversy and the recent litigation at, and ostensibly inconsistent findings of, the Supreme Court of Canada regarding the patentability of non-human higher life. Rather than provide clarity and certainty as Supreme Court decisions ought to do, the Canadian decisions leave the law in this jurisdiction and the extent of patent rights conferred disturbingly unsettled. The decisions simultaneously highlight the importance of coherent regulatory policies regarding biopatenting. But, the judiciary has *had* to act even while calling on governments to respond to fill the existing lacunae of law and policy to better address the challenges of biotechnology. The courts have generally been responsive to arguments expanding the scope of rights as well as the subject matter to which they attach for a host of reasons including

5 Ibid.

6 See <http://www.seedsectorreview.com/index-e.html>. See also National Farmers Union, 'Nine things farmers need to know about the Seed Sector Review' (13 May 2004), <http://www.nfu.ca/seedsector.pdf> and Stephen Leahy, 'Monsanto Victory Plants Seed of Privatization', Inter Press Service (5 October 2004).

incomplete understanding of the technical science, industrial pressure for greater protections, assumption of rights or utility based rhetoric, a desire for convergence in standards with more protective jurisdictions (such as in the USA), and with regard to international trade obligations. Yet, life has not always been patentable.

Advances in molecular genetics created the impetus to apply for life patents by allowing the necessary fragmentation of the body to its chemical composition and thereby enabled its manipulation, commodification, and commercialization. The geneticization of living matter created the ability to ignore the whole of a living organism as at *least* the sum of its parts. Addressing the question 'What is life?', the SCC in *Harvard* confirmed that

> there is a good deal of debate about what constitutes 'life' but some consensus about a few of its characteristics. These include the capacity to grow and develop (including reproduction) i.e., a metabolism, the ability of an organism to draw energy from its environment for this purpose, and the ability to respond to stimuli … Life is no less wondrous at the microscopic level, and to think of 'life' primarily in terms of dolphins, chimpanzees and blue whales (examples urged by the appellant Commissioner in the oral hearing) is something of an oversimplification.[7]

Philosophers and ethicists are also concerned with the inherent difference between life and traditional patentable technology given the former's unique capacity to self propagate, self-replicate, reproduce and mutate – all evolutionary indicia of 'life' – even at the molecular level. Biotechnological 'invention' is in fact biological *intervention* and is accordingly necessarily collaborative with the nature we dispose over:

> [i]n dealing with dead matter, the producer, confronted with a passive material, is the only one to act. In dealing with organisms, activity is confronted with activity: biotechnology is collaborative with the auto-activity of the active material, the biological system in its natural functioning into which a new determinant has to be incorporated … The mode of the technological act is intervention, not building.[8]

Biotechnological intervention is characterized by human conceit because the intervention, according to Hans Jonas, is irreversible in a self-regulated

---

7 *Harvard College v. Canada (Commissioner of Patents)* [2002] S.C.J. No. 77, 4 S.C.R. 45, File No. 28155 [*Harvard SCC*] para. 66.

8 Hans Jonas, 'Lasst uns einen Menschen klonieren', in Jonas, *Technik, Medizin und Eugenik*. Zur Praxis des Prinzips Verantwortung (Frankfurt am Main: Suhrkamp, 1985) at 165, cited in Jürgen Habermas, *The Future of Human Nature* (Cambridge: Polity Press, 2003) at 47.

process, and will lead to consequences we cannot control: 'To 'produce', here, means to commit something to the stream of evolution in which the producer himself is carried along'.[9] Jürgen Habermas discusses the philosophical writings of Jonas and concludes that the Jonas thesis 'resituates genetic engineering in the context of a self-destructive dialectics of enlightenment, according to which the species reverts itself from the domination of nature to servitude of nature'.[10] Molecular interventions commit future generations to their consequences and in binding the living to the dead, serve as a form of power asserted, according to Jonas, by the living over their heirs. Environmental protectionists and opponents of GMOs and GM food emphasize this point for the purpose of arguing the precautionary principle for risk assessments. Richard Gold also suggests that this feature means 'patent law needs to evolve to recognize that biotechnology is *different* from all other technologies'.[11]

Patenting life has the anomalous effect of granting proprietary rights that recognize the contribution and intervention of others to the exclusion of the contribution (auto-activity) of the living 'self'. There is said to be no 'inventive' quality to this latter endowment other than through our procreation and self-maintenance, neither of which occurs in the lab or is legally credited, but can be ascribed to our evolutionary communal 'authorship' jointly with nature. To challenge the patent system's asymmetrical recognition of labour and creation, in an act of protest on the proliferation of gene patents, Donna MacLean, a disgruntled British waitress and poet, filed a patent application (GB0000180.0) on herself claiming that she had reinvented herself, was new, useful and not obvious.[12]

Morris Cohen long ago argued that '*dominium* over things is also *imperium* over our fellow human beings'.[13] Because property confers a right establishing a legal relation between the person not in relation to the object but in respect of it as against the world, its dominant feature of exclusion creates a form of *sovereignty* over others affecting many human rights and in particular the right to (bio) self-determination.[14] The conceptual distinction between sovereignty and property becomes significantly blurred as the latter necessarily imposes on the former through the non-consensual and coercive use of state power.

9 Ibid.

10 Ibid. at 48.

11 Emphasis added, Stuart Laidlaw, 'Reckoning with Technology law in court: Monsanto decision hurts equity, innovation: Expert' *Toronto Star* (21 June 2004) D1 at 1.

12 See discussion in Amani and Coombe, 'The Human Genome Diversity Project', op. cit.

13 Morris R. Cohen 'Property and Sovereignty' (1927) 13 Cornell L. Q. 8 at 13.

14 Such politics play out in other forms of IPRs too. See e.g. Graham Dutfield, 'Case Study 6.1', in *IPRS, Trade and Biodiversity*, op. cit. at 87–8; Bita Amani 'A Penchant for Persian Rugs Over Palatable Products: The Use of Geographic Appellations as Trade-Marks' (2000) 14.2 *I.P.J.* 185; 14.3 *I.P.J.* 313.

Here, biocolonialism follows the mappings of colonialism before it and life, traditionally thought of as sacred and bountiful, now becomes a patentable 'invention' and therefore scarce. Patented life is largely privately controlled but simultaneously exceeds the control of its so-called inventor because of its generative and creative nature.[15]

Non-Patentable Subject Matter in the USA: Products of Nature

The evolution of the patented species finds its origins in the United States. The US patent code was silent on the issue of patenting life and products of nature. An early precedent was established in 1889 upon the Commissioner of Patents' rejection of an application over a fibre in the needles of a pine tree because, it was found, the ascertainment of the composition of the trees in the forest was not an 'invention' 'recognized by the statute, any more than to find a new gem or jewel in the earth would entitle the discoverer to patent all gems which should be subsequently found'.[16] And while the fibre was not known in the prior art, it was taken from a natural source. This precedent created the 'product-of-nature' doctrine that, although judicially eroding, theoretically continues to distinguish unpatentable objects of discovery from patentable processes created to extract what is found in nature. The Commissioner found that to allow patents on trees of the forest and the plants of the earth would be 'unreasonable and impossible'. That DNA sequences should be considered unpatentable subject matter is rooted in the idea that these should be treated as *discoveries* of products of nature or natural phenomena rather than *inventions*.

Patent law does not allow patents to be granted for discoveries. The legal basis for this in Canada relies in the interpretation of s. 27(8) of the CPA, which provides that 'No patent shall be granted for any mere scientific principle or abstract theorem'. The Federal Court of Appeal (FCA) in Canada has interpreted this provision to indicate a division between 'a discovery of a natural phenomenon' and something involving 'the application of inventive ingenuity'.[17] However, defining an invention as something involving ingenuity is not a helpful way to distinguish a discovery from an invention as '[i]t is axiomatic in discussions of patent law that $E=mc^2$ is not patentable because it is a discovery, and yet this clearly involves the application of human ingenuity'.[18]

15 See e.g. William Y. Brown, 'Promise and Peril' (2001) 18:5 *The Environmental Forum* 30 for the unanticipated ecological effects of GMOs.

16 *Ex Parte Latimer,* 1889 Dec. Com. Pat. 123.

17 *Harvard SCC, supra* note 7 at paras 46–7.

18 Austin and Amani, 'Patents on Genes', op. cit.

Austin and Amani offer another explanation to understand the invention/discovery dichotomy based on the United States Supreme Court decision in *Gottschalk v. Benson*.[19] The Court there reasoned that '[p]henomena of nature, though just discovered, mental processes, and abstract intellectual concepts are not patentable, as they are the *basic tools* of scientific and technological work'.[20] By this account, the bargain struck by patent legislation is that incentives for inventions are provided in return for the dissemination of knowledge and useful things but these policy considerations fail to compel in relation to 'the basic tools of scientific and technological work' because they would hamper, rather than help, the very progress that is sought. By analogy to the s. 27(8) exclusions, genetic mapping projects constitute basic research of natural phenomena and discoveries of scientific principles; their results should be excluded from patentability. Once the DNA structure became known, molecular manipulation became possible, legal claims inevitable, and the product of nature doctrine perverted. Ananda Chakrabarty, a biochemist at the General Electric Company, genetically modified a bacterium to break down crude oil spills by degrading hydrocarbons. In the 1980 decision of the USSC in *Diamond* v. *Chakrabarty*[21] (*Chakrabarty*) a narrowly divided court (5:4 majority) held that a patent could be granted for living micro-organisms as 'nonnaturally occurring manufacture or composition of matter – a product of human ingenuity'.[22] The focus of the court was not whether life or living matter *per se* was patentable subject matter, as the court found that 'anything under the sun made by man' is patentable.[23] Rather, the issue was whether *this* was an 'invention' within the statutory definition or a discovery of a product of nature. The majority ultimately found that the terms 'any manufacture' and 'composition of matter' were intended to be interpreted broadly and that Congress thus 'recognized that the relevant distinction was not between living and inanimate things, but between products of nature, whether living or not, and human-made inventions. Here, respondent's micro-organism is the result

19 409 U.S. 63. This case is referred to in *Parker v. Flook*, 437 U.S. 584 in which it was held that 'Even though a phenomenon of nature or mathematical formula may be well known, an inventive application of the principle may be patented. Conversely, *the discovery of such a phenomenon* cannot support a patent unless there is some *other* inventive concept in its application' at 594, emphasis added. *Flook* was relied upon in *Diamond* v. *Chakrabarty*, 447 U.S. 303 (1980) [*Chakrabarty*], which was in turn relied on by the FCA in *Harvard FCA, supra* note 61.

20 *Gottschalk v. Benson*, 409 U.S. 63 (1972). 409 U.S. 63 at 67.

21 *Chakrabarty*, *supra* n. 19.

22 Ibid. at 309–10. For a more current account of the legal treatment of micro-organisms, see Paul Oldham, *Global Status and Trends in IP Claims: Micro-organisms* (Lancaster: Centre for Economic and Social Aspects in Genomics, 2004), <http://www.cesagen.lancs.ac.uk>.

23 *Chakrabarty*, *supra* note 19 at 309 (citing S.Rep.No. 1979, 82d Cong., 2d Sess., 5 (1952); H.R. Rep. No. 1923, 82d Cong., 2d Sess., 6 (1952)).

of human ingenuity and research'. It was also found that Chakrabarty's new bacterium had 'markedly different characteristics from any found in nature … [and that] his discovery is not nature's handiwork, but his own; accordingly it is patentable subject matter under § 101'.[24] In the United States, what is patentable is set out in §.101 of Title 35 of the United States Code governing patent grants, which provides:

> Whoever invents or discovers any new and useful process, machine, manufacture, or composition of matter, or any new and useful improvement thereof, may obtain a patent therefore …[25]

This would not be the first time that the patentability of living matter was decided by a narrow majority of the Supreme Court of a country. Nor would this be the only occasion in which the court's finding would subsequently be used to obfuscate rather than clarify a most fundamental legal doctrine. One would have thought that with the majority's qualification, DNA sequences found in nature would not be held patentable even if engineered life, such as *Chakrabarty*'s waste management bacteria, was.

In 1985, the scope of patent protection in the United States was extended to higher life forms (man-made sexually reproducing plants). A species of corn engineered to produce an abundant amount of the amino acid trytophan was considered not to be a product of nature in *Ex Parte Hibberd*[26] because it did not occur *naturally*. *Ex Parte Hibberd* was an important decision because in 1930, both Congressional houses had rejected the idea of including sexually reproducing plants within the American Plant Patent Act (PPA)[27] and a 1968 proposed amendment to the PPA to extend patent protection to sexually reproducing plants was also defeated. In order to offer some form of legislated protection, Congress in 1970 enacted the Plant Variety Protection Act (PVA) as an alternative which extended to sexually reproducing plants (including most crops and seed varieties). The PVA conferred a 20-year term of protection with very expansive exclusive rights to the owner:

> Every certificate of plant variety protection shall certify that the breeder has the right, during the term of the plant variety protection, to exclude others from selling the variety, or offering it for sale, or reproducing it, or

24 Ibid. at 313.

25 35 U.S.C.A. §.101, <http://www4.law.cornell.edu/uscode/35/>. This is essentially the language used in the 1793 Thomas Jefferson draft except for the word 'process' which replaced the word 'art' in a 1952 Congressional reform of patent law.

26 227 United States Patents Quarterly 443 (Board of Patent Appeals and Interferences 1985).

27 Plant Patent Act, 35 U.S.C. §§ 161–3 (1930).

> importing it, or exporting it, or using it in producing a hybrid or different variety therefrom, to the extent provided by this Act.[28]

While the PVA enabled protection without the need to demonstrate utility and non-obviousness, or to make disclosure, the PVA also had two important legislated exemptions lacking in patent law: the PVPA allowed researchers to use the protected varieties for free exchange within the research community and also for farmers to save patented seed for replanting. After *Ex Parte Hibberd*, plants reproduced from seeds were patentable without these attendant safeguards as the USPTO changed its practice to reflect the court decision. By the time the result was confirmed in the Supreme Court decision of *J.E. M. Ag Supply Inc. v, Pioneer Hi-Bred Int'l, Inc.*, some 1800 utility patents for plants, plant parts and seeds had issued, 17 belonged to Pioneer Hi-Bred International, Inc.[29] Pioneer's patents covered the manufacturer, use, sale, and offer for sale of their corn seed products which, under the patent, extended to protect both the seeds and plants of the inbred line and as well as the hybrids produced in crossing the patented line with other corn lines. Since inbreeds are often weak and with low yields, their primary utility is in making hybrids produced from crossing two inbred corn plants. Hybrids on the other hand produce highly valued characteristics and are 'strong and vibrant plants' but they 'generally do not reproduce true-to-type, i.e. seeds produced by a hybrid plant do not reliably yield plants with the same hybrid characteristics. Thus a farmer who wishes to continue growing hybrid plants generally needs to buy more hybrid seed'.[30] But beyond practical quality based considerations, the licence under which Pioneer was selling its patented hybrid seeds was a limited one that only permitted use to produce grain and/or forage but did not extend to the use of seed from such crop or its progeny for propagation or seed multiplication or for the production or development of a hybrid or different variety seed, though these would constitute patent infringement if the Court were to find the patents valid as it did. *J.E.M.* confirms that utility patents under 35 U.S.C. § 101 could be extended to plants and sexually reproducing seeds based on a written description of the plant and a deposit of the seed if all of the other requirements of patentability were met. The court rejected the argument that since Congress had two alternative regimes for plant protection (the PPA and PVA) the case for utility patents was foreclosed and instead concluded that since Congress failed to explicitly exclude plants in the provision of the Patent Act providing for utility patents, extending patents to plants cannot be viewed

28 Plant Variety Protection Act, 7 U.S.C. §2483. For a detailed discussion of how US patent law was expanded to cover plant genetic resources, see generally Keith Aoki, *Seed Wars* (Durham, NC: Carolina Academic Press, 2008).

29 Ibid.

30 534 U.S. 124 (2001), 124–5.

as contrary to congressional intent. By the time of judgement, it had already been patent practice in the USA for at least 16 years:

> In the face of these developments, Congress has not only failed to pass legislation indicating that it disagrees with the PTO's interpretation of §101, it has even recognized the availability of utility patents for plants. In a 1999 amendment … Congress provided: 'Applications for plant breeder's rights filed in a WTO member country … shall have the same effect for the purpose of the right of priority … as applications for patents, subject to the same conditions and requirements of this section as apply to applications for patents'. … which suggests a recognition on the part of Congress that plants are patentable under §101.[31]

It seemed clear that without congressional intervention, the traditional role of farmers as both producers and consumers of seeds had been permanently shifted by the courts. In summarizing the impact of the decision, Keith Aoki argues that demands for strong and uniform international IPRs protections are betrayed by hypocritically inconsistent practices:

> *J.E.M. Ag Supply* finally, if problematically, clarifies the post-*Chakrabarty* lacunae … [holding] that the PVPA and §101 general utility patents are not mutually exclusive avenues for protection of property rights in germplasm, even if the legislative history underlying both the PPA and the PVPA appeared to contradict an interpretation of such protection as concurrent … It is time to consider the following proposition: the germplasm that provided the foundation for germplasm's commodification originated as a resource obtained for 'free' via voraciously collected seed and plant samples appropriated from regions and peoples around the world. This 'free' resource was then selected and distributed to U.S farmers 'free' of charge, first by the U.S. Patent Office and then by the USDA until 1924. Indeed there is an irony that the U.S. Patent Office which in the 19th century was responsible for starting germplasm collection, propagation, and distribution until the time of the USDA's creation, is today a key institutional actor in promoting laws and practices that prohibit seed saving.[32]

Higher animal life had already been considered as patentable subject matter years earlier in *Ex Parte Allen*.[33] The USPTO had rejected the patent over a polyplod Pacific Coast oyster genetically modified for year round consumption on the basis that it was both naturally occurring and obvious. On appeal to

31 Ibid. at 143.

32 Aoki, *supra* note 28 at 60.

33 2 U.S.P.Q. 2d 1425 (P.T.O. Bd. App. and Int. 1987); (1987), 1077 Official Gazette of the U.S. Patent and Trademark Office announced the new policy.

the Patent Board of Appeals and Interferences, the Board recognized that the subject matter was a patentable 'invention' provided the other requirements (novelty, utility and non-obviousness) were met. But in this case, the oysters though bigger and more resilient, were found to be not *significantly* different from those found in nature and thus were obvious despite the level of ingenuity in their making. In April 1987, immediately following the Board's decision, the USPTO issued the following statement on the subject matter it considered patentable under US law and confirmed that higher life that was not a product of nature *could* be patentable:

> The Patent and Trademark Office now considers non-naturally occurring non-human multicellular living organisms, including animals, to be patentable subject matter within the scope of 35 U.S.C. 101. A claim directed to or including within its scope a human being will not be considered patentable subject matter under 35 U.S.C. 101. The grant of a limited, but exclusive property right in a human being is prohibited by the Constitution … any claim directed to a non-plant multicellular organism which would include a human being within its scope [should] include the limitation 'non-human' to avoid this ground of rejection.[34]

The Manual for Patent Examining Procedure provides the guidelines on patentability and states that 'if the broadest reasonable interpretation of the claimed invention as a whole encompasses a human being, then a rejection under 35 U.S.C. 101 must be made indicating that the claimed invention is directed to nonstatutory subject matter'.[35] While the statement attempted to quell public concern over human 'ownership', significant public protest and moral outrage ensued on the patentability of higher 'non-human' life but did little to prevent Harvard University from obtaining the first animal patent in 1988.

In the interim, *Moore v. Regents of the University of California*[36] resulted in the Supreme Court of California extending patents to *human* genetic material while denying Mr Moore a property interest in the commercial sale of his cell line. In the 1970s John Moore suffered from hairy cell leukaemia. He had his spleen removed at the University of California, Los Angeles Medical Center during the course of which Dr David W. Golde discovered that Moore's spleen tissue produced a rare protein that appeared to be of some cancer-

---

34 D.J. Quigg (Commissioner of the United States Patent and Trademark Office, Policy Statement on Patentability of Animals, 1077 Off. Gaz. Pat. Office 24 (7 April 1987). See also 'Notice: Animals – Patentability', 1077 Official Gazette U.S. Pat. and Trademark Off. 8 (April 21, 1987).

35 Manual for Patent Examining Procedure, <http://www.uspto.gov/web/offices/pac/mpep/21.txt>.

36 *Moore v. Regents of the University of California*, op. cit.

fighting utility. Without Moore's consent or knowledge, a cell line, the Mo cell, was made and patented as 'invention' (No. 4,438,032) by Dr Golde and was subsequently sold to a drug company for $15 million with a reported current value in excess of $3 billion for the derivative drugs and therapies developed from the patented product. In 1984, upon learning of the facts, Moore sued the university and drug company on 11 causes of action including claims for conversion as 'owner' of the cells, breach of fiduciary duty, and lack of informed consent. The California Supreme Court eventually held (in 1990) that the doctors' conduct was tortious and that they had also breached fiduciary duties by failing to reveal to Moore their research and financial interest in his cells. Moore was granted compensatory damages for such wrongs but the Court also ruled that there were no grounds for Moore's property claim – once the cells were removed from his body – for policy reasons: 'research on human cells plays a critical role in medical research … [and granting the patient proprietary rights would threaten to] hinder research by restricting access to the necessary raw materials.'[37]

Moore's cells were considered to be raw materials in the *public domain* and thus available for the creative and 'inventive' activities of others. The irony of the decision in granting property rights to the patentees while denying them to Moore was highlighted in Justice Broussard's dissent: 'the majority's analysis cannot rest on the broad proposition that a removed part is not property, but … on the proposition that a *patient* retains no ownership interest in a body part once the body part has been removed.'[38] But the majority felt comfortable granting Dr Golde a property right in the cells on the basis that the patented cell-line was substantially different from Moore's own cells. Paradoxically, the majority did not recognize Moore's property right in his own cells and ignored the auto-collaboration of the living 'self' to which Habermas and Jonas refer. Moore's deliberate 'labour' in maintaining his living cells by eating, sleeping, and generally preserving his health were insufficient to bestow on him sovereignty over himself. Commentators have criticized the *Moore* decision because it denies the source of the valued material (Mr Moore) a property interest in his excised parts but simultaneously recognizes third party (commercial) interests in his living material as commodity.[39] Mason and McCall Smith appropriately

37 Ibid. at para. 96.

38 Ibid. at paras 133–4. For a detailed discussion, see Amani and Coombe, *supra* note 12.

39 '[b]ecause the courts tend to ignore sociological, psychological and anthropological attitudes towards the body in their decisions and, instead, reduce everything to economics, they ignore signs of people's ambiguity about technological advancement … [I]t is accepted that the encouragement of commercial profit-making by users, primarily by the biotech industry, can only bring benefits to the community in the guise of more therapies and cures. But … giving property right to biotechnological researchers focuses the medical industry primarily on cures and shifts attention away

ask, 'does not such an example demonstrate precisely the kind of value which individuals could retain in their excised parts? The properties of the patient's cell are a *sine qua non* of the ultimate invention'.[40] Cultural anthropologist and legal scholar Rosemary Coombe offers this insight:

> [j]ust as the authorship and ownership of others was not recognized in the nineteenth century pillage and plunder of their material artifacts ... today the authorship of others in shaping the biological 'specimens' that surround them (or the genes within them) is unacknowledged. 'They' do not understand their global significance, it would seem, but only their local and parochial meanings. Once again, 'our' appreciation enables 'our' appropriation on behalf of a global 'we' that classifies and maps, not the globe, but the biosphere (or the genome).[41]

The lesson from *Moore* was clear: while possession remains nine-tenths of the law, there is no substitute for legal title. And legal title is, as biocolonialism now and colonialism before it attest, a political enterprise; the underlying norms are themselves coercively imposed:

> The very tropes of discovery, invention, naming, and originality that animate modern intellectual property laws emerge from a historical era in which Europeans mapped the world in their own image – ignoring the human ecologies of others and denying the value to the pre-existing worlds of meaning in which such phenomena figured ontologically and spiritually ... We can see the same processes at work in contemporary political debates – in the so-called 'New World Order' – as authorial tropes are deployed to legitimate new forms of social domination. Emergent elites naturalize their claims to represent the global 'we' – protecting 'our' biodiversity and preserving 'our' gene pool – while attaching their own signatures to the mappings they effect.[42]

---

from the equally, or more important pursuit of determining – and eradicating – the underlying causes of illness and disease'. See J.K. Mason and R.A. McCall Smith, *Law and Medical Ethics* (London: Butterworths, 1999) at 489 referring to E.R Gold, *Body Parts: Ownership of Human Biological Materials* (Washington: Georgetown University Press, 1996).

40 Ibid. at 492–3.

41 See Coombe, *supra* note 1 at 1362

42 Ibid. at 1351.

Non-Patentable Subject Matter in Canada: The Disclosure Dilemma

In Canada the legal approach to the patenting of life has been somewhat different. Prior to 1982, living matter was perceived to be a product of nature and was not within the definition of 'invention' such that claims encompassing living matter were refused until *Re Application of Abitibi Co.*[43] when, shadowed by developments in American jurisprudence, the Patent Appeal Board found that lower life were patentable. In *Abitibi*, the application was for both process and product claims over a new mixed fungal yeast culture system that was useful for absorbing foaming spent-sulphite liquor generated in papermaking. The fungi were isolated and subjected to increasing concentrations of foaming sulphites and nutrients in a water medium; those that survived had functionally mutated (adapted) to consume the foaming effluent and became the subject of the product claims. Initially, the patent examiner had disallowed the product claims on the basis that microbial culture systems were living matter and thus not a patentable 'invention' under s. 2. The Board considered the similarity of our s. 2 definition of 'invention' with the US Patent Code, and the shared silence on living subject matter. Having the 'benefit' of the USSC precedent in *Chakrabarty*, the Board noted that the Canadian exclusion of patenting living matter had been based on UK precedents and that 'the Chakrabarty decision casts doubt on the correctness of that practice',[44] marking an early display of preference for judicial convergence with American based standards for more inclusive patentability. The Board found that '[i]f an inventor creates a new and unobvious insect which did not exist before (and thus was not a product of nature) *and can recreate it uniformly and at will,* and it is useful ... then it is every bit as much a new tool of man as a micro-organism'.[45] The Board suggested that the Patent Commissioner interpret the terms 'manufacture' and 'composition of matter' to incorporate lower life. The Commissioner accepted the Board's recommendation to allow the claims over micro-organisms and issued the patent. As a result of considerations very narrowly decided and extrinsic to our judicial system (with the 5:4 split in *Chakrabarty*) and without account for historical, social, political, and regulatory differences between the American and Canadian contexts, *Abitibi* found the yeast culture patentable.[46] The Canadian patent office listed these items as patentable 'invention' so long as the product claimed could be recreated uniformly on large scale and at will:

43 *Abitibi,* op. cit.

44 Ibid. at 87.

45 Ibid. at 90, emphasis added.

46 Ibid. at 88, the Court also considers German and Japanese law, both of which allowed for the patentability of micro-organisims. See also discussion in Fecenko, supra note 24 at 84.

> all micro-organisms, yeast, moulds, fungi, bacteria, actinomycetes, unicellular algae, cell lines, viruses or protozoa; in fact … all new life forms which are produced *en masse* as chemical compounds are prepared, and are formed in such large numbers that any measurable quantity will possess uniform properties and characteristics.[47]

The requirement of re-creation at will (*en masse*) and with uniformity is an issue that resurfaces before the Supreme Court of Canada in the Harvard mouse decision discussed below because it is a proposition that technically defies guarantee. It is *degree,* rather than principle, however, that seemed to matter more to the Patent Board:

> With still higher life it is of course less likely that the inventor will be able to produce it at will and consistently, as more complex life forms tend to vary more from individual to individual. But if it eventually becomes possible to achieve such a result, and all of the requirements of patentability are met, we do not see why it should be treated differently … However a micro-organism, being living matter, will reproduce itself on proper culture medium, so as the inventor can maintain his supply indefinitely. If he places samples of the organism in a culture collection to which others have access, they too will be able to reproduce the organism, and thus have access to his invention, and use it once the patent expires.[48]

Habermas and Jonas make clear that *all* organisms vary. The degree of genetic variation for the entire human family is only 0.1 per cent and for many diseases only a single genetic letter. On whether a deposition of the invention in the culture collection is sufficient to satisfy the disclosure requirement, the Board was again influenced by American case law which had found that it was sufficient disclosure for *process* claims since the ordinary skilled person could put the invention into practice with a sample of the deposit.[49] But the Board made a pivotal error in extending the sufficiency of a deposition in *process* patent applications to *product* patent claims:

> If deposition of a micro-organism in a culture collection is sufficient disclosure of it when an applicant claims a process utilizing that organism, it seems strange indeed to hold it is inadequate when the organism itself

47 *Abitibi*, *supra* 43 at 82.

48 Ibid. at 91.

49 *American Cyanamid Co. (Dann's) Patent,* [1971] R.P.C. 42 cited in *Abitibi supra* note 43 at 91.

> is claimed. In both instances the public needs the organism to work the invention, and in both instances it has it ...[50]

This finding renders moot the important distinction between process and patent claims.

Depositions are and should be sufficient for disclosure of *process* patents because what has been deposited when reproduced will operate to perform the process function for which it has been patented. However, if the claim being made is to the cultured bacteria itself, then a deposition should not suffice to meet the disclosure requirements for the bacteria *product* because of the variability inherent in life; even if the cultured bacteria grown from the deposition displays, in the aggregate, the dominant functional feature (the ability to eliminate waste contaminants – a process) claimed.[51]

The next major decision, *Pioneer Hi-Bred Ltd v. Canada (Commissioner of Patents)*,[52] recognized the disclosure dilemma as it pertained to higher life and was the first case to raise the issue of a multicellular organism in Canada based on an application for a new soybean variety developed from artificial crossbreeding and selection but cultivated naturally and without molecular manipulation. The issue before the Court was the patentability of plant life. The SCC upheld the Commissioner's decision denying the application over plants and its seed, not on the basis of patentable subject matter but rather on the basis that filing a seed sample was *insufficient disclosure*.

Disclosure requires the inventor to 'set forth clearly the various steps required to make the "composition of matter" in such full, clear, concise and exact terms as to enable any person skilled in the art or science to which it

---

50 Ibid. See also s. 17.05 of the Canadian Manual of Patent Office Practice (MOPOP) which provides that the examiner has the discretion to determine whether a written description is sufficient to satisfy the specification and disclosure requirements of s. 27(3) of the CPA or if a deposit of biological material is also needed to work the invention, [hereafter CIPO, MOPOP]. CIPO, <http://strategis.ic.gc.ca/sc_mrksv/cipo/patents/mopop/chap16-e.html>.

51 Cloning technology has extended the possibility of reproducing 'en masse', and arguably *at will*, to higher (plant and animal) life but the level of genetic variation in the clone from the original is not insignificant and affects the significantly lower survival rate and the statistically higher risks of developing health problems. See L. Paterson, P. DeSousa, W. Ritchie, T. King, and I. Wilmt, 'Application of Reproductive Biotechnology in Animals: Implications and Potential Applications of Reproductive Cloning' (2003) 79(3–4) *Animal Reproduction Science* 137–43; W.M. Rideout, K. Eggan and R. Jaenisch, 'Nuclear Cloning Epigenetic Reprogramming of the Genome' (2001) 293 *Science* 1093–8; and G. Schatten, R. Prather, and I. Wilmut, 'Cloning Claim is Science Fiction, Not Science' (2003) 299 *Science* 344. See also 'Sorcerer's Apprenticeship', *The Economist* (15 March 2007).

52 *Pioneer Hi-Bred Ltd v. Canada (Commissioner of Patents)* [1989] 1 S.C.R. 1623, [1989] S.C.J. No.72 [*Pioneer*].

pertains, or with which it is most closely connected, to make it'.[53] In this case, it was held that the disclosure failed to *teach* others skilled in the art how to *reproduce* and work the invention. The SCC expressly declined to consider whether new soybean varieties are patentable subject matter under the s. 2 definition of 'invention', while the FCA had found that the plant derived from traditional cross-breeding could not be construed as 'manufacture' or 'composition of matter' within the s. 2 language in part because plant breeding, an established practice prior to the enactment of the CPA, was not alluded to at all within the definition as it could have been were the intention otherwise with reference to words such as 'strain', 'variety' or 'hybrid'.[54] This was a rare instance of judicial finding directly contrary to the line of reasoning adopted by the US Supreme Court in *J.E.M. Ag Supply Inc.* regarding Congressional intent and demonstrates how two highest courts may approach the issue of legislative intent from divergent perspectives.

Pioneer was the last major case on the patentability of higher life in Canada before two recent SCC decisions; it was also the closest one to recognizing the auto-contribution of the living subject in deciding what may be a patentable object. The Court's reasoning recognized that higher multi-cellular life was not patentable because the so-called 'invention' could not be *fully* disclosed in a manner that would allow the full replication of the invention. When given proper consideration, while it is well established that deposits can legally meet the disclosure requirement of living matter, technically, it is not a small one to overcome for genetic inventions because the self-propagation and mutation of living matter means that inventors do not have full control over all of the characteristics of the resulting product and therefore technically can *never* teach it in their patent application. The legal acceptability of deposits for living product claims fails to account for epigenomic mutations (based on chemical tags that determine how a gene interacts with environmental factors) to which entire scientific consortia are dedicated. For example, the private-public collaboration of the Human Epigenome Project (HEP)

> aims to identify, catalogue and interpret genome-wide DNA methylation patterns of all human genes in all major tissues. Methylation is the only flexible genomic parameter that can change genome function under exogenous influence. Hence it constitutes the main and so far missing link between genetics, disease and the environment that is widely thought to place a decisive role in the aetiology of virtually all human pathologies.[55]

---

53 See Binnie J's discussion of *Pioneer* in *Harvard SCC*, *supra* note 7 at para. 27.

54 Fecencko, *Biotechnology Law*, op. cit. at 85.

55 See http://www.epigenome.org/index.php?page=project.

That we have accepted the practice of deposits of living matter as a means to the end of conferring patents over life does not mean that to do so is not in error given that all living matter at its most basic molecular level is affected by environmental stimuli. From this perspective, normatively speaking, product patents on life will always be disingenuous as they will technically fail to meet the conditions of the 'bargain' with the crown even if a court holds otherwise. And, because of the organism's own 'reproducibility', any rights granted to the invention claimed *cannot* be commensurate with the 'invention' made.

Justine Pila, commenting on the insufficiency of written forms of description to ensure 'a third party's ability to repeat an invention involving biological material', writes that '[i]t followed from the requirements of enablement and disclosure that if biotech patenting were to be allowed, traditional conceptions of reproducibility would need to be revised or an alternative means of description introduced'.[56] The 1977 Budapest Treaty on the International Recognition of the Deposit of Microorganisms for the Purposes of Patent Procedure[57] worked around this problem and allowed for the deposit of micro-organisms in lieu of – or as a complement to – written descriptions. However, as Pila notes, some nations refused to accept these samples as fully meeting the disclosure requirement. A new, useful and non-obvious cultivation method may be disclosed and patented as *process* claims so long as the living matter generated from the deposits continues to function to achieve the process disclosed. But, granting a claim for a 20-year monopoly in the underlying *product* sanctions a misrepresentation on the public as inventors simply cannot account for the variable contribution associated with life. *Pioneer* was used by the patent office to reject claims for higher animal and plant life on this basis until the case of the Harvard mouse, which needs to be followed through three jurisdictions in order to understand its significance to the international patenting life debate.

Comparative Treatment: Of Mice …

The Harvard mouse case focused on the legally distinct issue of whether non-human higher life (plants and animals) were patentable subject matter within the s. 2 statutory definition of 'invention' in Canada. The 'oncomouse' epic began in a Harvard lab in the early 1980's with two scientists, Philip Leder and Tim Stewart, successfully engineering a mouse to carry an oncogene (a tumour-causing gene tied to a promoter that activates the gene in the mammary glands of mice). The scientists did this by introducing transgenic material into mice DNA. The oncogene increases the probability that the mice carrying it will develop malignant tumours without the traditional requisite degree of

56 Pila, 'Bound Futures', op. cit. at 336.

57 Apr. 28, 1977, 32 U.S.T. 1241, 1861 U.N.T.S. 361.

exposure to carcinogens but does not ensure cancer for all oncogene-carrying mice. Still, the oncomouse is useful for facilitating carcinogenic studies. Daniel Kevles observes that contrary to popular belief, the

> work had not been done for the sake of devising a patentable product, but once it was accomplished, Leder recognized that it might have commercial possibilities. In 1983, he brought his mice to the attention of the Office of Technology Licencing and Industry Sponsored Research, the recently established patents arm of the Harvard Medical School.[58]

Leder and Stewart were motivated by the scientific drive to facilitate their research but the *Bayh Dole University and Small Business Patent Procedures Act*[59] enacted by the US Congress in 1980 enabled the patenting of inventions produced at federally funded public universities. The oncomouse was made in the course of employment; any rights that these scientists might enjoy would be assigned to Harvard. After legal consultation with outside counsel, Harvard decided the mouse would have to be patented as an invented product in order to ensure maximum protection (and profit) under the law. DuPont, as Harvard's exclusive licensee, was to oversee the commercialization of the oncomouse increasing the market value of the patented invention as licenced technology to other labs conducting cancer research.[60]

A patent application over animal life was unprecedented but the subject matter was not expected to create a legal hurdle given that the mouse was new, involved an inventive step (it was not-obvious), and had significant research-based utility. Harvard's product claims were excessively broad extending beyond the transgenic *oncomouse* to cover a significantly broader 'invention': *all* transgenic non-human mammals. Claim 1 was for:

> A transgenic non-human mammal whose germ cells and somatic cells contain an activated oncogene sequence introduced into said mammal, or an ancestor of said mammal, at an embryonic stage. 'Germ cells' are reproductive cells, and 'somatic cells are all other cells of the organism other than reproductive cells. It is understood that the transgenic non-human

58 See Kevles, *supra* note 3 on the history of patenting life (at 44) based on the author's interviews with Philip Leder, 21 June 1988 and 6 June 1991. Leder confirms this in a more recent interview, see *IDEAS*, CBC Radio One (June 2003) archived at <http://www.cbc.ca/ideas/calendar/2003/10_October.html>.

59 35 U.S.C. §200–212 (2000). See generally Arti K. Rai and Rebecca Eisenberg, 'Bayh-Dole Reform and the Progress of Biomedicine' (2003) 66 *Law and Contemp. Probs.* 289.

60 Kevles, *supra* note 3.

mammals which are sought to be patented will have the oncogene present in all reproductive and somatic cells.[61]

In 1984, Leder and Steward's US patent application was assigned to Harvard creating the catalyst for the ensuing controversy over animal patents; the first animal patent was formally granted to Harvard by the USPTO in 1988 for the oncomouse (US Patent 4,736,866). Once again it was administratively confirmed that higher life is patentable subject matter – a finding that established the precedent for the range of life patented in the USA since. Genetically engineered animals include fish, cows, mice and pigs and there are currently more than 190 of these varieties waiting to be patented.[62] Moreover, Harvard succeeded in securing rights to its reach-through claim which meant that Harvard would enjoy an entirely separate revenue stream from licensing any transgenic *mammal* covered within the claim as well as its *oncomice* (which it licences to labs at $1/per) even though disclosure for what was 'invented' was only for a mouse and not, say, an *oncomonkey*.

Once patenting life forms becomes common practice, legally restricting it becomes harder to justify by the courts, if for no other reason than the principle of *stare decisis*. Moreover, the search for greater commercial profit means that biotechnology may move beyond R&D in health and food and expand into markets where the 'anarchic whims of consumers and clients'[63] will translate into demand for novelty products from designer pets to designer children. These offer greater potential profit as 'luxury goods' but create unforeseeable consequences for future regulation. Even if we are to put aside the issue of designer babies, new ethical questions are raised where the transgenic animal is designed exclusively for 'pet ownership'. If the technology is there, there is a profitable market, and no prohibitive regulation to the contrary, corporate firms are likely to develop and commercially exploit such products *anyway*; are patents as incentives *ex ante* really needed or do they simply provide *ex post* rewards as in the case of Harvard's oncomouse?

Glow mice, glow fish, and luminescent rabbits[64] have all been 'invented' and may foreshadow the market for GM designer cats and dogs. The Glofish is a genetically modified zebrafish with a coral gene that makes it fluoresce red. It has been selling well in states such as Florida according to Michael LePage, and is rivalled only by its darker cousin, Night Pearl – a fish that

61 *Harvard FCA*, op. cit. at fn 5.

62 See Shiva, *Protect or Plunder*, op. cit. at 112.

63 Habermas, *supra* note 8 at 48.

64 Performance art has also embraced biotechnology as part of its protest. San Francisco artist Marlilyn Donahue commenced a 'genome certification program' and prompts members of the public to 'become a certified original human. Copyright your DNA'. She recommends fixing 'authorship' by licking a stamp and posting it on a self-addressed envelope, <http://www.mudhaus.com/marilyn/donahue>.

glows in the dark due to an added jellyfish gene that has been selling in Taiwan since June 2002. California state fish and game commissioner Sam Schuchat told reporters, in response to the new Glofish, 'I just don't think it's right to produce a new organism just to be a pet … To me, this seems like an abuse of the power we have over life'.[65]

Patenting, creating, and manufacturing for commercial sale of biological 'products' are all distinct matters but the issues they raise are common and many are ethical. Many of these concerns are tied to alarmist apprehensions of 'playing god' and public interest unease at privatizing and commoditizing life.[66] Public welfare arguments that support biotech innovation (and even their patenting) in health and food industries are less compelling in cases where pets are consumable goods. However, the debate necessary to determine what qualifications, if any, need to be made to IPRs in the life sciences becomes even less imperative after-the-fact. Industry proponents will assert, as they have with gene patenting and plant variety protection extended to sexually reproducing seeds, that what is already being done hardly begs further debate. If the *J.E.M.* experience teaches us anything, the Courts will agree.

Would transgenic animal research cease without patent protection? Likely not given what we now know of what drives scientists and what actually drives markets. But granting IPRs would further the *commercialization* incentive that drives industry to innovate genetically modified higher life, whether research promoting or as a consumable good, even while opponents object on numerous grounds such as the survival of the family farm; undesirable distributional consequences; commodification and devaluation of life; increase in animal suffering; interference with the natural world; the inviolability and inalienability of the body; the loss of control and creation of unknown risks; and the penetration of undesirable commercial imperatives in the organization and priorities of academic research.[67] Additionally, there are concerns over animal and human health and welfare effects with xenotransplants (which rely on the commercial production of genetically modified donor animals).[68]

65 Michael LePage, 'They Came, They Glowed … They Could Conquer the Pet Trade' *New Scientist*, 20 December 2003–9 January 2004, at 24, available at <http://www.newscientist.com>.

66 For a discussion of the ethical issues, see Schrecker et al. 'Ethical Issues Associated with the Patenting of Higher Life Forms', 17 May 1997, Westminister Institute for Ethics and Human Values McGill Centre for Medicine, Ethics and Law, Strategis.gc.ca, <http://strategis.ic.gc.ca/epic/internet/inippd-dppi.nsf/en/ip00095e.html> [Schrecker et al.]. See generally, Fecenko, *supra* note 54.

67 Rebecca Dresser, 'Ethical and Legal Issues in Patenting New Animal Life' (Summer 1988) *Jurimetrics Journal* 399–435.

68 Report of the National Forum on Xenotransplantation: Clinical, Ethical and Regulatory Issues, Ottawa, 6–8 November 1997 (Ottawa: Health Canada, 1998), <http://www.hc-sc.gc.ca/hpb-dgps/therapeut/zfiles/ennglish/btox/reports/frmrptx_e.pdf> discussed in Ted Schrecker and Alex Wellington, 'Patenting of Biotechnological

Proponents may celebrate the experiments underway for the creation of human/animal chimeras because of the utility of these hybrids to harvest cells genetically identical to those of humans that in turn would enable damaged organ repair and ultimately transplantation of human organs cultivated in animal hosts. But critics are concerned that retroviruses dormant in animal DNA could mutate or activate into infectious forms for humans. While some clinical trials have commenced in the United States, the technique is expensive to develop. And what about the ethics? 'There is the moral issue. Some people oppose the creation of all human-animal chimeras on religious grounds, and many more would join them if there were the slightest chance that sheep with human brain cells might be more than just sheep'.[69] The potential reward of a patent would encourage this research, and possibly even research *races* to secure the grant and recoup expenses, but ignores some deeply fundamental and inherently ethical issues. At the end of the day, it is not clear whether the public in the United States, Europe, Canada or any other jurisdiction in fact unequivocally believe that such commercial activity merits further encouragement by letters patent, particularly when it arises from judicial fiat without public participation that can and should inform necessary legislated oversights.

American patent policy ignores all issues of morality in its consideration of patentable subject matter. In Europe, national patent law of signatories must conform to the European Patent Convention (EPC)[70] which contains a moral clause. The EPC is a treaty that was created to harmonize the laws of signatory European nations and to create collaboration and convergence regarding patentable subject matter. Administered by the European Patent Office (EPO), the EPC allows inventors to process their application through one office, the EPO, to obtain multiple European national patent rights. Neither the EPC nor the laws of its signatories specifically deal with biotechnology patents. Biotechnology patents have been considered under Article 53(a) of the EPC which prohibits patents on inventions *contrary to ordre public or morality;* although these terms, which also appear in Article 27(2) of TRIPS discussed in text below, are undefined in both contexts.

In 1984, Harvard filed its patent application in Europe. It was the first case to invoke Article 53 of the EPC. The application was initially refused in 1989 by the European Patent Office (EPO), not because of *ordre public*

Innovations Concerning Animals and Human Beings', prepared for the CBAC Project Steering Committee on IP and the Patenting of Higher Life Forms, 31 March 1999; the CBAC reports are available online, Grover, op. cit. See also Sylvia Pagan Westphal 'Growing Human Organs on the Farm' New Science (20 December 2003–9 January 2004) at 4.

69 Ibid.

70 European Patent Convention, October 1977 [EPC], <http://www.european-patent-office.org/legal/epc/e/mal.html>.

or morality, but on the basis that the patent was prohibited by Article 53(b) of the EPC which expressly excludes 'plant or animal *varieties* or essentially biological processes for the production of plants or animals'.[71] Article 53(b) was drafted in order to prevent interference with the international system for protecting plant breeders' rights under UPOV created in 1961.[72] Despite the ongoing public outcry against the patenting of life both in the United States and Europe, the *ordre public* and *morality* provision was surprisingly not referenced.[73] Instead, the oncomouse was considered the product of natural biological processes – a different breed constituting a *new* variety – making it ineligible for a patent under the EPC.

Harvard appealed the decision to the EPO's Technical Board of Appeal (TBA) on the basis that this mouse was not a *variety* to be excluded but a new type of animal 'transcending varietal classification, and that [it] was not a natural biological product but – echoing Chakrabarty's claim – a biological entity made by man'.[74] Unlike the Canadian patent system, which limits third party filings to prior art documents for re-examination after the patent has issued (under ss. 48.1–48.5 CPA), the EPC allows interested third parties to file comments opposing or in support of applications pending or on appeal on broader grounds. The Harvard appeal was mired with such filings from various environmental and animal welfare interest groups on moral grounds opposing the patent grant; they had an impact. The TBA reversed the EPO decision in 1990 after examining the French and German language equivalents of this text to reach the conclusion that the Examining Division was wrong to interpret Article 53(b) as excluding from patentability animals per se. The oncomouse application had been for 'mammals' and 'rodents'. It was remitted to the original examiners for reconsideration to determine whether the patent would be excluded based on Article 53(a) as contrary to *ordre public* and

---

71 European Patent Office Press Release No. 10/89, 'EPO Refuses Patent Application for Oncogenic Mouse'. See generally Kevles, *supra* note 3 at 58.

72 The International Union for the Protection of New Varieties of Plants (UPOV) is an international agreement providing the prevailing model for plant breeders' rights. The 1978 UPOV Act was presumed to permit farmers to re-use proprietary seeds. UPOV 1991 was considered to prevent farmer plant-back. See The Crucible II Group, *Seeding Solutions: Policy Options for Genetic Resources – People, Plants and Patents Revisited*, vol. 1 (Ottawa: International Development Research Centre, 2000) [Crucible II, *Seeding Solutions*], IDRC, <http://web.idrc.ca/en/ev-31619-201-1-DO_TOPIC.html> at 40–71. Plant variety protection provides a different form of protection than patents for plants, see Mark D. Janis, 'Sustainable Agriculture, Patent Rights, and Plant Innovation' 9.1 *Ind. J. Global Legal Stud.* 91–119.

73 C.M. Ho, 'Patent Law and Policy Symposium: Re-engineering Patent Law: The Challenge of New Technologies: Part III: International and Comparative Law Issues: Splicing Morality and Patent Law: Issues Arising from Mixing Mice and Men' (2000) 2 *Wash. U. J. L. and Pol'y* 247 at 257–8.

74 Kevles, *supra* note 3 at 58.

morality with the direction that a balancing test be undertaken to consider three main concerns: the potential suffering of the animal due to cancer; the potential risks posed to the environment if a GM mouse escaped; and the human benefit of expedited cancer research.[75]

Harvard's reply discharged the burden by asserting that if patented, the oncomouse would be widely available under licence – this would reduce the aggregate number of mice needed for cancer research. In addition, using mice as research subjects has long been an acceptable practice for society due to the research benefits. Finally, the probability of a mouse escaping and causing environmental disaster was negligible and outside of the contemplation of the patent system, although possibly subject to external regulations. This argument is consistent with what patent law in all jurisdictions confers: a right to *exclude* and not a right to *commercialize*. Other regulatory compliance requirements could be imposed for the sake of commercialization, Harvard argued, but agencies charged with control, over hazardous materials for example, would have better jurisdiction over such issues.[76] On 4 October 1991, the Examining Division issued a communication under Rule 51(4) of the EPC that it would grant Harvard its patent. After a period of further opposition pursuant to the EPC, the mouse emerged triumphant under Patent No. EP0169672.[77] The Board took the added rare measure of publishing in the Bulletin its comments on the fundamental patentability issues arising from the case.[78] The ruling provoked ten more years of controversy and the 'no patents on life' campaign that continued helped secure the Article 27.2 and 27.3 exemptions to patentability into the final text of TRIPS in Uruguay.

Amani and Coombe observe that in this case, the potential adverse effects of a technology were held sufficient reason to regulate its use in Europe but not to prevent a patent from issuing in the first instance. Concerns over the need to protect the environment from the unwanted spread of genes and animals from cruelty were outweighed by the overwhelming consensus in the scientific community that the oncomouse would be instrumental in furthering cancer research. The authors note, '[o]nce again, the Western conception of the scientific research process was reified and valorized over and above all

75 EPO, 'Decision of the Technical Board of Appeal 3.3.2 of 3 October 1990'.

76 'European Patent Application No. 86 304490.7, President and Fellows of Harvard College, Response to the Official Letter of 11 December, cited in Kevles, *supra* note 3 at 60; Ho, *supra* note 73 at 260.

77 The patent would take effect later in May once published in the European Patent Bulletin (European Patent Office, Press Release 3/92, 'European Patent for Harvard's Mouse').

78 For a timeline, see Official Journal of the EPO, 10/1992, <http://www.european-patent-office.org/news/pressrel/pdf/oj1992_10_p588_593.pdf>.

other social considerations while the costs of monopoly rights to the actual practice of research remained unaddressed'.[79]

The European Harvard mouse decision signalled that ethical concerns were of some relevance and that patentees in European patent applications must discharge the Article 53(a) requirements. However, it also demonstrated the Division's determination that an animal (mice), found to be in a higher taxonomic classification than a species, would not be captured by the exclusion for *varieties* in Article 53(b), which means that animals as subject matter are patentable if 'invention' is demonstrated.[80] In the long term, however, the efficacy of the *ordre public* and *morality* provision to moderate the future of biotechnology animal patents will only be as good as the policy in determining what the substantive standards enlivening these terms are. How the Article 53 exclusions are operationalized determines their utility in addressing policy concerns and for balancing economic interests with policies geared to protecting human rights. The terms *ordre public* capture important features underlying the need for regulatory diversity and when translated from French are different than mere *public order*. Daniel Gervais explains:

> [w]hile public order may be defined as the maintenance of public safety, *ordre public* concerns the fundaments from which one cannot derogate without endangering the institutions of a given society. A typical application of this rule in civil law jurisdictions is that while private persons are free to contract in or out of any obligation, that freedom is limited by *ordre public*. Covenants ... require approval of the State ... where interests of society (represented by the relevant state institutions) are involved.[81]

'Morality' writes Daniel Gervais, 'is a different concept ... It seems to correspond to the French concept of "*bonnes moeurs*". It naturally depends to a certain degree on the particular culture of a country or region'.[82]

The patenting life debate continued unabated and in 1998 the EU issued the Biotechnology Directive (EBD), now part of European Community law, to try and clarify the role of ethical considerations in EU patent jurisprudence

79 Amani and Coombe, *supra* note 12.

80 Harvard/onco-mouse [1991] E.P.O.R. 525. The Examining Division avoids interpreting 'animal varieties' but concludes at 526 it is 'not entirely clear, in particular in view of the differing wording in the three equally binding languages of the EPC, it nevertheless could be stated with certainty that rodents, or even mammals, constituted a taxonomic classification unit much higher than species ("Tierart"). An "animal variety" or "race animale" is a sub-unit of a species and therefore of even lower ranking than a species'.

81 See Daniel Gervais, *The TRIPS Agreement: Drafting History and Analysis* (London: Sweet and Maxwell, 2003) at 223.

82 Ibid. at 222–3. See also Philip C. Mendes da Costa, 'Patent Harmonization through GATT: TRIP or Trap' (1992) 45 *Patent World* 23 at 25.

and introduce harmonized standards that assure biotech patentability. The EBD was opposed by a number of EU member states. Members were to implement the Directive into national law by the year 2000 but at least nine members refused to do so because of strong opposition to the patenting of genetic material of all life forms – a decision which can subject them to the European Court of Justice (ECJ) for infringement on referral by the European Commission.

The EBD prohibits the patenting of human clones and of innovations that involve industrial use of embryos as well as 'parts of the human body ... at various stages of its formation and development, and the simple discovery of one of its elements'. Nevertheless, Article 5 provides that 'elements isolated from the body, including genes, will be patentable if their [industrial] utility can be established'. After the EBD was adopted, the Netherlands, with the support of Italy and Norway challenged the legality of the EBD at the ECJ (ECJ, Case C-377/98) with the Dutch government arguing that the EBD was in breach of the principles of subsidiarity and legal certainty of international obligations, that it undermined the inalienable nature of human beings, and conflicted with the fundamental right to human dignity and integrity.

In 2001 the ECJ found otherwise and in confirming that the EBD was legal, held that there were sufficient safeguards within the text (such as Article 6 and the 38th recital to the preamble relating to *ordre public* and *morality*) to protect these fundamental rights given that the human body as such cannot be patented. Additionally, if a process offends human dignity, the ECJ ruled, it *must* be excluded from patentability.[83] Netherlands was ordered to pay costs for the European Parliament and of the Council. But despite the bioindustries' celebration of this decision, more than a year after the deadline, eleven out of the fifteen member states had yet to become compliant. The prolonged delay in implementing the EBD into national law of many EU member states suggested cultural resistance and continued controversy over biopatenting despite industrial practice and the law.

The prospects of using the *ordre public* and *morality* provisions in Article 53 of the EPC to limit domestic patents on human genetic materials appears increasingly remote now given the apparent high threshold for offending human dignity. The successive narrow interpretation by the EPO of Article 53 has served to restrict its potential to inventions that threaten national security or an individual's physical integrity in society; presumably biochemical agents for bioterrorism would qualify, but one cannot be sure.

In 1995, *Howard Florey Institute/Relaxin*[84] determined that the 'morality' exception applied only 'to ensure that patents would not be granted for

83 ECJ decision, <http://curia.europa.eu/en/actu/activites/act01/0125en.htm>.

84 *Re Howard Florey Institute/Relaxin, EPO Opposition Division* [1995] E.P.O. R. 541 [*Howard*]. See also Neil Jenkins, 'In Brief: The Impact of the EU Biotechnology Directive on the Patenting of Biotechnology' (2001) *Patent World* 128.

inventions which would *universally* be regarded as *outrageous*', establishing a standard that is arguably impossible to achieve. Moreover, it was held that the invention – human DNA fragments with a particular encoding – did not involve the patenting of human *life* per se. The capacity of a patent granting body to determine the meaning and boundaries of human life was questionable to many but, even on appeal, the Technical Board of Appeal affirmed that '... DNA is not 'life' but a chemical substance which carries genetic information and can be used as an intermediate in the production of proteins ...'.[85] But DNA engages in the same kind of organic auto-activity we attribute to life: it replicates, mutates, and creates; it is sufficiently personal that it requires consent to extract, helps forensically identify criminals, and tells us not only about ourselves, but our pasts and futures An artificially narrow interpretation as merely chemical perverts the actual (legal) consequences it gives rise to.

In Canada a rare thing happened: Harvard did not succeed despite the substantial similarity in the legal definition of 'invention' under s. 2 of the CPA with the American statute (on which it was modelled) and the general proclivity for judicial convergence. The patent application for Harvard's mouse took 17 years to resolve in Canada during which time patent protection for transgenic non-human higher life remained uncertain. But this did not stifle innovative activity in this field. Perhaps Canadian inventors are greater puzzle seekers or simply they derive their incentive to innovate from sources external to the Canadian patent regime. The prospect of patenting in more viable and lucrative commercial markets, like those of the United States and Europe, may be a stronger incentive for Canadian inventors than domestic patent law and policy; what is optimal protection as an incentive in each specific jurisdiction may vary.

Harvard's application in Canada covered 26 claims and was made on 21 June 1985. It came to rest with the December 5, 2002 judgment of the Supreme Court of Canada. The application claimed patent protection for both the *process* of creating transgenic mice and *product* claims over an array of related goods including the transgenic material (oncogene sequence), plasmids, oncoeggs, and of course the GM mouse itself. The Canadian patent examiner originally rejected 18 of the 24 claims submitted. Upon Harvard's request for review and an amended application for 26 claims, the examiner in 1993 allowed the process claims 13–26 but again rejected the product claims 1–12 as non-statutory subject matter. In 1994, the Patent Appeal Board heard the appeal from the decision of the Examiner as endorsed by the Patent Commissioner endorsed. Section 40 of the CPA gives the Commissioner discretion to refuse a patent which the applicant is not entitled to by law.[86] Accordingly, the Commissioner refused to extend 'manufacture' or 'composition of matter'

85 *Howard*, op. cit. at 541–2.

86 *President and Fellows of Harvard College v. Canada (Commissioner of Patents) (T.D)* [Harvard Trial], <http://reports.fja.gc.ca/fc/1998/pub/v3/1998fc22348.html>.

in the definition of s. 2 'invention' to include non-human higher life forms. Harvard's appeal to the Federal Court Trial Division in 1998 was dismissed by Justice Nadon but was allowed by the Federal Court of Appeal in 2000.[87] Rothstein J.A. for the majority (Linden J.A. concurring, Isaac J.A. dissenting), quashed the decision of the Commissioner and trial judge and remitted the matter to the Patent Commissioner with the direction that the patent granted cover the product claims 1–12 set out in the application.

Canada's heavy investments in the biotechnology sector and its political support for stronger IPRs led many legal and business commentators to assume what was later confirmed by the Federal Court of Appeal (FCA): that Canada's industrial needs would trump to ensure the patentability of genetically modified life forms. The SCC decision was expected to affect approximately 600 patent applications on GM animals and plants.[88] But the refusal of the SCC to follow the United States and European precedent recognizing the validity of Harvard's product claims suggested that critical scrutiny of these patents within relevant industrial sectors as well as growing public and NGO opposition was finally having some impact. While Canada formally lacks a morality provision in its patent laws, issues of morality and environmental risk were raised by the numerous intervenors before the Supreme Court. The majority for the Court made sure, however, that they would not be perceived as engaging in making judicial patent policy. They did this by restricting their reasoning to avoid obscuring the matter of statutory interpretation with which they were charged with the issues of morality for which there was no legislative authority. Per Justice Bastarache,

> [t]he sole question in this appeal is whether the words 'manufacture' and 'composition of matter', in the context of the PA, are sufficiently broad to include higher life forms. If these words are not sufficiently broad to include higher life forms, it is irrelevant whether this Court believes that higher life forms such as the oncomouse ought to be patentable ... In my view, whether higher life forms such as the oncomouse ought to be patentable is a matter for Parliament to determine. This Court's view as to the utility or propriety of patenting non-human life forms such as the oncomouse are wholly irrelevant.[89]

Harvard advocated a strict application of the CPA by the Supreme Court in order to preserve the status quo and strongly argued that moral and policy judgments pertaining to an invented technology were not within the jurisdiction of patent law to contemplate, given that statutory interpretation

87 *Harvard FCA, supra* note 61.

88 Phillip B.C. Jones, 'Will the Oncomouse Squeak Through the Supreme Court of Canada?' (July 2002), <http://www.isb.vt.edu/articles/jul0206.htm>.

89 *Harvard SCC, supra* note 7 at para. 153.

must inherently be amoral. In addition, Harvard's counsel asserted that the issues raised by the intervenors were better addressed by external regulation, as they successfully had in Europe. This would make patentability the default subject to political will to legislate to the contrary. The mere grant of a patent on living subject matter or our conception of IPRs, however, are not entirely value free or morally ambiguous as Mason and McCall Smith have found:

> The idea of property is primarily a legal one. It is a construct which allows us to order our society according to a chosen value system, which in turn greatly facilitates the achievement of certain of our social goals – in our case, those of most Western states which encourage commerce and sanction commodification.[90]

The authors insightfully add, that '[t]he role of ethics lies not in grounding a property right but in determining whether it is appropriate to commodify something such as the human body which … has a particular moral status'. Granting patent rights to life is a policy choice itself reflecting a particular set of values in a given society; one which normatively prescribes and reinforces the judicially endorsed ideology that patents can be granted for 'anything under the sun made by man'[91] and, in relation to DNA sequences, even found in nature. But the relationship between patent law and morality has been inconsistent and dependent on domestic economic and political conditions. Justice Binnie's dissent found that the 1993 repeal of the s. 27(3) prohibition in Canada against patenting 'an invention that has an illicit object in view' made clear

> that granting a patent is not an expression of approval or disapproval. At that time, Parliament did not add a provision, present in the European Patent Convention and in many civil law systems and international agreements, that patents will not be granted for inventions whose use of exploitation would be inconsistent with *ordre public*, public morality or environmental or health protection. That type of provision would open the door to value judgments in assessing patentability … Parliament thereby signaled, however passively, that these important aspects of public policy would continue to be dealt with by regulatory regimes outside the Patent Act.[92]

For the SCC, while various references were made to international law, references to *ordre public*, *morality,* and patent policy objectives proved

90 Mason and McCall Smith, *supra* note 39 at 486.

91 *Chakrabarty, supra* note 19.

92 *Harvard SCC, supra* note 7 at 12. The provision was CPA, *supra* note 57 s. 27(3), prior to amendment by the North American Free Trade Agreement Implementation Act, S.C. 1993, c. 44, s. 192 [*NAFTA I A*].

unnecessary in the end but also indicate judicial reluctance to engage in what were believed to be determinations better suited to the legislative branch of government. The majority upheld the Commissioner's decision distinguishing the patentability between Harvard's *product* and *process* claims.

The Court's decision was grounded in a common sense reading of the statute. The words used in the definition of 'invention' are broad but it is an exhaustive definition that signals on its plain meaning a clear intention of parliament to include certain subject matter as patentable while excluding others. Bastarache J. writing for the majority held that the oncomouse did not fit within the s. 2 definition of 'invention' because it could not accord with common statutory interpretation rules for interpreting the 'imprecise' and 'ambiguous' words of 'manufacture' or 'composition of matter' within the definition. Moreover, the issue of whether something falls within the s. 2 statutory language is, according to the Court, a pure determination of law,[93] supporting a jurisdictional approach.

Another reason for the majority's decision was its deference to the democratic process and parliament's role to settle these complicated issues legislatively. In so doing, the SCC recognized that before the one-size-fits-all patent is extended to all biotech innovations, it should be decided whether the scientific developments claimed are ones which we would want to otherwise regulate outside of patent legislation. This is a sensible anticipatory approach that seeks to minimize legal inconsistencies that would otherwise bring the administration of justice into disrepute. It is inefficient and a paradox to provide legal incentives, for example, for that which is duly criminalized to be deterred. The balance of interests in aggregate social goals should occur *a priori* in the development of coherent industrial policy and its coordination with other public policy areas. A recent Bill to amend the Canadian Criminal Code provisions on cruelty to animals under examination by the Canadian Parliament's Standing Senate Committee on Legal and Constitutional Affairs reportedly 'has researchers in the biotechnology sector worried'.[94] The concern is over animal experimentation for testing new drugs since the provisions against animal cruelty have been removed from Part XI of the Criminal Code (Willful and Forbidden Acts in Respect of Certain Property) to a new Part V.1 of the Code; this change may effectively remove access to the specific defences of colour of right and legal justification available under the former property-based provision and would signal a departure from the vestiges of such ownership in animals. If animals are no longer conceptualized as tangible personal property should they now be treated instead as intangible objects commoditized by IPRs?

93 *Harvard SCC*, *supra* note 7 paras. 144–66.

94 Paul Marcotte, 'The Biotechnology Industry and Animal Experimentation in Canada', Biotechnology Bulletin, Fasken Martineau (April 2003), <http://www.fasken.com>.

In Canada, having found plants and animals to constitute non-human *higher* life, the Supreme Court relied on the existence of the *Plant Breeders' Rights Act*[95] as an expression of legislative intent to exclude higher life from patentability. The 1990 legislation specifically provides protection to breeders of plant varieties and while like its US equivalent it provides a lower burden for obtaining rights, it offers a correspondingly narrower monopoly than that available under the Patent Act. Unlike the Supreme Court of the United States, the SCC took the provision of plant variety protection in separate legislation as an expression of parliamentary intent to preclude plants from patentability. The absence of similar separate legislation for animals led to a judicial inference that parliament did not mean to provide such protection for animals and supported the finding against extending patent protection to them. The majority found that 'although Parliament passed "special legislation" to provide protection for plant breeders, it made no move to amend the Patent Act or to adopt other special legislation to provide for the protection of forms of animal life'.[96] According to Bastarache J., '[g]iven the unique concerns associated with the grant of a monopoly right over higher life forms, it is my view that Parliament would not likely choose the Patent Act as it currently exists as the appropriate vehicle to protect the rights of inventors of this type of subject matter'.[97]

Policy analysts across sectors and legislatures across jurisdictions must work together in consultation with their public to make certain value based decisions on the incentives to be provided, the rewards to be given, and the costs that are willing to be borne in determining whether life is patentable. Undoubtedly, Harvard's oncomouse facilitates cancer research but how should national governments cooperate in resolving the arising practical and legal issues? In Canada, the delivery of health typically falls within provincial jurisdiction over property and civil rights[98] but the provinces are restrained by federal objectives conditioned on funding transfers (the Canada Health

95 *Plant Breeders' Rights Act*, S.C. 1990, c. 20 [PBRA]. See discussion in *Harvard SCC*, *supra* note 7 at para. 188–90.

96 Ibid. at para. 189. See also PBRA, op. cit.

97 Ibid. at para.120.

98 Section 92(7) grants the provinces authority over hospitals. Under section 92(13), power over health is included in the provincial property and civil rights, and under section 92(16) as a matter of local and private nature. However, health can also fall under the federal power under s. 91(27) for criminal law aspects of health (criminal regulation of food and drugs) punishing conduct dangerous to health, within the general jurisdiction of the peace, order and good government federal power if the health problem has national dimension (such as the spread of epidemics, water pollution, and pestilence) and in relation to labour relation standards, is under federal power; jurisdiction depends on 'the purpose and effect of the particular health measure in issue'. Peter W. Hogg, *Constitutional Law of Canada*, 3rd edn (Toronto: Carswell, 1992) at 476.

and Social Transfer) as part of the shared-cost programmes within each province. This unique relationship makes inter-governmental cooperation and collaboration imperative for the viability of the Canadian health care system and its internal coherence between provinces.[99] IP regulation, on the other hand, remains within federal jurisdiction[100] making policy coordination between two jurisdictions and two separate ministries – provincial health and federal industrial – that much more difficult. Not surprisingly, the majority of the SCC found that the patenting of life,

> is a policy issue that raises questions of great significance and importance and that would appear to require a dramatic expansion of the traditional patent regime. Absent explicit legislative direction, the Court should not order the Commissioner to grant a patent on a higher life form.[101]

Next, the Court found that the patenting of higher life raised 'unique concerns' that do not accord with the scheme of the CPA and are not shared with respect to traditional invention. Here, the Court was referring to the ability of such inventions to self-replicate; what they call *reproducibility* and Habermas refers to as *auto-activity*.[102] The Court found the line drawn in *Abitibi* regarding reproducibility *en masse* of micro-organism arbitrary as are other legal lines drawn between higher and lower life: 'All of them are policy driven and, if they are to be introduced at all, should be introduced by Parliament'.[103] The majority recognized that life is qualitatively different subject matter for the purpose of determining patentability and adopting the view expressed in the June 2002 Canadian Biotechnology Advisory Committee's (CBAC) final report, the Court found:

> [b]ecause higher life forms reproduce by themselves, the grant of a patent covers not only the particular plant, seed or animal sold, but also all of its progeny containing the patented invention. In the CBAC's view, this represents a significant increase in the scope of the rights offered to patent

99 See National Forum on Health, *Canada Health Action: Building on the Legacy – Final Report of the National Forum on Health* (Ottawa: Minister of Public Works and Government Services, 1997) at 20; Auditor General of Canada, *Report of the Auditor General of Canada to the House of Commons, 1999* (Ottawa; Minister of Public Works and Government Services Canada, 1999) c. 29 at 19; Martha Jackman, 'Constitutional Jurisdiction Over Health in Canada' (2000) 8 *Health L. J.* 95, University of Alberta <http://www.law.ualberta.ca/centres/hli/pdfs/hlj/v8/jackmanfrm.pdf>.

100 Patent regulation is expressly granted to federal parliament under s. 91(22) of the *Constitution Act, 1867* (U.K.), 30 and 31 Vict. c. 3, reprinted in R.S.C. 1985, App II, No. 5.

101 *Harvard SCC, supra* note 7 para. 155.

102 Ibid. at para. 170.

103 Ibid. at para. 52.

> holders that is not in line with the scope of patent rights provided in other fields.[104]

Harvard's counsel highlighted that lower life (bacteria, yeast and moulds) have long been patentable in Canada as a 'composition of matter' within the s. 2 definition and that there was no justifiable or justiciable reason for excluding higher life.[105] The Patent Office's distinction between higher and lower life, Harvard argued, was ill-founded and an arbitrary exercise in line drawing. The majority was wise to recognize that all judicial decisions and legislative ones too are about line drawing and that with such bright lines, Parliament has the better authority. Any critique offered of the majority's decision to discern a difference between higher and lower life is a criticism of the process of judicial decision making in general which necessarily draws lines. *Abitibi* drew a line as did *Chakrabarty*. Each line must be reviewed for the balance it achieves between the interests of various parties affected and competing interpretation of statutory provisions. The decision to allow biopatents for DNA and lower life is also a bright line the courts have drawn; the majority of the SCC was not willing to extend it to non-human higher life.

Like many of the Supreme Court decisions on biopatents, *Harvard* was derived from a narrow difference of opinion in which the majority recognized that expansion of 'invention' under s. 2 of the CPA in such circumstances would be a policy decision best left for Parliament while approving the line drawn by the PO and courts in the past distinguishing between higher and lower life for three main reasons: (1) micro-organisms are produced en masse as are chemical compounds. Any measurable quantity can possess uniform properties and characteristics which cannot be said of plants and animals. This premise may become moot once cloning technology becomes effective for en masse production of higher human and non-human life, particularly for organ harvesting. Still, on this point, the more important finding by SCC was that the disclosure of the invention would be insufficiently enabling to allow one skilled in the art to duplicate as the capacity to self-propagate and reproduce creates natural variability that cannot be disclosed by the patentee nor repeated by a PHOSITA; (2) there are significant distinctions between higher and lower life which include variations in responses to stimuli; and that (3) the *Agreement on Trade Related Aspects of Intellectual Property Rights* (TRIPS) and the *North American Free Trade Agreement* (NAFTA), each contain an article whereby members can 'exclude from patentability' specifically provided-for subject matter which, while including plant and animal life, does not extend to micro-organisms. The SCC's final reason reflects a desire to reduce friction with international obligations and notes that the majority's decision is consistent with permissible regulatory scope under

104 Ibid. at para. 171.
105 Ibid. at para. 198.

TRIPS and would not compromise Canada's relations as a result of trade obligations. In short, the Patent Office was justified in its denial of the product claims according to Justice Bastarache since:

> [o]nly Parliament has the institutional competence to extend patent rights or another form of intellectual property protection to *plants and animals* and to attach appropriate conditions to the right that is granted. In the interim, I see no reason to alter the line drawn by the Patent Office. The distinction between lower and higher life forms, though not explicit in the Act, is nonetheless defensible on the basis of common sense differences between the two. Perhaps more importantly, there appears to be consensus that human life is not patentable; yet this distinction is also not explicit in the Act. If the line between lower and higher life forms is defensible and arbitrary, so too is the line between human beings and other higher life forms.[106]

… And Men: Schmeiser and the Definition of 'Use'

The Case and its Challenges

The problem with line drawing is that the line can always be redrawn. The SCC in *Harvard* conferred the responsibility on parliament and yet, with the changing bench, the SCC reclaimed jurisdiction to determine the patentability of higher life – this time in relation to plants. The *Harvard* Court was cognizant of the leave to appeal to the SCC from the FCA decision in *Monsanto Canada Inc. v. Schmeiser*[107] (*Schmeiser*) upholding a genetically modified plant patent as valid and infringed. As the case before the SCC in *Harvard* was only about animals, presumably the consistent reference to plants *and* animals, and the specific finding in *obiter* that existing plant breeder legislation suggested that plants were not patentable (in Canada), was an effort by the majority in *Harvard* to prescribe a legal standard for treating the issue of patents on higher life forms as a regulatory class inclusive of plants *and* animals with the pending *Schmeiser* appeal in mind. However, *Harvard*'s majority became *Schmeiser*'s minority opinion and the deference to political processes to legislatively determine the patentability of higher life was quickly abandoned only two years later by a reconstituted Supreme Court. Amongst industry participants, some speculated that the restructuring of the Bench would dictate a different outcome for Monsanto's appeal, particularly given the potential influence of the political fallout of *Harvard* and that it was narrowly decided with a 5:4 split. Politicians are adversely motivated to respond to the fallout because

---

106 *Harvard SCC, supra* note 7 para. 199.

107 *Monsanto Canada Inc. v. Schmeiser* (2001), 12 C.P.R. (4th) 204 (F.C.T.S.), aff'd. [2003] F.C. 165 (F.C.A.) [*Schmeiser FCA*].

the issues are significantly charged and likely to polarize constituents. They would rather have the matter settled by the courts than in the court of public opinion. The result is a very closely divided Supreme Court of Canada on the question of what may be patented and what constitutes infringement even if *not* patented.[108]

In *Schmeiser*, Monsanto had obtained a Canadian Patent (No. 1,313,830) for an invention relating to glyphosate resistant plants.[109] The patent covered claims to a chimeric plant gene that encoded for an enzyme which conferred resistance to such glyphosate herbicide as Roundup®, which Monsanto produces. The claims extended to plant cells that contained the chimeric gene. Monsanto Canada licenses Canadian commercial seed growers to grow their Roundup Ready Canola (RRC). RRC is engineered to express a particular gene that makes the plant resistant to Roundup. When Roundup is sprayed, it will kill all other plants but the GM RRC and thereby makes it easier for farmers to control weeds. Licensees sign an undertaking not to save seed from one year's crop for future replanting or inventory and not to share (sell or give) the seed to third parties.[110] Mr Schmeiser, a farmer of 50 years, never purchased the canola, never obtained a licence, and claims not to have taken advantage of the patented feature of the canola plants. Yet, after an anonymous tip

108 In *Harvard* the 5:4 majority decision against patentability of higher life (animals) was delivered by Bastarache J, with justices L'Heureux-Dubé, Gonthier, Iacobucci, and Lebel concurring. The dissent was written by Binnie J. with McLachlin, Major, and Arbour concurring. In *Schmeiser*, the 5:4 majority opinion effectively supporting patent protection for higher life (plants) was delivered by Chief Justice McLachlin and Fish J., with Major, Binnie and Deschamps concurring. The dissent was delivered by Arbour J (dissenting in part) with reasons of Iacobucci, Bastarache, and LeBel.

109 Claim 1: a chimeric plant gene which comprises: a promoter sequence which functions in plant cells; a coding sequence which causes the production of RNA, encoding a chloroplast transit peptide/5-enolpyruvylshikimate-3-phosphate synthase (EPSPS0 fusion polypeptide, which chloroplast transit peptide permits the fusion polypeptide to be imported into a chloroplast of a plant cell; and a 3′ non-translated region which encodes a polyadenylation signal which functions in plant cells to cause the addition to polyadenylate nucleotides to the 3′ end of the RNA; the promoter being heterologous with respect to the coding sequence and adapted to cause sufficient expression of the fusion polypeptide to enhance the glyphosate resistance of a plant cell transformed with the gene.

110 Michael Blakeney reveals, however, that the 'conspicuous success of RoundUp Ready® seed is explained because the increase in seed price to cover the price of genetically engineered herbicide tolerance is offset by profits from enhanced yields. On the other hand, increased yields might result in glutted markets, with a commensurate decline in profits'. Michael Blakeney, 'Stimulating Agricultural Innovation' in K.E. Maskus and J.H. Reichman, eds, *International Public Goods and Transfer of Technology under a Globalized Intellectual Property Regime* (Cambridge, MA: Cambridge University Press, 2005) [*IPG*].

sent investigators to his Saskatchewan farm in 1997, it was discovered that the vast majority of his 1,000 acres (95–98 per cent by some accounts) of canola crop was made from Roundup Ready plants although the origin of the plants remained speculative.[111] During trial, Schmeiser claimed that the canola plants were the result of cross-field contamination (by wind or insects), making him at most an *accidental* infringer, if at all. Trial Judge W. Andrew Mackay was not persuaded that the innocent bystander claims put forward could explain the high degree of concentration of RRC '"of a commercial quality" ultimately present in Schmeiser's crop'. In his defence, Mr Schmeiser argued that 'the subject matter claimed in the patent is unpatentable. While acknowledging that Monsanto claims protection only over a gene and a cell, Schmeiser contends that the result of extending such protection is to restrict use of a plant and a seed'. Both Mr Schmeiser and the Court agreed that seeds, plants, and animals are not patentable in Canada pursuant to *Harvard*[112] but disagreed that the patentability of plants or seeds was the issue before the Court in *Schmeiser*. How the legal issue is framed is always imperative to its outcome:

> [t]his case is different from *Harvard Mouse*, where the patent refused was for a mammal ... The Patent Commissioner had allowed other claims, which were not at issue before the Court in that case, notably a plasmid and a somatic cell culture. The claims at issue in this case, for a gene and a cell, are somewhat analogous, suggesting that to find a gene and a cell to be patentable is in fact consistent with both the majority and the minority holdings in *Harvard Mouse*.[113]

Moreover, the majority in *Schmeiser* found that whether the scope of patent protection extends from gene and cell claims to provide rights to activities in relation to a plant was not relevant for the determination of *validity* but *infringement*.[114] On the issue of patentable subject matter, the majority did not focus on whether the case raised the issue of patent protection over plants, which is contested post-*Harvard*, but on the narrower issue of the validity of the patent claims to genetic material. They referred to *Kirin Amgen Inc v. Hoechst Marion Roussel Ltd:*

> That case dealt with a protein useful in the diagnosis and treatment of blood disorders. The English court construed the claims to exclude the naturally occurring form of the DNA sequence in a human cell. However, this was done to accord with the provisions of a regulatory scheme that has no

111 See *Schmeiser* FCA, *supra* note 107.

112 Ibid. at para. 21.

113 Ibid. at para. 22.

114 Ibid. at para. 24.

> parallel in Canada: Article 5 of the European Parliament's Directive 98/44/EC, which regulates the patentability of biotechnological inventions ... The legislature has not enacted a comparable statutory scheme in Canada to narrow the scope of patent construction. Thus *Kirin Amgen* is not applicable to the case before the Court.[115]

Consequently, Monsanto's gene and cell product claims were held to be valid.

The Trouble with 'Use'

The Federal Court of Appeal (FCA) affirmed the trial judge's finding of patent validity and Schmeiser's infringing 'use'. This encouraged the Canadian Intellectual Property bar to advise their clients to pursue claims to cells, sequence listings and so forth because the appellate decision broadly interpreted patent rights through the definition of 'use' and effectively established backdoor protection for higher (plant) life through exclusive rights claimed over the genetic and cellular material.[116] When the case reached the SCC, the majority stated that 'it is uncontested that Monsanto's patented claim is only for the gene and cell that it developed ... The more difficult question – and the nub of this case – is whether, by cultivating plants *containing the cell and gene*, the appellants *used* the patented components of those plants'.[117]

In a decision delivered in May 2004, the majority affirmed that while Schmeiser did not infringe Monsanto's exclusive rights under s. 42 of the CPA by *making* or *constructing* the invention, the collection, saving, and planting of seeds containing Monsanto's patented gene and cell constituted an infringing 'use' of the invention, and contravened the patentee's guarantee of exclusive rights.[118] The error in the majority's reasoning to support this interpretation of 'use' is apparent from the following passage which starts with a quote from a leading international IP academic but leads to an inappropriate analogy and consequently the wrong conclusion by the majority:

> As Professor Vaver states ...:

---

115 E.W.J. No. 3792 (QL), [2002] EWCA Civ. 1096 (C.A.), in *Schmeiser* SCC, *supra* note 121, para. 89.

116 See Torys Newsletter, No. 2002-29T, 6 December 2002, <http://www.torys.com>.

117 Emphasis added. See *Schmeiser SCC*, *supra* note 121 at paras 77–80; the court found that 'provided the patented invention is a significant aspect of the defendant's activity, the defendant will be held to have 'used the invention and violated the patent ... Infringement does not require use of the gene or cell in isolation'.

118 For a summary of the majorities propositions, see *Schmeiser SCC*, ibid. at para. 58.

> 'Use' applies both to patented products and processes, and also to their output. A patent that covers a zipper-making machine or method extends to zippers made by the machine or method. Each zipper sold without authority infringes the patent, even if the zippers themselves are unpatented. This expansive doctrine applies, however, only if the patent plays an important part in production'. By analogy, then, the law holds that a defendant infringes a patent when the defendant manufactures, seeks to use, or uses a patented part that is contained within something that is not patented, provided the patented part is significant or important. In the case at bar, the patented genes and cells are not merely a 'part' of the plant; rather, the patented genes are present throughout the genetically modified plant and the patented cells compose its entire physical structure. In that sense, the cells are somewhat analogous to Lego blocks: if an infringement use were alleged in building a structure with patented Lego blocks, it would be no bar to a finding of infringement that only the blocks were patented and not the entire structure. If anything, the fact that the Lego structure could not exist independently of the patented blocks would strengthen the claim, underlining the significance of the patented invention to the whole product, object or process. Infringement through use is thus possible even where the patented invention is part of, or composes, a broader unpatented structure or process.[119]

With due respect to the Court, the majority makes the fatal though common error of collapsing the essential distinction in patent law that weighed so heavily in *Harvard,* between *product* versus *process* claims just as the Board had before it in *Abitibi.*

Recall, Harvard wanted a patent product claim over the oncomouse and was not satisfied with the CPO's grant of claims extending to genetic and cellular material (products) as well as to the process used to make such material and a transgenic organism. This was a legitimate pursuit since if a rogue mouse escaped, was stolen or was created by some other processes not within Harvard's patent and went on to create more mice with the patented oncogene through natural breeding, there would be no infringement. That a run-away, stolen, or otherwise manufactured mouse (through an unpatented process) prominently featured the patented qualities of the oncogene would be insufficient to extend Harvard rights to the mouse unless Harvard secured a product patent. Otherwise the effort to secure a product claim over the mouse would effectively be redundant with claims granted over the transgenic matter. Vaver's 'zipper' analogy is correct but improperly applied by the Court. It is *only* if the zipper is itself patentable that a zipper made by some *other* process and not made 'by that machine' or 'by that method' would constitute an infringement of the patentee's rights. Similarly, if a plant or animal carries patented genetic or cellular matter, but does so because it is

119 Ibid. at paras 41–3.

the result of an unpatentable (natural) process, such as independent mating or cross-pollination, then arguably the resulting progeny or plant should *not* constitute an infringement unless these plants or animals are themselves patented *products*. If liability were to be imposed, then it would have to be based on some fault-based principle not in the CPA that would render the acts of the defendant nevertheless culpable for dirty hands or wrongful conduct – i.e., that a mouse was stolen and procreated through *breeding* rather than independent mating.

GM contamination in nature is not uncommon.[120] The SCC concedes that this may occur and Monsanto asserted that it would not bother with litigating such cases. But here was Mr Schmeiser making these same 'innocent by-stander' arguments, contending that any occurrence of the GM crop occurred through wind-blown contamination of his fields, over which he had no control (an alternate process)[121] and yet Monsanto *was* litigating it. Whether the RRC appeared innocently on Schmeiser's farm through no act of his was also held to be irrelevant to the determination of *infringement* which, the court affirmed, is based not on intent but on whether the defendant has *made*, *used*, or *sold* the patented invention.[122] Strangely, 'use' was strictly determined by this SCC. Chief Justice McLachlin and Justice Fish, writing for the majority identify the 'salient facts' to include that Schmeiser did not purchase the GM crop or obtain a licence to plant it,

> [y]et, in 1998, tests revealed that 95 to 98 percent of his 1,000 acres of canola crop was made up of Roundup Ready plants. The origin of the plants is unclear ... The trial judge found that 'none of the suggested sources [proposed by Schmeiser] could reasonably explain the concentration or extent of Roundup Ready canola of a commercial quality' ultimately present in Schmeiser's crop ((2001), 202 F.T.R. 78, at para. 118).[123]

Schmeiser was shortsighted in its implications but indicates that the majority was influenced by the specific facts of the case and two factors in particular. First, the Court could not get around the facts which, they seem to imply, suggest that there may have been some bad faith on the part of the appellants, supported by suspicious findings that most of Mr Schmeiser's corn was RRC

120 For a discussion on transgene contamination in Mexico, see the NGO, Action Group on Erosion, technology and concentration (ETC Group), report 'Maize Rage in Mexico: GM Maize Contamination in Mexico – 2 Years Later', 10 October 2003. The ETC report provides that a massive recall of tainted food products in the US lead Aventis to withdraw StarLink from the US market. Transgene contamination has reportedly affected a number of crops worldwide, see <http://www.etcgroup.>

121 *Schmeiser* SCC, *supra* note 121, para. 6.

122 Ibid. at para. 49.

123 Ibid. at para. 6.

and that he actively cultivated and harvested RRC seeds. Although his fields of canola were not sprayed with Roundup Ready®, so, technically, he derived no use, he did end up with 1,030 acres of RRC which would have cost him $15,000.[124] This led the majority to conclude that '[i]n these circumstances, the presumption of use flowing from possession stands unrebutted'.[125] Second, the Court was compelled by the 'stand-by or insurance utility' of the properties of the patented genes and cells. Regardless of whether a farmer actually sprays the Roundup herbicide on the RRC, '[t]he farmer benefits from that advantage from the outset: if there is reason to spray in the future, the farmer may proceed to do so'.[126] The Court also found that 'cultivating RRC also presents future revenue opportunities to 'brown-bag' the product to other farmers unwilling to pay the licence fee, thus depriving Monsanto of the full enjoyment of their monopoly'.[127] But, to conclude that cross-contamination with the patented crop invariably is a benefit to its owner is to reveal the normative value placed on the patent regardless of the nuisance and harm that such unintended GM contamination may cause in diminishing the integrity of Mr Schmeiser's crop (important in an age where consumer niche markets are developing around 'organic' and 'GM free' labels). The potential liability of such companies in trespass to real property, nuisance, and negligence law may have been given due consideration had the facts not led to the uncertainty regarding the cause of contamination of Mr Schmeiser's fields. Subsequent litigation to this effect has failed, placing the burden of GM contamination on the farmer rather than the corporate polluter.[128]

David Morrow, who was Harvard's counsel at the SCC and acted as co-counsel for the intervener, the Canadian Seed Trade Association in *Schmeiser*, summarizes the impact of the *Schmeiser* decision for biopatenting:

> The result of this decision provides a fairly strong confirmation that the making, using or selling of a 'macrobiological article such as a plant or seed, even when made by a macrobiological method such as planting seeds, infringes a claim for a microbiological article such as a gene or cell that is

124 Ibid. paras 60–70, and para. 87. Incidentally, there are current efforts underway for devising a 'synthetic' gene coding for glyphosate resistance in plants that would achieve the same result as Monsanto's patented product but without infringement. See e.g. Todd Funke et al., 'Molecular Basis for the Herbicide Resistance of Round Up Ready Crops' (August 2006) 103.35 PNAS 13010–13015, <http://www.pnas.org/cgi/content/abstract/103/35/13010?ck=nck>.

125 Ibid. at para. 87.

126 Ibid. at para. 84.

127 Ibid. at para. 85.

128 See *Hoffman v. Monsanto Canada Inc.*, Sask. Q.B. 10 April 2003, 2003 Sask. D.J. Lexis 130, certification of class action on organic farmers denied. See also Etc Group (20 May 2004) 'Canada's Supreme Court rules on "David and Goliath" tomorrow: Tell Monsanto where to go!', <http://www.etcgroup.org/article.asp?newsid=452>.

> included within the macrobiological article. In other words, even though there was no patent on the 'Roundup Ready' canola plant, the plaintiff Monsanto was able to get remedies for infringement as though there had been a patent on the actual plant. As a result, this decision may provide means for patentees of microbiological articles to have effective protection for plants even in the absence of a claim to the patent.[129]

The majority of the Supreme Court allowed Mr Schmeiser's appeal in part, finding that on the facts the appellants had made no actual profits as a result of his use of the patented invention (that is, he made no additional profits beyond that which he would have made in any event through the sale of his crop). The award for account of profits was set aside but in all other respects the trial Judge's decision was confirmed.[130] Like the SCC in *Harvard*, the Court in *Schmeiser* also invited a legislative response to settle patentability.[131]

The dissent casts the issue very differently, noting that the lower courts' decisions in *Schmeiser* were without the benefit of the SCC *Harvard* decision: 'The heart of the issue is whether the Federal Court of Appeal's decision can stand in light of our decision in that case [*Harvard*] ... that higher life forms, including plants are not patentable'.[132] The majority's decision conflicts with the Patent Office's long standing policy not to grant exclusive rights to higher life forms (defined as plants and animals) as products.[133] Finally, the majority's emphasis on the commercial value of the exclusive rights to the patentee as the primary consideration for extracting the 'essential elements' of the patent claims ignores three other considerations that the dissent considered relevant to a

---

129 See A. David Morrow and Sandee Smordin, 'Examination of Biotech Patent Issues: The Harvard Mouse Case', International Conference on Intellectual Property, University of British Columbia, 19 September 2003 at 18 discussing the F.C.A. decision which was upheld by the majority of the SCC.

130 *Schmeiser SCC, supra* note 121 at para. 106.

131 Ibid. at para. 94: 'Where Parliament has not seen fit to distinguish between inventions concerning plants and other inventions, neither should the courts'; para. 95: 'Again, if Parliament wishes to respond legislatively to biotechnology inventions concerning plants, it is free to do so. Thus far it has not chosen to do so.'

132 Ibid. at paras 107–11. The court quotes from E.R. Gold and W.A. Adams, 'The Monsanto Decision: The Edge or the Wedge' (2001), 19 *Nat. Biotechnol.* 587, asserting that both lower court decisions 'allo[w] Monsanto to do indirectly what Canadian patent law has not allowed them to do directly: namely, to acquire patent protection over whole plants' at 108.

133 See Industry Canada, Canadian Intellectual Property Office, *Manual of Patent Office Practice* (1998) at para. 16.05 which states: 'Higher life forms are not patentable subject matter. However, a process for producing a higher life form may be patentable provided the process requires significant technical intervention by man and is not essentially a natural biological process which occurs according to the laws of nature.'

purposive construction of the claims: namely (1) fairness and predictability;[134] (2) what is not claimed is disclaimed;[135] (3) the person skilled in the art.[136] The dissent found that 'a person skilled in the art, upon filing of Monsanto's patent, could not reasonably have expected that the exclusive rights for gene, cell, vector, and method claims extended exclusive rights over unpatentable plants and their offspring'.[137] Using the same three principles of statutory interpretation that the majority applied, the minority reached a completely different conclusion, defining an infringing 'use' as one linked to the invention and therefore limited by the subject matter and the construction of claims. As a result, the dissenting opinion concludes that use of the chimeric gene in its isolated form to create an expression or cloning vector or a transformation vector which is used to create a transgenic plant cell, or the use of the cell to regenerate a 'founder plant' but not its offspring, would all be infringing uses. That would mean that other herbicide producers could not use the transgenic material or processes disclosed in the patent to create a different herbicide resistant plant but Mr Schmeiser's use would be permitted. This is consistent, Arbour J. asserts, with TRIPS Article 27.1, which obligates Canada to make patents available without discrimination to the field of technology, and Article 27.3 which specifically allows the exclusion from patentability of plant and animal life (although if not a patent, an effective *sui generis* system is required for the protection of plant varieties; plant breeders' legislation satisfies this requirement).

Reflections on the SCC Decisions

The two SCC decisions on the patentability of higher life allow us to draw certain conclusions. First, the issue of patenting life remains highly contentious and more important today than it was before these two decisions. Its significance can be evinced by the number of interveners who sought and were granted

134 Where predictability is achieved by tying the patentee to its claims and fairness is achieved by interpreting the claims in an informed and purposive manner. See *Schmeiser SCC*, *supra* note 121 at para. 122, fn omitted.

135 Ibid. at para. 124. This is the 'classic rule' that an inventor may not get exclusive rights to an invention that was not part of the public disclosure and feeds into the predictability that the public must derive from what is claimed in order to decide whether their activities will infringe on the exclusive rights granted. If the patentee has limited the claims (as Monsanto did by explicitly disclaiming rights over the RRC plant), then the public is entitled to rely on that limitation.

136 While patent claims must be interpreted from the view of a hypothetical worker skilled in the art, a reasonable person skilled in the art must also be taken to know the state of the law as it relates to the subject matter of his invention. See *Schmeiser SCC*, *supra* note 121, paras 125–6.

137 Ibid. at para. 126.

leave to participate in the proceedings; in *Harvard*, no fewer than 12[138] and in *Schmeiser* 11 including the Attorney General of Ontario.[139] Second, the highest court has adopted an inconsistent approach to protecting proprietary interests in the subject matter and demonstrated ambivalence to what the law should be. Both courts left unchallenged the CPO practice of issuing patents on genes, cells, hormones, etc. But *Harvard* was decided by a 5:4 split against the patentability of higher life where an animal was at issue and *Schmeiser* was decided by a 5:4 split effectively favouring an interpretation that extends patent rights to unpatentable higher (plant) life through the definition of 'use'. The result is that the status of the patentability of life is murkier today than it was prior to these decisions, with findings that have in tandem obfuscated patent law and Patent Office practice. Third, both decisions invite parliamentary response through legislative reform, which has a number of apparent advantages over judicial fiat. Most notably, courts can decide only the issues presented in the case before them, are limited to the collation of facts given, and may be swayed by such extrinsic emotional factors as the desire to correct against piracy and bad faith. As well, they may be affected by such political influences as a desire for a convergence with patent law in other jurisdictions perceived to be more optimal to economic growth or in better compliance with international obligations. But, the precedents established extend beyond the narrow facts of a given case to all fields of technology and areas of patent practice, often conflicting with other areas of public policy. Consequently, patent law develops incrementally without any consideration of the technology, a centralized coherent underlying policy or the need for its coordination with other sectors of vital importance.

This approach to patent policy means that with time patent law becomes internally incoherent. For example, there are exemptions in Canada for abstract theorems and methods of medical treatment[140] but not for basic research

138 They were: Canadian Council of Churches, Evangelical Fellowship of Canada, Canadian Environmental Law Association, Greenpeace Canada, Canadian Association of Physicians for the Environment, Action Group on Erosion, Technology, and Concentration, Canadian Institute for Environmental Law and Policy, Sierra Club of Canada, Animal Alliance of Canada, International Fund for Animal Welfare Inc. and Zoocheck Canada Inc.

139 They were: Attorney General of Ontario, Canadian Canola Growers' Association (CCGA), Ag-West Biotech Inc., BIOTECanada, Canadian Seed Trade Association, Council of Canadians, Action Group on Erosion, Technology and Concentration, Sierra Club of Canada, National Farmers Union, Research Foundation for Science, Technology and Ecology and International Centre for Technology Assessment.

140 In *Tennessee Eastmen Co. v. Commissioner of Patents*, [1974] S.C.R. 111 (1974), 8 C.P.R. 92d) 202, the Commissioner of patents refused to grant a patent for a surgical method of applying specified new uses for known compounds. The Court found that a method of bonding incisions and wounds was not an 'art' or 'process'. It should be noted that while this common law precedent continues to exist in Canada

(DNA sequences) or diagnostic methods of testing even though they do not seem fundamentally different from both a policy and doctrinal perspective. The Canadian Manual of Patent Office Practice (MOPOP) provides under s. 12.04.02:

> A method or process of surgery or therapy on living humans or animals is not considered to be within the scope of 'invention' as defined by section 2 of the *Patent Act,* because such methods do not produce an essentially economic result in relation to trade, industry, or commerce ... Methods of diagnosing a physical disease or physical medical condition in a human being, provided that the methods do not contain any step of surgery or therapy, may be patentable.[141]

The 'methods of medical treatment' exclusion is very important for a country which prides itself on its public health system. More than eighty countries around the world reportedly have this public policy exclusion.[142] Other jurisdictions suffer from even less coherence. In the United States, it was judicially decided that methods of medical treatment *are* patentable[143] but to deal with the health policy issue, in 1996 a patent law amendment (s. 287 (c) of the 35 US Code) made such patents *unenforceable* against medical practitioners, performers of a medical activity, or against some other related health care entity. As a precedent, this model offers great possibilities for law makers regarding biotechnological inventions for health, particularly as they relate to gene therapies and regenerative medicine (such as methods of cultivating skin or processing cells). But the exception amended into US patent law explicitly excludes application to biotechnology patents (s. 287(c) 2 (A) (iii)). Neverthless, the variation in legal approaches is apparent and these emerge as remedial responses to policy issues: normally, a system strives to generate patented inventions that are held valid for infringement. Litigation may prove some patent claims valid but not infringed, or that the patent is not valid. To add to the confusion, we now have examples where some inventions are not patentable (as with the RRC plant) but are nevertheless infringed and some that are patentable but not enforceable as infringement; such confusion only adds to the costs of biopatenting. Herder and Gold wisely posit that any assessment of the relative value of IPRs to a given society should be consequentialist:

---

as the basis for the exclusion, the statutory provision on which it was based has since been repealed.

141 CIPO, MOPOP, *supra* note 50.

142 O. Mitnoveski and D. Nicol, 'Are patents for methods of medical treatment contrary to the *ordre public* and morality or 'generally inconvenient'?' (2004) 30 *J. Med. Ethics* 470–75, <http://jme.bmjjournals.com/cgi/content/full/30/5/470>.

143 See *Ex Parte Scherer*, 103 U.S.P.Q. 107 (Bd. Pat, App. 1954).

> … [P]atents are justified to the extent that the direct and indirect consequences of their existence increase wealth, health, and industrial practices more than they harm these … *What is important to note … is that effects (positive and negative) are not those that arise from a reading of applicable patent legislation, but rather from how real people actually react and deal with IP. Therefore if IP rights are so confusing that people stay away from some research or if there is a miscommunication that leads people to not conduct certain work, that negative effect is, from a consequentialist perspective accurately attributable to IP.*[144]

To add to the jumble, the United States after leading the world down a path towards greater *property fundamentalism* is on the brink of considering a new proposal to legislatively change course with a *ban* on patents for all human nucleotides (and possibly polynucleotide) and their functions. On 9 February 2007, California Congressman Xavier Becerra and Mr Weldon proposed a new Bill (US, Bill H.R. 977, *Genomic Research and Accessibility Act*);[145] in his introductory speech, Becerra implored:

> I rise today with the hope of fixing what I believe to be a regulatory mistake – a mistake that at first glance may seem minor in scope, but upon further examination has dramatic, costly and harmful implications for every American. I speak of the practice of gene patenting, where private corporations, universities and even the Federal Government are granted a monopoly by the United States Patent and Trademark Office on significant sections of the human genome … It is my belief that this practice is wrong, ill-conceived and stunts scientific advancement … My legislation, the Genomic Research and Accessibility Act, is straightforward: it ends the practice of gene patenting. It gives guidance to the United States Patent and Trademark Office (PTO) on what is not patentable-in this case, genetic material, naturally-occurring or modified. It is not retroactive – it does not rescind the patents already issued … We have overstepped our bounds. We have made a regulatory mistake. We have allowed the patenting of a product of nature. Fortunately, we have the power to end the practice expeditiously and for the benefit of all.[146]

In Canada, it is clear that neither of the SCC decisions affect the practice or legality of gene patenting, but taken together they significantly expand

144 Emphasis added. Herder and Gold, *supra* note 53 at 8.

145 110th Cong., 2007 to Amend Title 35, United States Code), <http://thomas.loc.gov/home/gpoxmlc110/h977_ih.xml>.

146 See US, *Cong. Rec.*, daily edn, at E315 (9 February 2007) (Rep. Becerra)), Library of Congress, <http://thomas.loc.gov/cgi-bin/query/D?r110:60:./temp/~r110KYJQHb::>.

the value and scope of gene patent rights after-the-fact. While the MOPOP confirms that plants, animals, seeds, and mushrooms are all examples of higher life and therefore not patentable,[147] standard patent practice now provides additional 'windfall' and unexpected licence generating streams for those who patent or have patented lower life and genetic material, producing consequences unanticipated by all parties including the issuing office when the grant was made. Human, plant, animal genetic sequences and micro-organisms continue to be patentable and patented, which means that we are on the horns of an interesting legal dilemma. On the one hand, the SCC in *Harvard* has found that higher life is not a patentable product. On the other hand, the SCC in *Schmeiser* has found that a use of an unpatentable product (e.g. plants, animals, and humans) which contains a patentable component (e.g. a micro-organism, virus, vector, DNA sequence, gene, cell, or plasmid) may be an infringing 'use' of the patent holder's exclusive monopoly rights in the subsumed patented invention. Not only are there major resulting policy concerns, but inequitable results will follow.

Imagine that Myriad manufactures a mouse with its patented BRCA1/2 gene using a process that does not encroach on Harvard's transgenic process claims. The BRCA1 mouse is useful for facilitating breast cancer research. In Canada, Myriad would effectively enjoy a patent right extending to the whole mouse as a product due to the *Schmeiser* definition of 'use' despite the finding in *Harvard* that higher life forms (transgenic mice) are not patentable. In fact, Myriad's rights would extend to the whole mouse or any other higher life that carried its patented gene(s), even if it did not manufacture or invent such a mouse or higher life. Disproportionately harmful impacts of existing DNA patents may be felt, such as on gene therapy.[148] While a patent does not give you a right to the actual tangible embodiment of the invention, the product patent was important to Harvard because it would have allowed Harvard to enjoin others from using such an oncomouse whether they independently created, reverse engineered, or came across it innocently. A product patent would have made its unlicensed manufacture, use, or sale an infringement of Harvard's rights in all contexts. The *Schmeiser* definition of 'use' has corrected for this 'gap' and in fact overextends protection not just to the animal as

147 CIPO MOPOP, *supra* note 50, s. 12.04.01.

148 Gene therapy is a means of correcting defective (faulty) genes responsible for disease development and entails one of several approaches to accomplish this including inserting a normal gene 'into a nonspecific location within the genome to replace a nonfunctional gene … An abnormal gene could be swapped for a normal gene through homologous recombination[,] the abnormal gene could be repaired through selective reverse mutation, which returns the gene to its normal function[,] the regulation (the degree to which a gene is turned on or off) of a particular gene could be altered'. Human Genome Project Information, <http://www.ornl.gov/sci/techresources/Human_Genome/medicine/genetherapy.shtml>.

product based on the fact that the mouse contains patented material (such as the plasmids) but to any potential embodiment. Considering the statistics on the significant portion of human, animal, and plant genomes patented, the implications of the SCC *Schmeiser* decision may be so dire they simply cannot be overestimated even if they cannot be measured empirically either.

The merging of mice and men, as poetic as it may be, is being contemplated by North American scientists considering the implantation of human embryonic stem cells into a mouse embryo. The experiment is to test whether human embryonic stem cells – on which research is controversial because a human embryo is destroyed during the procedure – have the capacity to become every human tissue type while in the mouse's body:

> Scientists have for years successfully inserted human genes into mouse DNA to study certain diseases. They have also fused mouse cells with human cells. But they have never created a mouse carrying human cells, each of which contains a full complement of human genes, said Ronald Worton, head of Canada's Stem Cell Network and the Ottawa Hospital Research Institute … The Canadian Institutes of Health Research, which distributes federal funds for medical research … president Dr Alan Bernstein noted … that no Canadian scientist receiving CIHR funds would be permitted to conduct such an experiment. 'We don't even really know what happens when you mix two different species like this', he said, 'Are you going to have a mouse walking around with human sperm – what would be the public reaction to that?[149]

In addition to raising the ethical and philosophical question of what is the essence of 'human' and the degree of genetic hybridization that any given society will tolerate within this definition, this experiment raises an interesting implication in light of the SCC's definition of 'use'. Effectively, patent protection would extend to the chimera product if it contained a patented component despite the fact that the Court in *Harvard* found that neither mice nor men are patentable.[150] Justice Binnie's dissent at the SCC emphasized that granting Harvard its patent is not equivalent with crediting it with inventing life. By Justice Binnie's account, what Harvard did was to 'modify every cell of

---

149 See Carolyn Abraham, 'Scientists look at Creating a Human-Mouse Embryo', *Globe and Mail*, <http://rtnews.globetechnology.com/servlet/ARticleNews/tech/RTGAM/20021128/wxmous11>. This report reaffirms three very important points. First, it highlights the ethical and socially contentious nature of such research. Second, that in the absence of public consultation and regulation there is no need to use the carrot of IPRs to create incentives for this research; in fact, the administrative decision to grant a patent should be stayed on such inventions until further consultation is made. Finally, it creates the quandary discussed in text.

150 *Harvard SCC*, *supra* note 7, para. 168.

a living creature in a new and useful way, and to the extent that modification is a valuable addition to the advancement of learning …'.[151] The majority, however, found that while that may be so, since the progeny of the mice were created through natural reproduction, to grant Harvard a patent in the mouse as product would be to over compensate Harvard for more than the scope of its invention, creating a windfall. Biotechnology '… can contain important characteristics that have nothing to do with the invention'. After *Schmeiser*, we can expect that in addition to the growing patenting trend there will be a significant shift in the type of biotech patents sought, away from macrobiological products (such as plants, animals, humans) the legal status for which is uncertain, to permitted microbiological products such as viruses, bacteria, fungi, and also genetic sequences of these, that will effectively provide protection to higher, possibly human, life. Private mapping projects have begun to shift to lower life, microorganisms, bacteria, and viruses for which patents can be obtained on the whole organism or simply its genetic sequences with broader enforceable rights.[152]

Finally, in the Harvard mouse case in Canada, a key issue was the distinction between higher and lower life given that since 1982, patents have routinely issued by the Commissioner in Canada on lower life forms as *composition of matter* or *manufacture* within the s. 2 definition of 'invention'. Justice Bastarache maintained the importance of the distinction for the Court's determination, writing for the *Harvard* majority:

> [t]he patentability of lower life forms is not at issue before this Court, and was in fact never litigated … [but] it is now accepted in Canada that lower life forms are patentable. Nonetheless, I agree with the appellant that this does not necessarily lead to the conclusion that higher life forms are patentable, at least for the reason that it is easier to conceptualize a lower life form as a 'composition of matter' or a 'manufacture' than it is to conceptualize a higher life form in these terms.[153]

Despite this finding, the majority's definition of 'use' in *Schmeiser* now renders this exercise moot. It is no longer necessary to obtain a patent for

151 Ibid. at para. 169.

152 See e.g. 'Craig Venter: Beyond BioBabble', *Business Week*, 2 June 2003, online: <http://www.businessweek.com/magazine/content/03_22/b3835715.htm>.

153 Ibid. at 198 and 201. Although the issue of whether lower life has ever been litigated remains contested by Justice Binnie in his dissent at para. 36. Binnie J. adopts an organic definition of life and states 'certain enzyme products (which are living matter) were held to be patentable by this Court 60 years ago in *Continental Soya Co. v. J.R. Short Milling Co.* [1942] S.C.R. 187, as were engineered micro-organisms used as an antibiotic in *Laboratoire Pentagone Ltée v. Parke Davis and Co.* [1968] S.C.R. 307'. The majority of the SCC omitted reference to these two decisions.

higher life when the patent to lower life and genes will effectively garner protection in essentially *anything* containing it. While the American maxim was '*anything* under the sun is patentable', the Canadian maxim is now '*everything* under the sun can be enforced as an infringement'. Such an outcome does not well anticipate the likely future of complex genetic manipulation of plants as biopharmaceuticals and nutriceuticals containing multiple patented subcomponents held by different patentees, such as a 'ready' variety vitamin-A enriched food crop.

Since TRIPS also makes a distinction between higher and lower life in the Article 27.3 exception, this means that the permissive language in TRIPS for excluding higher life from patentability is now also moot, emasculated in Canada by the judicial finding that lower life and genetic patents provide higher life protections. This expands IPRs well beyond what has been invented and disclosed, to reward (without qualification) that which contains it based on the liberal definition of 'use'. The SCC has effectively extended patent protection to *all* living matter that contains the patented invention. This would allow findings of infringement not only for matter not demarked by the patentee's claim, but possibly even those actually *disclaimed* in the patent application (such as human embodiments of the activated oncogene in Harvard's Claim No. 1).

The Sincerity of 'Invention': An Imperative for State Agency

Reviewing modern patent practice at administrative offices, in the market, and before the courts where they are asserted, tested, and enforced helps to identify less-than-sincere inventions in the biopatent field while warning of the consequences of an incremental judicial expansion of patents to life without consideration or commitment to coherent patent policy. While marked by continued controversy, the determination that life is patentable subject matter has nevertheless occurred without *political* debate and continues to erode critical regulatory space. Industry influences these often closely split judgments, urging for convergence in Canadian patent law with legal outcomes in the United States, where a more powerful corporate lobby exerts economic and political pressure. In *Schmeiser*, it proved effective. But there are important differences in the regulatory climate, in public, cultural and social values, and in national commitments to social and economic rights under international instruments.[154] Patent offices and the judiciary experience similar pressures to harmonize of patent laws and to converge enforceability protections with those in the US. They can play a more effective role, however, as active gatekeepers by demanding high standards of novelty, utility, and non-obviousness are met,

---

154 Unlike Canada, the United States has not ratified the International Covenant on Economic, Social and Cultural Rights under which health rights are located.

full disclosure made, construing claims narrowly and resetting the level of skill for the PHOSITA, taking into account worldwide public knowledge in their analyses of prior art.[155] A more cautious and conservative approach to grants should be preferred over revocation and compulsory licensing not only since a patent may be used to secure priority rights in other jurisdictions but also because national standards invariably inform trends for judicial convergence and international harmonization.

The existence of bilateral (NAFTA) and multilateral (TRIPS) agreement has become more important to the SCC in recent years and is usually seen as a basis for strengthened IPRs. Sometimes national agents point to these international instruments as proof that the present level of protection they contain is somehow an *optimal* standard according to which domestic courts should rule and patent offices operate. But, we cannot overlook the exceptions that riddle such international agreements. Neither can we ignore the lack of empirical evidence that might substantiate patent law's primary assumption, that stronger regimes spur greater innovation. There is growing evidence of the harm associated with such assumptions.[156] The judiciary is responsive to industry pressure and more often than not is influenced by the compelling force of economic arguments premised on private property rights. Some jurists may view regulatory deviation from other key protective jurisdictions as something that would discourage direct foreign investment in biotech or as a disadvantage to local industry and an impediment to Canada's competitiveness in the international market. Currently, there is no provision for affordable proper scrutiny of the merits of the invention claimed, moral consideration of its privatization, or policy considerations outside of those raised by the litigants in a private dispute. In this context, one might argue that the role of the court as the conscience of society and the keeper of the public's interest in balanced patent law becomes even more critical. But, it does not require judicial activism; rather, it requires judicial restraint and a conservative approach to invasive IPRs. The jurisprudential review demonstrates the piecemeal basis by which patent law has developed and the dangers of this. Instead of following legal developments in other jurisdictions, Canada and other countries, would do well to carve their own niche, possibly as beneficiaries of stronger IP regimes in such larger markets as the USA, in which patentees are predominantly interested, without necessarily committing to provide similar overprotection when they fashion their own laws. Free market competition in upstream basic research in Canada would reduce transaction costs for applied research and

155 Some of these suggestions were made by Austin and Amani, *supra* note 18, to the Ontario Advisory Committee on Predictive Genetic Technologies in 2001.

156 See e.g. Lee G. Branstetter, 'Do Stronger Patents Induce More Local Innovation' (2004) 7.2 *J. Int'l Econ. L.* 359-370. See also Jay P. Kesan (2002) 'Carrots and Sticks to Create a Better Patent System' 17.2 *Berkeley Technology Law Journal*, 763–97.

likely attract foreign direct investment to Canadian applied R&D, where the true innovation can be said to exist. Our patent policy need not be the same as our neighbour's for its own sake, given the significant difference in cultural values and social programmes. Any patent regime that is an *imitation* risks repeating the emulated model's errors; extension of private rights should be held to strict doctrinal scrutiny and public policy objectives; as instrumental tools, the strength of a patent regime has relative worth for reasons well explained by Vandana Shiva:

> Due to the inherent conflict between private and public interest, patent laws that are strong for protecting the private interest are thus weak for protecting the public interest. However there is no 'strong' or 'weak' patent law in an absolute sense. Strength and weakness are basically relative to the interest being protected. The one-sided reference to 'strong systems' in the debate on IPRs in GATT has an underlying, tacit assumption that only corporate rights count.[157]

Neglected is the fact that the system must be customized because 'patents have differential effects depending on a country's level of development. IP is simply so embedded in the culture and economy in which it operates, the general conclusions about the costs and benefits it brings are meaningless'.[158] The role of IPRs in public policy objectives is complex but whatever the level of protection, it can only be in the public's interest if the integrity of the system is retained, the bar for patentability is high enough to justify the monopoly, and safeguards are in place to curtail anti-competitive behaviour. These three suggestions would strengthen any IPR system according to merit, unite patent theory with practice, and situate patent policy in terms of distributive effects in light of total gains and losses for a society. Benefit to the public cannot be measured purely on the basis of the greater *number* of patented inventions disclosed or the degree to which Canadian patent law is congruent, for example, with American jurisprudence. Rather, the public must be educated and encouraged to participate in a uniquely Canadian dialogue on patent policy and the implications of monopolies legitimately created, enjoyed, and tolerated; other countries should take note. Given the numerous existing patents over DNA sequences of various species, relying on challenges to patent validity through litigation is a poor form of quality control and policy development; it constitutes a form of taxation on the public.[159]

---

157 Shiva, *supra* note 49 at 6.

158 Herder and Gold, *supra* note 53 at 35.

159 See Rebecca S. Eisenberg and Robert P. Merges, 'Opinion Letter as to the Patentability of Certain Inventions Associated with the Identification of Partial cDNA Sequences, 23 *AIPLA Q.J.* 1 (1995). Compare with Melanie J. Howlett and Andrew Christie, 'An Analysis of the Approach of the European, Japanese and United States

Strong advocacy and fierce resistance by developing and least developed countries, during contentious and politicized negotiations, led to significant exclusions from the requirements of national patentability provided under TRIPS. As demonstrated in the next chapter, these gains were achieved by matching trade-offs and compromises, in IP and other areas of negotiation.[160] Such hard won sovereignty should not be so readily surrendered to greater protection than TRIPS requires. Judicial agency has more than compensated for the state's failure to regulate industrial policy in light of biotechnologies' formidable challenges, rather than maintain the strict integrity of the patent system through a conservative exercise of its discretion. If the court is to err in adjudging, it should err on the side of caution by recognizing the inherent harm of monopolies and the instrumental goals of patent regimes. In a perverse sense, the legal evolution of life has become more material than the scientific one and reflects a remarkable conceptual shift away from *creation* to *invention,* as though all species were *invented*. In 1959, Fritz Machlup famously concluded he had just enough confidence in the patent system that he could not recommend it be abandoned to the US Senate; however, had it not existed he would not recommend adopting such a system.[161] Having a system and its universal minimum requirements is now prescribed for all WTO members but TRIPS retains sufficient flexibility to allow the instrumental potential of IPRs and human rights obligations to be realized through appropriate agency of a state and its agents.

Patent Offices to Patenting Partial DNA Sequences (ESTs)' University of Melbourne *Legal Studies* Research Paper No. 82.

160 See Gervais's drafting history, *supra* note 81; see also Trebilcock and Howse, *International Trade Regulation*, op. cit.

161 F. Machlup, *An Economic Review of the Patent System* (1958) Study No. 15, US Senate Committee on the Judiciary, Sub-Committee on Patents, Trademarks, and Copyrights.

Chapter 4
The Global Governance of Trade: Trading in Political Ethics

> [I]ntellectual property rights are rule-governed privileges that regulate the ownership and exploitation of abstract objects in many fields of human activity ... [I]ntellectual property rights are *liberty-intruding* privileges of a special kind ... [T]hey promote factionalism and a dangerous level of private power. From the viewpoint of distributive justice, their scope should be limited.[1]

Introduction

Our institutions of global governance, according to Errol Mendes and Ozay Mehmet, like us display hubris as their 'tragic flaw':

> The aspirations of humankind to eradicate the conditions that led to the Second World War and the evils that occurred ... were soon overwhelmed by the tragic flaw within the nature of humankind ... [an] urge in human nature, which is then reflected in the institutions of global governance, to seek the supremacy of territorial integrity over human integrity and dignity in the pursuit of perceived collective power and self-interest.[2]

Their characterization does not augur well for the survival of the WTO and its authority in an age of 'global restructuring of capital'[3] with its noticeable

1 Drahos, *A Philosophy*, op. cit. 105 at 5.

2 Errol Mendes and Ozay Mehmet, *Global Governance, Economy and Law: Waiting for Justice* (New York: Routledge, 2003) at 1.

3 Defined by Rosemary Coombe as an 'opaque phrase that attempts to encompass a multiplicity of phenomena – the emergence of a global interconnected economy, the dispersion of manufacturing production to ever-shifting sites around the globe – largely from so-called First World to so-called Third World areas, the proliferation of export processing zones in indebted areas facing World Bank and IMF pressure, the growth of international finance markets, the increasing feminization of the global manufacturing labour force, new migration patterns and the development of a global network of factories, service outlets, and capital investments'. See R.J. Coombe, 'The Cultural Life of Things: Anthropological Approaches to Law and Society in Conditions of Globalization' (1995) 10 *Am. U.J. Int'l and Pol'y* 791 [Coombe, 'Anthropological Approaches'] at 797–8.

absence of a 'global political ethics'.[4] Nicolaidis and Howse contend that while

> one might not be able to have democracy at the global level, but one can certainly have democrats, people who operate, advocate and decide, informed by a political ethics inspired by democratic ideals. *Such a political ethics – defined by values such as inclusion, participation, transparency, attentiveness to distributive effects, tolerance of diversity and of their levels of legitimate governance* – may already be implicit in the utopian projection of the European elites; even if their natural instincts as elites often cause them to act at odds with that utopian projection. If this is the case, it may be easier to describe this utopia in a mirror as it were, as the image that could inspire the rest of the world.[5]

Many scholars suggest that changes in the past 50 years in world politics, sources and centres of power, emerging technology, decolonization, globalization, and the changing dynamics of international organizations (IOs) has created a 'legitimacy deficit'[6] to which the existence and efficacy of the WTO – its power in governance – is tied. WTO legitimacy is not only contested amongst member states but equally at the level of state action in committing to *technocratically* determined trade policy[7] – without broader consideration of the interests and preferences of civilian constituent populations or the moral force of their human rights claims. Kalypso Nicolaidis and Robert Howse comment on the economic and technocratic sensibility informing regime management and the absence of political and juridical processes. '[T]rade issues' they write, 'were largely left, domestically, to a single part of the government, and to trade specialists – the trade policy elite', the sensibility of whom

> was fundamentally that of the Anglo-Saxon petit bourgeois … [and] a mentality that, generally speaking, likes to respond to pressure for change by readjusting bargains rather than building or rebuilding institutions, that likes rules for their regularity but resists their transformation into a

4 Kalypso Nicolaidis and Robert Howse, '"This is my EUtopia …": Narrative as Power' (2002) 40.4 JCMS 767 at 775–6, citation omitted ['EUtopia'].

5 Ibid.

6 See essays in Jean-Marc Coicaud and Veijo Heiskanen, eds, *The Legitimacy of International Organizations* (New York: United Nations University Press, 2001) [*Legitimacy of IOs*]; Daniel C. Esty, 'The World Trade Organization's Legitimacy Crisis' (2002) 1 World Trade Review 7; Robert Howse, 'The Legitimacy of the World Trade Organization' in *Legitimacy of IOs* at 355.

7 Robert Howse, 'From Politics to Technocracy – and Back Again: The Fate of the Multilateral Trading Regime' (Symposium: the Boundaries of the WTO) (2002) 96 AJIL 94 [Howse, 'Technocracy'], <http://www.asil.org/ajil/wto6.pdf>.

> normative system that constrains the messy methods of 'ad hoc' and 'muddling through'. Public policies are viewed as sets of alternative means to fixed, single objectives or ends, not as expressions of social and cultural norms.[8]

The turmoil in the multilateral trading regime is, according to Jeffrey Dunoff, 'in part over the issue of public participation … a story about litigation, frustration, agitation, and legitimation'.[9] Others present even more alarmed portrayals:

> [T]he WTO currently face[s] crucial challenges – dare we say crises? – of legitimacy. One speaks in both cases of the 'democratic deficit' and the need for legitimate governance; … political elites are accused of letting the social fabric fall victim to the forces of markets and globalization.[10]

Civil society is subject to this palpably democratically deficient legalization of trade policies which is said to be 'evidence of the ultimate powerlessness of progressive legal politics'.[11] The internationalization of IPRs in the multilateral trade regime represents a creeping and unaccountable encroachment of global politics in all facets of civilian day-to-day lives. It compromises the state's ability to respond to demands for social justice, cultural diversity, and human rights while bypassing the meaningful participation of various publics in devising strategic policy and optimal regulatory outcomes. Rosemary Coombe had well anticipated these problems over a decade ago with observations that have even greater resonance today:

> The global restructuring of capital and the intensified flows of capital, goods, imagery, people, and ideas has shaken the authority of nation states, cast cultural differences into sharp relief, and undermined the capacity of governments to deal with social welfare concerns. This raises new questions about the loci of power, the nature of accountability, and the authority of traditional communities and leaders, and creates crises of legitimacy and representation of unprecedented scope. In such circumstances, we might ask whether and to what extent any singular legal regime is 'constitutive' ….If diverse laws govern worlds of their own creation, people may occupy a number of juridically mediated worlds simultaneously.[12]

---

8 'EUtopia', *supra* note 4 at 776.

9 Jeffrey Dunoff, 'Public Participation in the Trade Regime: Of Litigation, Frustration, Agitation, and Legitimation' (2004) 56 *Rutgers L. Rev.* 961.

10 Nicolaidis and Howse, 'EUtopia', *supra* note 4 at 769.

11 Denise Reaume, 'The Politics of Globalization' *Ultra Vires* 13–14, March 2001.

12 Coombe, 'Anthropological Approaches', *supra* note 3 at 798.

Stratified publics share their global identity as subjects of multiple governance norms and institutions derived from multiple sources of law. The issue is whether the cultural and political demands of such *consumers* of governance – in neoliberal terms – may be realized when they are no longer treated as purely subjects of representative governments within their respective communities. There is only one tangible and accessible government in the absence of a supra-national global sovereign: the government of the nation state in which the individual resides; it is what makes citizenship, an increasingly contested status, so critical. Bound by multiple governance norms, the individuals are only able to formally participate in norm formation at the national level. But the scope of participation is circumscribed by international obligations committed to by the executive or legislative branches of government and exceeded by the judiciary. With growing legal fragmentation, how to hold any external source accountable to obligations arising from a failing 'Club Model' of trade policy-making seems insurmountable:

> A relatively small number of economists and diplomats from like-minded states worked quietly to make trade policy without much public input or oversight … it was successful in the sense that it oversaw dramatic decreases in tariffs and other trade barriers, and a corresponding increase in global trade and prosperity … Paradoxically, however, the advantages of the Club Model of trade policy-making contained the seeds of their own destruction … First, increasing trade liberalization caused citizens to be more sensitive to further liberalization … complicat[ing] future efforts at liberalization. In addition, the Club Model was not sustainable in a context where developing states began to demand a greater role in trade negotiations and policy-making. Developing states distrusted a club that excluded them from a seat at the table, and generated rules that seemed to help rich states more than poorer states … [13]

Unlike human rights instruments, since the multilateral trading system is founded conceptually on a barter process of negotiation and contract, those states with more limited resources and fewer 'trade elites' are disparately affected by the process and therefore its outcomes.[14] Consequently, legitimacy

13 Dunoff, *supra* note 9 at 969–70, footnotes omitted. See also Robert O. Keohane and Joseph S. Nye, Jr, 'The Club Model of Multilateral Cooperation and Problems of Democratic Legitimacy', in Roger B. Porter et al., eds, *Efficiency, Equity and Legitimacy: The Multilateral Trading System at the Millennium* (Washington, DC: Brookings Institution Press, 2001).

14 The inequality of bargaining power and coercion that governed the TRIPS negotiation process is documented by several scholars, see e.g. Susan K. Sell, *Private Power, Public Law: The Globalizing of Intellectual Property Rights* (Cambridge, Cambridge University Press, 2003), 108–20 describing bargaining strategies [Sell,

becomes a paramount concern for less privileged member states and their adversely affected citizens, who suffer the consequences of state agency as a measure of the paucity of their governmental resources, technocratic 'know-how' and experience with the regime's architecture. The history of TRIPS is a testament to this, shaping perceptions of fairness, justice, and equity in form (the processes through which the individual has legal recourse for complaint) and substance (the content of the *juridification* process affecting people in their daily lives) but also in practice (the relative number of complaints made by the developed members as compared to the LDCs and DCs). Juridification is usually accompanied by pluralization and privatization of authority relations[15] and remains a problem for national constitutions to solve because it presents 'the possibility of legal domination in the face of institutions that cannot organize a singular and unified popular sovereignty'.[16] Furthermore, it has been stated that '[s]uch legal domination is not simply tyranny, but rather the imposition of a cooperative scheme upon others who cannot influence or revise its terms'.[17] WTO reform and just dispute settlement may be more than simply matters of social good.

Lack of capital informational resources in the trade negotiation process for developing countries (DCs) has resulted in initiatives for a joint capacity building project on IP and development launched by the UN Conference on Trade and Development (UNCTAD) and the International Centre for Trade and Sustainable Development (ICTSD) in 2001. The Project's resource book on TRIPS and development is intended to assist with WTO and other IP negotiations for officials from developing countries.[18] These remedial efforts are useful but do not allay public concern. Nor do they mitigate the consequences of obligations that came to exist when power relations came to be dominated by international law and by the politics that included TRIPS in the multilateral system. The result is an institutional lack of attention to democratic processes and the perpetuation of social injustices as domestic

---

Private Power]; Peter Drahos, 'Developing Countries and International Intellectual Property Standard Settings' ['Standard Setting'], 5 J. *World Intell. Prop* 765 critiquing the 'trade-off' theory of TRIPS as part of a larger package of concessions; and Keith Maskus, *Intellectual Property Rights in the Global Economy* (Washington, DC: Institute for International Economics, 2000), for an empirical analysis.

15 A. Claire Cutler, *Private Power and Global Authority: Transnational Merchant Law in the Global Political Economy* (Cambridge: Cambridge University Press, 2003) [*Transnational Merchant*].

16 James Bohman, 'Constitution Making and Democratic Innovation: The European Union and Transnational Governance' (2004) 3 *Eur. J. Pol. Theory* 315–37.

17 Ibid.

18 See UNCTAD-ICTSD *Capacity Building Project on Intellectual Property Rights*, <http://www.IPRsonline.org/unctadictsd/description.htm> and *Resource Book on TRIPS and Development: An Authoritative and Practical Guide to the TRIPS Agreement* at <http://www.IPRsonline.org/unctadictsd/ResourceBookIndex.htm>.

democratic governance in ways inconsistent with the history and ideals of the multilateral trading system and the *law of nations*.

The Multilateral Trading System

Trade Theory's Contradiction: Institutionalizing IPRs

Mercantilism, dominating trade in the seventeenth and eighteenth century. It favoured government regulation restricting imports and liberal export policies to promote a strong domestic manufacturing industry but was successfully displaced by moral philosopher Adam Smith's Theory of Absolute Advantage presented in his seminal treatise, *The Wealth of Nations*.[19] Revered by economists for his classic trade theory, Smith posited that states would benefit from realizable gains from even unilateral liberal trade where there was specialization and division of labour. Critics, however, questioned whether liberal trade and economic integration would remain a desirable policy for countries naturally poorly endowed (in all respects) when compared to trading partners, and argued that trade liberalization, while good for rich countries with established economies, was damaging for developing countries with emerging ones. David Ricardo's Theory of Comparative Advantage[20] continues to enjoy preeminence amongst trade proponents for recognizing that even if a country is less efficient in *all* sectors of domestic production and enjoys no absolute advantage, it could still realize gains from trade by specializing in production for export in the area in which its *comparative advantage* is *most* and disadvantage is least. All states are encouraged to trade; 'history tells us clearly that isolationism and policies aimed at economic self-sufficiency are in fact a prescription for poverty'.[21]

Scholars, economists, Western diplomats and agents of TNCs rely on this theory of trade to support global efforts towards greater market liberalization with the view that changing market structure and encouraging competition is

19 Adam Smith, *The Wealth of Nations*, <http://www.econlib.org/library/Smith/smWN.html>.

20 See David Ricardo, *The Principles of Political Economy (1817)*, discussed in Trebilcock and Howse, *International Trade Regulation*, op. cit. at 3–6, with additional commentary on periodic modifications to this theory.

21 Ross P. Buckley, 'The Changing Face of World Trade and the Greatest Challenge Facing the WTO and the World Today' in Ross P. Buckley (ed.) *The WTO and the Doha Round: The Changing Face of World Trade*, Global Trade and Finance Series (New York: Kluwer Law International, 2004) [*Changing Face*] at 4 with examples offered in accompanying footnote 9. See also the United Nations Conference on Trade and Development, 'Linking International Trade with Poverty Reduction' Least Developed Countries Report 2004, United Nations, UNCTAD, <http://www.unctad.org/en/docs/ldc2004_en.pdf>.

a panacea for poverty. It will, they argue, stimulate development by increasing domestic production for export, and will thereby increase capital flows and personal income. Financially empowered, the poor can further participate in growth and improve their standard of living. Enhancing living standards and life expectancy improves political stability and personal security, as a corollary. This creates a social feedback loop through trade which, by reducing costs for consumers through liberal markets, promotes further gains in health, education, innovation and personal dignity, enabling further innovation and economic development. For free trade proponents, the question is not *whether* the 'deal' will make everyone better off but rather *how* to ensure that the promises of trade theory translate into experienced realities.

Still other scholars, such as Ralph Gomory and William Baumol, argue that the classic trade model's theory, that all who participate in free trade will benefit from it, depends on assumptions that may no longer be true in today's society. When classicists wrote, they argue, the agricultural industry dominated production which supported the then prevailing economic principle of *diminishing returns to scale*. Today's world is marked with significant *economies of scale* which has ushered in the age of large multinational corporations that challenge the classicists' conclusion of only a single stable positive outcome ('equilibrium') driven by the invisible hand of the free market. The various equilibria, from a world-output perspective, undermine the efficiency claims of the comparative-advantage model because if a particular equilibrium is sustained by the market, it may serve to perpetuate distributional inequalities based on pre-existing factors (for example, fortuitous or disastrous conditions such as famines and wars, migration patterns, political orientations of prior governments etc.). These factors influence market forces and determine *which* equilibrium is selected. A country that is producing more than its population's share of world commodities has, they argue, more to trade, more to consume, and is better able to compete. The particular equilibrium of the market helps retain this advantage and reinforces a state's advanced position while freezing out other countries from most industries, rendering them unable to enter the market:

> [W]hich country makes what product is generally uniquely determined in the classical economic model of trade. And the outcome always serves the economic interests of the general public in all the countries involved because a country can be the prime supplier in an industry only if it is the lower-cost supplier of the item at a fixed quality level, or alternatively, that at a given cost, it is the higher–quality supplier ....But today's world of industry contrasts sharply ....[I]n many of today's industries, with large-scale operations required with difficulties of entry, and with acquired advantages

> rather than natural ones playing a more decisive role, the situation is basically different … there *is no single clear cut and natural outcome*.[22]

Knowledge asymmetries between negotiating parties arise from and also exacerbate economic factors, with the result that developed countries secure more favourable terms for themselves in trade negotiations. *Harmonization* efforts by the West strive 'to integrate the economies of non-Western societies into a global economic system which it dominates'[23] and are often labelled by civil society as *homogenization* efforts. At the 2001 G8 summit in Genoa, Italy, the Group of Eight leading industrialized nations conceded that the poorest countries of the world had yet to benefit from the multilateral trade regime and that greater effort was needed to increase access to markets of the developed world for developing countries' exports. Yet these same 'richer' parties are intransigent in trade negotiations over agriculture and textiles, two areas in which poor countries can nevertheless compete due to cheap labour and low regulatory compliance costs. The institutionalization of the trade-IP linkage under the WTO, by contrast, clearly exemplifies asymmetric benefits garnered by developed countries of the 'North' through agency capture by industry lobbyists within their boundaries. TRIPS illustrates how international (trade) law can translate the ideology of the powerful within powerful nations into a social construction that strives through politics, economics, and foreign and domestic policy to perpetuate itself as paradigmatically normative despite its bias towards the developed world's market. In fact, despite common claims that strong IPRs promote domestic economic growth by protecting new innovations and attracting foreign direct investment, the correlation is only empirically supported when IP protections are linked with liberal trade policies. Keith Maskus reports on the findings of a 1996 Gould and Gruben study:

> They found no strong direct effects of patents on growth, but there was a significantly positive effect when patents interacted with a measure of openness to trade … stronger patents in open economies raised growth rates by 0.66 percent on average, suggesting that market liberalization in combination with stronger IPRs tends to increase growth.[24]

TRIPS scholars criticize its coercive nature as contradicting proponents' claims that the universally adopted strong standards prescribed will benefit all member economies and therefore their respective publics. These critics

---

22 Ralph E. Gomory and William J. Baumol, *Global Trade and Conflicting National Interests* (Cambridge, MA: MIT Press, 2000) at 5–6.

23 Lloyd Geering, 'The World to Come: From Christian Past to Global Future', http://www.religion-online.org/showchapter.asp?title=2735andC=2467: '*What is universalism to the West is imperialism to the rest*' (at 184, emphasis added).

24 Maskus, *IPRs in the Global Economy*, op. cit. at 168.

intimate that TRIPS may become 'one of the most effective vehicles of Western imperliasm in history'.[25] Conclusions to the contrary ignore distributional realities: members 'differ in terms of their levels of wealth, economic structures, technological capabilities, political systems, and cultural traditions. They have different needs and aspirations and require different intellectual property systems'.[26] From an economic perspective, any additional monopolistic rights provided above the base amount necessary for innovation in a given state is *inefficient* and fails to achieve optimal patent policy. TRIPS' 'one size fits all' regime turns out to be ill fit for *any* according to David Vaver, who writes:

> [l]egal standardization may be beneficial in general but is not so for intellectual property in either the developed or developing world. The law in developed countries is currently incoherent and itself requires major reconsideration. The imposition of such a defective law on the developing world is helpful to neither side.[27]

The impact of TRIPS is greater than that which is textually provided, due to the scope *imputed* to the obligations, i.e., that they are both more restrictive than the language providing for them implies (furthering the optics of a repressive regime for regulatory governments) and more emancipating (the claim that IPRs are necessarily good for national development) than practice supports. Paul J. Heald suggests a country-specific approach to devising rational IP policy for the purposes of economic growth and development. He points out that stronger protections can benefit only countries sufficiently advanced to be developing their own IP and not those in which most people lack the basic necessities of life, effective government and adequate or effective law enforcement.[28]

---

25 Marci A. Hamilton (1996) 'The TRIPS Agreement: Imperialistic, Outdated, and Overprotective', 29.3 Vand. J. Transnat'l L 613 at 614. See also Ruth L. Gana, 'Has Creativity Died in the Third World? Some Implications of the Internationalization of Intellectual Property' (1995) 24 *Denv. J. Int'L. and Pol'y* 109, at 112 examining TRIPS 'as a form of passive coercion' and J.H. Reichman, 'Intellectual Property in International Trade: Opportunities and Risks of a GATT Connection' (1989) 22 *Vand. J. Transnat'l L.* 747 at 813 contending that 'Imposition of foreign legal standards on unwilling states in the name of 'harmonization' remains ... namely, a polite form of economic imperialism' (footnote omitted).

26 Peter K. Yu, 'Four Common Misconceptions About Copyright Piracy' (2003) 26.1 *Loy. L.A. Int'l and Comp. L.J.* 127 at 135.

27 David Vaver, 'Need Intellectual Property Be Everywhere? Against Ubiquity and Uniformity' (2002) 25 Dal. L.J. 1 at 5 [Vaver, 'Uniformity'].

28 See Paul J. Heald, 'Mowing the Playing Field: Addressing Information Distortion and Asymmetry in the TRIPS Game' (2002) Vanderbilt Law and Economics Research Paper No. 02-21: Vanderbilt Public Law Research Paper 02-15, <http://papers.ssrn.com/sol13/papers.cfm?abstract_id=319301>.

The lack of basic institutions and public infrastructure for some WTO members translates into comparative advantage differences. IP continues to be a contentious negotiating issue as DCs and LDCs complain that they are not domestically in a position to generate intellectual capital necessary to benefit equally from the IP-trade linkage. Michael Trebilcock and Robert Howse rephrase the problem in trade terms:

> [i]n terms of neoclassical trade theory, whether a particular country will want stronger or weaker intellectual property protection will depend on whether its *comparative advantage* lies more in innovation or in the imitation and adaptation of others' innovations. This is simply an extension of the argument concerning the allocation of resources domestically between imitation and innovation. More precisely still, a rational country would have different levels of protection for different industries, representing different trade-offs between innovation and imitation in each industry depending upon where its comparative advantage lies.[29]

Some DCs enjoy a comparative advantage in *imitation* due to cheap labour and low regulatory compliance costs rather than in *innovation*. Innovation capacity is stronger in industrialized and industrializing economies with liberal democracies and supporting public institutions and infrastructure in health and education; these publicly funded systems effectively serve as export subsidies for IP products. Lacking in comparable domestic contexts favouring knowledge production, DCs now more vocally object to the diminishing value of their comparative advantage as evidenced by the troubled Doha Round of negotiations. Ross P. Buckley contends that the Doha Round presents a particular challenge for 'redressing this sense of broken promises and unfinished business'.[30]

In short, the inclusion of TRIPS contradicts the very tenets central to trade theory on two counts. First, it *requires* that members of the WTO *create* market monopolies by necessitating domestic IPRs legislation and enforcement mechanisms, whereas free trade agreements generally try to reduce monopolies by increasing access to markets and opening them up for competition to create a positive effect on prices. Second, derogating from first principles in trade theory with the TRIPS instrument not only renders that deviation suspect but also undermines the legitimacy of trade policy and thereby threatens to undermine further potential benefits the system is to serve. Sampord et al. write:

> [b]arriers to entry foster conditions where domestic companies become monopolies. This environment interferes with the workings of market

29 Trebilcock and Howse, *supra* note 20 at 400, emphasis added.

30 See *Changing Face*, *supra* note 21.

> mechanisms and leads to an inefficient use of resources. The opening of domestic markets weakens the possibility of monopolistic power.[31]

Private economic activities of traders that reflect their different resources, endowments, and skills, are facilitated by the increased contract opportunities and reduced transaction costs that free international trade enables. Adam Smith not only argued for free trade as a means of increasing the wealth of nations but also denounced the evil of monopolies[32] because liberal trade and free competition are linked; any gains to be garnered by states and their consumers through increased global competition and reduced government intervention (such as tariffs, duties and other barriers) in markets cannot be fully realized if replaced with substantive standards justifying anti-competitive monopolies.

The international protection of IPRs under threat of sanctions has led the critics of TRIPS to conclude that its substantive requirements are unjustifiable incursions on regulatory governments and the intrusions on public goods and individual rights make the welfare promoting function of trade disingenuous. *Global* public goods, such as health, usually demand a trade-off between efficiency and equity. The nature of global public goods, or their opposite, 'public bads', is that their respective benefits or costs cut across borders;[33] particularly given that patent rights secured in one jurisdiction can be used to secure priority rights in others. Strong, internationally mandated IPRs convert knowledge, an essential global public good for the attainment of other GPG, into a global public *bad* with potentially significant adverse human impacts that raise their costs. GPG are integral to multilateralism according to former UN Secretary General, Kofi A. Annan:

> Whether we are talking about *preserving biodiversity*, preventing climate change, *fighting the spread of communicable diseases*, establishing rules for trade and aviation, or *setting global standards of human rights*, it is impossible for any single state to secure such goods on its own. Quite the contrary, *global public goods can only be attained if countries work together, and globalization has only increased this fundamental interdependence*.[34]

31 Charles Sampford et al., 'Living up to the Promises of Global Trade' ibid. at 11.

32 Robert D. Anderson and Peter Holmes, 'Competition Policy and the Future of the Multilateral Trading System' (2002) 5.2 *J. Int'l Econ. L.* 531 at 535.

33 Inge Kaul et al., 'Why Do Global Public Goods Matter Today?' in Kaul et al., op. cit. at 2.

34 'Endorsements' in Kaul et al., ibid. at ii, emphasis added. See also *WTO Agreements and Public Health: A joint study by the WHO and the WTO Secretariat*, Geneva (August 2002), World Trade Organization, <http://www.wto.org/english/res_e/booksp_e/who_wto_e.pdf>.

The trade-related substantive code of TRIPS is anomalous with the free-market promoting regime in which it is found and creates barriers for cooperation and provision of public goods; it is inimical to trade-and-competition policy and has proven objectionable to free-market proponents as well as social justice seekers given the:

> wide acknowledgement that competition policy and trade liberalization share common objectives relating to the promotion of economic efficiency, consumer welfare, and that a lack of effective competition policies can impede the realization of the gains from liberalization. Furthermore, there is a growing recognition that anti-competitive practices impact directly on the welfare and prospects of developing countries, and hence are no longer (if they ever were) a concern exclusively for the developed world.[35]

TRIPS is not only inconsistent with the historical territoriality of patents, but contradicts heuristic utilitarian and economic rationales. The principle of Pareto-efficiency is inapt in justifying the imposition of higher IP protection through trade obligations because strengthened protection may improve welfare in some countries but would reduce it in others, making it Pareto-inferior. The dissonance of stronger trade-related protections with neoclassical trade theory is often justified as part of the *bargaining* process that created 'fair' trade-offs between diverse ('competing or conflicting economic') interests of states during trade negotiations.[36] But the diminution of the *comparative advantage* in imitation over the artificial inflation of that in innovation should be attributed, at least in part, to the system's normative ascription of a comparative advantage in *process dominators* and *realpolitik*.[37]

---

35 As a result, some are calling for WTO rules on competition policy. See Anderson and Holmes, *supra* note 32 at 532 and n. 4 reference to World Trade Organization, 'Special Study on Trade and Competition Policy', in *Annual Report for 1997* (Geneva: WTO, 1997), ch. IV; UNCTAD, *World Investment Report: Transnational Corporations, Market Structure and Competition Policy* (Geneva: 1997) and the OECD, *Trade and Competition Policies – Options for a Greater Coherence* (Paris: 2001). Some commentators argue against the WTO's extensive involvement in the area of competition policy based on compatibility concerns over operational modalities, the implications of an overly rigid codification of the principles of competition policy, and the propriety of applying the WTO Dispute Settlement Understanding (DSU) to competition law and policy. See Trebilcock and Howse, *supra* note 20 at 588–611.

36 See e.g. ibid. at 401.

37 By comparison, 'idealpolitik [is] the formulation of policy choices based on an idealistic view of world politics ... Idealpolitik presumes a linkage between domestic and international politics and assigns to international law an overarching mission of maintaining world order conducive to the survival of constitutional and representative forms of government ... Although seemingly irreconcilable, the positivism of realpolitik and the naturalism of idealpolitik do have a point of congruence in that

Realpolitik is the most widely held theory of international politics and refers to realistic calculations of state power,[38] helping to explain the rupture between the aspirations and realities of the WTO as implemented in policies. 'We have to be careful', warn Sampford et al., 'that our own actions do not undermine our promises by securing for ourselves trading arrangements so favourable that we benefit at the expense of those to whom we have promised beneficial trade liberalization. We would be making liars of ourselves!'[39]

The differences in negotiated outcomes arising from disparity of resources, inequality of bargaining power, and power politics at the WTO continue to undermine classic trade theory in their consequences. Member states without an *absolute* advantage in any area of trade cannot exploit the area in which they do have a *comparative* advantage, because the international rules were abruptly changed at the last moment of a heavily invested and very exhausting 'all or nothing' Uruguay Round of negotiations to include never before established uniform substantive standards.[40] With their comparative advantage in imitation neutralized by TRIPS, DCs and LDCs are wary of making concessions in traditional areas of trade negotiation during the most recent troubled Doha Round. Sylvia Ostry, a former Canadian trade negotiator offers us additional insights explaining why:

> The degree of intrusiveness into domestic sovereignty bears little resemblance to the shallow integration of the GATT with its focus on border barriers … The WTO has shifted from the GATT model of negative regulation – what governments must not do – to positive regulations, or what governments *must* do.[41]

---

both schools of thought recognize the utility of international law in providing for the reciprocity so necessary for the orderly relations of states'. See James Wolfe, *Modern International Law: An Introduction to the Law of Nations* (Upper Saddle River, NJ: Pearson Education, Inc., 2002) at 4.

38 See generally Peter Drahos, 'Where the Weak Bargain with the Strong: Negotiations in the World Trade Organization' (2003) 8 International Negotiation 79–109 [Drahos, 'Weak Bargain']. See also Sarah Anderson, ed., *Views from the South: The Effects of Globalization and the WTO on Third World Countries* (Chicago: Food First Books and International Forum on Globalization, 2000).

39 Sampford et al. *supra* note 31 at 9.

40 See Gervais, *The TRIPS Agreement*, op. cit. at 3–26 and discussing resistance to the 'TRIPS package' and the all or nothing nature of its adoption as part of broader negotiations in other sectors.

41 Sylvia Ostry, 'The WTO: Institutional Design for Better Governance' (paper presented at the Conference on Efficiency, Equity, and Legitimacy: The Multilateral Trading System at the Millennium, Kennedy School of Government, Harvard University, 1–2 June 2000) 6, cited in Claude E. Barfield, *Free Trade, Sovereignty, Democracy: The Future of the World Trade Organization* (Washington, DC: The AEI Press, 2001) at 5-6.

History of the GATT

With an end in sight to the atrocities of the Second World War, an important idea formed: world peace could be attained through international law. An institutional structure for the multilateral trade system emerged to increase global security by fostering economic interdependence. The First World War had been accompanied by an ensuing retreat from world trade and a return to protectionist polices giving rise to the depression of 1920s and eventually the Great Depression beginning in 1929.[42] The Second World War (1939–1945) again diverted the state's focus to national concerns and isolationist policies but as it ended, enthusiasm renewed to rebuild relations and improve conditions for stable economic growth. It became apparent that a cooperative reconstruction strategy would be needed if the devastating economic conditions following WWI were to be avoided. A series of conferences was organized to discuss this goal and the mutual gains of mobilizing efforts at increasing market access, discouraging and reducing monopolistic behaviours, and promoting domestic and international policies to support economic growth, integration, and interdependence. The Bretton Woods Agreement was reached at the New Hampshire UN Monetary and Financial Conference of July 1944; 'At this stage, it was evident that trade, economic stability, peace, international security and human rights *were clearly linked*'.[43]

There were three pillars to the Bretton Woods System: the International Monetary Fund (IMF), the International Bank for Reconstruction and Development (IBRD referred to as the World Bank (WB)) along with the International Development Association (IDA) and finally, the ill-fated ITO (International Trade Organization) (the intended institution for negotiating and administering the envisaged multilateral trade regime). The IMF was agreed to by forty-five government representatives as a framework for international economic cooperation designed to ensure the stability of financial markets in order to avoid the disastrous economic policies that led to the Great Depression. The IMF would oversee exchange rate stability and assist in the establishment of a multilateral system of payments. The IMF would also provide resources with safeguarding conditions – so that members experiencing fiscal hardship would have alternative responses than to resort to restrictive trade measures. IMF membership was a required pre-condition for WB membership. The role of the WB was to

42 Post war, the international system unhinged as world trade shrank from $30.3 billion in 1929 to $20.3 billion in 1931and by '1932, the United States, the major provider of capital … had virtually stopped investing its funds oversees'. Akira Iriye, *The Cambridge History of American Foreign Relations, Volume III: The Globalizing of America, 1912–1945* (Cambridge: Cambridge University Press: 1993) at 119.

43 Mendez and Mehmet, *supra* note 2 at 68, emphasis added.

> provide reconstruction capital from countries like the USA whose economies had not been devastated by the war to the shattered economies of Europe and Japan – the success of the Marshall Plan that the USA subsequently adopted in promoting this objective meant that the World Bank was able quickly to redefine its focus as providing development capital to less developed countries ...[44]

In February 1946, the newly forged United Nations Economic and Social Council (ECOSOC) accepted US initiated proposals for establishing the International Trade Organization (ITO) for the negotiation of an international trade agreement within the context of a proposed United Nations Conference on Trade and Employment in Havana. The United States' original proposal for the ITO carried provisions to address the anti-competitive harms of private combines and cartels, on the basis that private agreement dividing markets amongst cartel members would hamper trade liberalization and the free flow of goods across borders. The proposals were incorporated into the 1948 Havana Charter (Chapter V). A transitional agreement for tariff reductions had been agreed to in the interim and by January 1, 1947, the *General Agreement on Tariffs and Trade* (GATT 1947) and its *Protocol of Provisional Application* (PAP) had come into effect with 23 signatories including Canada. GATT 1947 was purposely a minimalist instrument that created few new obligations other than preliminary tariff reductions and non-discrimination principles cornerstone to pre-existing bilateral agreements. The GATT 1947 had no expiry date but for further clarification the PAP provided that the GATT was only a temporary agreement. To facilitate American support without the need to first obtain Congressional approval, the GATT 1947 only applied to future legislation and did not mandate amendment to existing domestic law. GATT 1947 also had provisions to ensure commitments would be observed pending the formation of the ITO to administer and enforce it. But the US administration announced that it would not submit the Havana Charter to Congress for approval, over concerns related to constraints on sovereignty, and the ITO was abandoned along with the Havana Charter. Without American

---

44 Trebilcock and Howse, *supra* note 20 at 23. But see Walden Bello, 'Building an Iron Cage: The Bretton Woods institutions, the WTO and the South' in Anderson, *supra* note 38 at 54–90 arguing that these institutions formed as part of the same continuum of 'a campaign of global economic containment of the legitimate aspirations to development on the part of Third World countries' marked by 'the reorientation of the World Bank toward managing development in the South in the late 1950s', the IMF's role 'as the watchdog of the Third World countries' external economic relations in the 1970s', the 'universalization of structural adjustments in the 1980s' and the 'unilateralist trade campaign waged against the Asian 'tiger economies' by Washington beginning in the early 1980s' (ibid. at 55).

participation, few of its other signatories were interested in the deal. The two financial pillars and the *provisional* GATT remained and,

> [b]y default, the GATT, a minimal code for trade relations, became the main game for the organization and coordination of international trade rules ... [W]hat also died along with the ITO charter was the intention to have a world trade regime that would be infused with the social dimensions of trade, as well as global values of justice and human rights ... [45]

Once the post-war urgency for full cooperation subsided, the significant compromise needed to achieve international agreement amongst stratified states with disparate economic and regulatory needs became less tenable. Further attempts to establish a permanent ITO-like institution failed, explain Mendez and Mehmet, and the multilateral trade regime evolved without the intended central framework of the Havana Charter through a series of formal arm-wrestling negotiations called *Rounds*. The eight Rounds[46] successfully concluded have achieved a substantial reduction in the tariff levels from an average of 47 per cent on manufactured products in the post war period to about 3 per cent after the full implementation of negotiated reductions from the Uruguay Round.[47] Each Round has been progressively lengthier, a reflection of increasing complexity and stated ambitions.

The first six rounds, ending with the Kennedy Round in 1967, focused on significantly reducing tariff barriers and import restrictions and were very successful in achieving negotiated reciprocal tariff concessions. Historically, tariffs had been negotiated on an 'item by item' bilateral basis, which meant that between two negotiating countries, one would bargain for a reduction of tariffs on a specific class of products which would be met with a counter request, but this proved too complex with the increase in participating parties. The Kennedy Round negotiations were on a 'linear basis'; countries agreed

---

45 Mendez and Mehmet, *supra* note 2 at 68–9: '[I]t is a moral, legal and indeed economic imperative that the world trade regime return to the original vision'. See also Jeffery S. Thomas and Michael A. Meyer, eds, *The New Rules of Global Trade: A Guide to the World Trade Organization* (Scarborough, ON: Carswell, 1997).

46 The Rounds are: 1) 1946, in parallel with a negotiation of the Agreement, in Geneva Switzerland; 2) 1949, Annecy, France; 3) 1950–1951, Torquay, England; 4) 1955–1956, Geneva, Switzerland; 5) 1960–1961, the 'Dillon Round' in Geneva, Switzerland, 6) 1964–67 the 'Kennedy Round' in Geneva, Switzerland; 7) 1973–1979, the 'Tokyo Round' in Geneva, Switzerland, 8) 1986 to 1994, the 'Uruguay Round' in Marrakesh, Uruguay.

47 Michael Trebilcock and Robert Howse, 'A Cautious View of International Harmonization: Implications from Breton's Theory of Competitive Governments' (unpublished paper presented 1 March 1999) at 1 [Trebilcock and Howse, 'Regulatory Diversity']. See also their 'Trade liberalization and Regulatory Diversity: Reconciling Competitive Markets with Competitive Politics' (1998) 6.1 Eur. J. L. and Econ. 5.

as a group to reduce their tariffs (an import or custom duty that acts as a border tax imposed on imported goods) by a specific percentage on large classes of products.[48] Sector specific discussions (i.e. for textiles) occurred for the first time. With marked success in tariff reductions underway, the focus of the negotiations expanded beyond lowering tariff barriers to include other trade distorting measures as non-tariff barriers to trade (NTBs).[49] NTBs encompass any trade restrictive measure including a 'law, regulation, policy, or practice of a government other than an import duty …'. The distinction between NTBs and tariff barriers is less valuable today due to the great success of the GATT 1947 in fulfilling its (provisional) mandate for lowering tariff levels and quantitative restrictions and because the NTBs are 'negatively defined' as anything that is not a tariff barrier; a definition so broad it lacks significance.[50] But historically, the GATT's prior success in reducing border measures came with a rise in new protectionism with domestic measures as permissible proxies for insulating such policies. NTBs, including measures 'such as quotas, voluntary export restraint agreements, orderly marketing agreements, industrial and agricultural subsidies and more aggressive unilateral invocation of trade remedy laws, particularly antidumping and countervailing duty laws',[51] provided the impetus for the shift in focus in the seventh (Tokyo) Round of negotiations, which launched at behest of the Americans. Linking trade with regulatory measures formed the foundation for the gradual invasive encroachment of trade law on sovereignty; it required, but failed to acquire, *public* consultation or approval.

Meanwhile, the LDCs that had banded together in 1964 to form the United Nations Commission on Trade and Development (UNCTAD) in response to GATT as a 'rich man's club', produced a report to address their special economic needs to be met through trade liberalization policies. As a result, Part IV was added to GATT 1947 to recognize the 'special and differential status' of least developed countries and to invite developed countries to provide unilateral and voluntary concessions on trade items of export interest to the LDCs, forming the basis for the subsequent Generalized System of Preferences (GSP). An understanding for dispute settlement (DSU)[52] and the 'Enabling

---

48 Thomas and Meyer, *supra* note 45 at 6.

49 Such as the Anti-Dumping Code of 1967, formally the *Agreement on Implementation of Article VI of the Agreement on Tariffs and Trade*, GATT BISD 15S/24 (1968) was the first code moving beyond tariffs.

50 See Federico Ortino, *Basic Legal Instruments for the Liberalization of Trade: A Comparative Analysis of EC and WTO Law* (Portland: Hart Publishing, 2003) at 7–16

51 Trebilcock and Howse, *International Trade Regulation*, op. cit. at 24.

52 *The Understanding Regarding Notification, Consultation, Dispute Settlement and Surveillance*, BISD 26S/210, which improved the existing dispute settlement process under GATT articles XXII and XXIII and incorporated certain customary practices.

Clause'[53] were other significant concessions obtained at the last minute under threat that the LDCs and DCs would abandon negotiations. These new arrangements helped soften the disparate impact of the international trading regime and thereby secured continued state participation even as the DCs saw the failure to establish a new textiles and clothing arrangement to replace the MFA (that restricted their trade in both) as a notable failure of the Tokyo Round and the unsettled issues in agriculture, safeguards, voluntary export restraints, and institutional improvements to the GATT remained.

The 'Enabling Clause' permits members to provide 'differential and more favourable treatment' to developing countries 'in spite of' the MFN obligation of Article I:1 of GATT 1947 which otherwise requires that *same* be extended to all members of the WTO 'immediately and unconditionally'. Because of this provision, GSPs are permissible and states may give better and preferential tariff treatment to DCs than the bound tariff rates provided to all other WTO members. Preferences are not binding but voluntarily given by the granting state and can be withdrawn. They require legal authority under the GATT nevertheless, because they would otherwise violate the MFN clause. Moreover, this provision enabled DCs to enter tariff reduction agreements amongst themselves without the strict application of GATT Article XXIV (on custom unions, territorial application, and free trade areas). These provisions of the GATT 1947 authorize preferential treatment aimed at promoting *development* through *trade* rather than *aid*.

The Uruguay Round and the WTO

The Uruguay Round, referred to as the most ambitious and lengthy, secured participation of 100 governments. From a highly contentious agenda, 26,000 pages of documents comprised of several instruments resulted.[54] The aims were to improve access to the world's markets, reform the rules and institutions of the international trade system, and establish an overseeing organization. The Round took four years to prepare and seven more to complete. It improved participation by the DCs[55] and

> dramatically expanded the traditional portfolio of trade interests. The World Trade Organization (WTO) regime now extends to services, intellectual

---

53 The enabling clause is in the *Differential and More Favourable Treatment, Reciprocity and Fuller Participation of Developing Countries*, Decision of the Contracting Parties L/4903, BISD 26S/203 (1979).

54 John Croome, *Reshaping the World Trading System: A History of the Uruguay Round*, 2nd and rev. edn, World Trade Organization (Boston: Kluwer Law International, 1999).

55 Jordana Hunter, 'Broken Promises: Trade, Agriculture and Development in the WTO' (2003) 4 *Melbourne Journal of International Trade Law* at 299.

> property, investment, national subsidies policies, and domestic regulation of product standards, food safety, and quarantine ... This expanded agenda has blurred the boundaries between trade and other subjects of national and international policy, creating the potential for tensions between competing policy goals'.[56]

The 8th Round Agreements (Agreements) signed in April 1994 in Marrakesh, Morocco were preceded by three very controversial Ministerial meetings in Punta del Este (1986), Montreal (1988), and Brussels (1990), and a Ministerial Declaration premised on the understanding that the issues and concessions arising from the IP discussions would *not* be directly linked to the other trade in goods issues.[57] Formal bodies established under the Declaration were charged with overseeing negotiations in different areas, such as sensitive sectors including agriculture, intellectual property, textiles and clothing, and safeguards, and for improving GATT as an institution for global economic policy coherence.[58] Some of the DCs, including agricultural exporting countries of Latin America, refused to deal in any area unless there was agreement on the agricultural issues addressing the domestic subsidy of such products, reducing tariffs, and eliminating quantitative restrictions. They were prepared to collectively veto any package that did not include progress in agricultural markets.[59] Negotiations in agriculture, however, remained intractable and threatened to undermine the successful conclusion of the Round. GATT Director General Arthur Dunkel adapted the negotiating structure and asked the Chairs of each negotiating group to submit 'consensus' achieved drafts and if none were reached, to provide 'arbitrated' texts based on their knowledge of the national objectives of the negotiating parties. The Draft Final Act (the 'Dunkel Draft') was compiled by the Secretariat for review and acceptance:

> The Director-General's hope was that, as a consolidated text, participants would be able to see a positive balance of concessions across the entire range of agreements for the first time. This, it was thought, might then persuade all participants to accept the Draft Final Act without alterations and conclude the Round.[60]

The US sought to amend specific areas and to abandon the proposed Multilateral Trade Organization (MTO), now the WTO, when Dunkel's

---

56 Jan McDonald 'It's Not Easy Being Green': Trade and Environment Linkages beyond Doha' in Buckley, *supra* note 21, 145–69 at 145.

57 Gervais, *supra* note 40; Thomas and Meyer, *supra* note 45.

58 Ibid. at 16–19.

59 Ibid. at 21.

60 Ibid. at 22.

successor Peter Sutherland re-launched negotiations to conclude the Round with 'consensus' on all major aspects.[61]

State signatories to the GATT 1947, labelled GATT 'Contracting Parties' in accordance to consent theories that treated international law as little more than a contract pursuant to a provisional agreement limited to tariff reduction, were now 'members' under GATT 1994 due to the formation of the WTO. While GATT 1947 encompassed rules for voting, all important decisions – except waiver and accession – were generally based on consensus and without a formal vote. Signatories to GATT 1947 not WTO members remain as 'contracting parties' and are not subject to GATT 1994 unless they sign on to the WTO Agreements, which carried an implementation deadline of 1 January 1995.[62] But the fate of the new WTO was only really secured once US Congress passed implementing legislation late December 1994; an institution that was not on the original negotiating agenda soon became known as the significant accomplishment heralded as the 'crown jewel' of the Round.[63] The WTO is now the permanent institution overseeing trade relations as prescribed by the legal instruments included in its Annexes; these constitute an integral part of the Agreement (as a group, the *Multilateral Trade Agreements* (MTA)).[64] Some 60 different instruments are now included under the WTO Agreement;[65] the most relevant for our discussion are the GATT 1947 (incorporated by GATT 1994), the DSU Agreement, and TRIPS.[66] The WTO oversees the application of its rules through a structured dispute resolution system – with

61 Ibid. at 29.

62 The Final Act Embodying the Results of the Uruguay Round of Multilateral Trade Negotiations, concluded in Marrakesh, Morocco, on 15 April 1994, published in WTO Secretariat, *The Results of the Uruguay Round of Multilateral Trade Negotiations, The Legal Texts* (Geneva, 1995), <http://www.wto.org/english/docs_e/legal_e/03-fa.doc>.

63 John M. Curtis, 'Trade and Civil Society: Towards Greater Transparency in the Policy Process', <http://www.international.gc.ca> at 302.

64 Article 1 of the WTO Agreement establishes the Organization and Article 2 its scope. The Multilateral Agreements on Trade in Goods (GATT 1947 and GATT 1994), the General Agreement on Trade and Services, and the Agreement on Trade Related Aspects of Intellectual Property Rights all are included in Annex 1 (A, B, C, respectively). Annex 2 is the Understanding on Rules and Procedures Governing the Settlement of Disputes (DSU). Annex 3 outlines the WTO's Trade Policy Review Mechanism (TPRM) and Annex 4 includes four additional agreements.

65 Joost Pauwelyn, *Conflict of Norms in Public International Law: How WTO Law Relates to Other Rules of International Law* (New York: Cambridge University Press, 2003). [Pauwelyn, *Conflict of Norms*] at 23.

66 Full text of all agreements available online at http://www.wto.org/.

panels and an Appellate Body (AB) – extending treaty based disciplines to formerly unilateral remedies.[67]

The current Round of trade negotations was to restore commitments to the DCs and LDCs. This 9th Round was launched with the agreement of 142 members at the 4th Ministerial meeting, on 14 November 2001 in Doha, Qatar and has been dubbed the Doha 'Development Round'. Former WTO Director-General Mike Moore has predicted: 'the meeting at Doha will be remembered as a turning point in the history of the WTO and the trading system and in relations between developed and developing countries within that system'.[68] The agenda for this Round indicates an attempt to remedy prior imbalances, to make future negotiations and obligations more equitable, to promote development, and to protect sufficient flexibility in trade agreements (and their interpretation) to allow for greater regulatory discretion towards development; these should provide positive gains for human rights as well. The Round was scheduled to conclude by 1 January 2005, collapsed in 2003, was revived and extended to December 2006, and continues still eight years after its launch with uncertain prospects for its conclusion after the failure to secure an agreement in July 2008. Despite important developments arising from this Round, some critics believe the WTO is not substantively or politically sustainable; there are too many broken promises and 'no real consensus among WTO members on many of the complex regulatory issues that the panels and the Appellate Body will be asked to rule upon'.[69]

Rise of a Corporate Republic

Prior to the Uruguay Round, IP was mentioned in only a few GATT articles. Article XX(d) of GATT 1947 created an IP *exception* to free trade which allowed for diverse regulatory measures.[70] While DCs and LDCs were being

---

67 Thomas and Meyer, *supra* note 45 at 259; Trebilcock and Howse, *supra* note 51 at 407–8.

68 28 November 2001, 14th General Meeting of the Pacific Economic Cooperation Council, as cited in Terence P. Stewart, *After Doha: The Changing Attitude and Ideas of the New WTO Round* (Ardsley, NY: Transnational Publishers Inc., 2002) at 2.

69 Barfield, *supra* note 41 at 7.

70 The exception was invoked in two GATT panel disputes. The panel's decision in *United-States-Imports of Certain Automotive Spring Assemblies* L/5333, adopted 26 May 1983, BISD 30S/107, the first patent infringement case, reaffirmed that patent protection was an area in which Parties could take measures that otherwise would be non-compliant with GATT obligations. The panel in *United States Section 337 of the Tariff Act of 1930*, L/6439, adopted 7 November 1989, BISD 36S/34 (EEC v. US), clarified that the Article XX(d) exception meant that the substantive patent law of a party was not challengeable under GATT but that parties remain charged with the obligation of ensuring that patent legislation was enforced in a manner not inconsistent with GATT. As a result, the

assured that stronger IP regimes would benefit their development objectives, they were not oblivious to the political pressure from the Japan, US, Swiss and EU alliance at the rent-seeking insistence of their respective information, entertainment, and pharmaceutical industries during the Uruguay Round. The politics of the process made the inclusion of TRIPS as part of the 'final package' questionable to the end.[71] Regionally, the United States pushed for the inclusion of substantive IP protection in its free trade agreement with Canada and Mexico in order to ensure its foothold as a major industry player in the knowledge economy but also to provide a precedent for the trade-related IPR regime being negotiated multilaterally.

Chapter 17 of the *North America Free Trade Agreement* (NAFTA) preceded TRIPS by a year and, as the first treaty to regulate and enforce substantive IP rights, became the model for TRIPS.[72] The shift to an 'integrated' Round at Uruguay required adoption of all WTO Agreements as a 'single package' undertaking[73] that linked IP to trade in goods and thereby made the trade-offs more agreeable based on a broader set of issues. Concessions were thus achieved where a member might otherwise have been economically or politically disinclined because 'somewhere in the package there should be something for everyone'.[74] This was not the structure, however, that members had agreed to when first committing to the Round.[75]

DCs and LDCs constitute over 85 per cent of WTO membership but given the degree of complexity and number of issues to be dealt with, the ability of DCs and LDCs to negotiate gains commensurate with those of developed countries is limited.[76] Some scholars write that TRIPS, for better or worse, was simply the result of this 'trade off' of concessions: what DCs and LDCs lost with IP standardization, they made up for in negotiations over agriculture and textiles.[77] There were some negotiated gains for DCs in the Round for trade

---

domestic provision that allowed the International Trade Commission (ITC) to suspend the importation of patented goods allegedly infringing a US patent were held to violate GATT since the infringer could not avail himself of the same procedural rights as the aggrieved national plaintiff. See Gervais, *supra* note 40 at 6. IP disputes predating TRIPS stress the importance of the non-discrimination principle.

71 Ibid. at 3–26.

72 NAFTA provides a high level of IP protection and mandates that parties to the agreement provide adequate and effective measures to protect and enforce IPR without the measures themselves becoming barriers to legitimate trade (Article 1701).

73 With some exceptions, e.g. government procurement, special and differential status of LDCs, longer implementation time frames for some members etc.

74 <http://www.wto.org/english/thewto_e/whatis_e/tif_e/fact4_e.htm>.

75 Gervais, *supra* note 40 at 3–26.

76 Drahos, 'Weak Bargain' *supra* note 38.

77 See Trebilcock and Howse, *supra* note 51. Note, the *Agreement on Textiles and Clothing* in the Uruguay Round provided for full phase out of quotas in the area by the end of 2004.

in cotton and textiles but these could hardly be considered a *trade off* against strong universal standards in IPRs. DCs, LDCs, and newly industrialized countries (NIC) have long enjoyed a comparative advantage in textiles due to the large pool of unskilled labour accessible and modest technology needed for production. Cotton and textiles were considered a desirable point of entry for further growth and development through manufacturing and exports, yet trade in textiles and clothing largely occurred outside of GATT because of an apprehension by developed countries that their domestic markets would be flooded and that the resulting market disruptions would cause harm to their producers. Any trade liberalization gains to be enjoyed by DCs were quickly offset by the 'voluntary' bilateral arrangements restraining cotton textile exports through quantitative restrictions that these competitive textile exporting countries were pressured into in the 1950s. The *Short-Term Arrangement on Cotton Textiles* in 1961 – 'in essence, a negotiated derogation from the then-existing GATT rules applicable to cotton fabrics' – formalized these bilateral agreements into an agreement under the GATT framework and within a year, was replaced by the *Long Term Arrangement Regarding International Trade in Cotton Textiles* 'which continued the negotiated derogation for trade in cotton textile' until it expired in 1973. It was then replaced by the *Multi-Fibre Arrangement* (MFA). With six extensions between the years of 1978–1993, LDCs and DC had been denied their comparative advantage in clothing and textiles for a near half century.[78] We can conclude that retrograde industrial pressures and policies commencing in the 1950s had succeeded in the inequity of keeping the textile and agricultural sectors largely outside the purview of GATT 1947; any *trade off* here was based on justice concerns for *restoring* old promises rather than bargaining anew. In the end, the politics of trade effectively dictated preferential first-world outcomes in Uruguay, creating 1) a legalized trade linkage in IP when other trade linkages (like trade and the environment or trade and human rights) had not been readily or expressly embraced[79] and 2) legally enforceable standards for minimally prescribed monopolies antithetical to trade theory and policy (which link development and prosperity to the exploitation of comparative advantage). Although TRIPS is incongruent with the tenets of trade liberalization, it borrowed generously

---

78 See Trebilcock and Howse, *supra* note 51 at 21; Thomas and Meyer, *supra* note 45 at 6–7, 95–104.

79 See Steve Charnovitz, *Trade Law and Global Governance* (London: Cameron May International Law and Policy, 2002) at 11. Charnovitz asks: 'How should the World Trade Organization ('WTO') and the World Intellectual Property Organization ('WIPO') work together? Governments, multinational corporations, non-governmental organizations ("NGOs"), and academics have offered answers to these questions, but often the answers seem internally inconsistent. For example, business groups have generally favoured linking trade policy to intellectual property rights, but have typically opposed linking trade and environmental policies.'

from the public benefit rationales of domestically strong IPRs of countries like the United States, Japan, Canada and Britain and the consumer benefit rationales of liberal trade, concealing at the time its heavy corporate impetus. During the ten years since TRIPS, a growing sentiment is that its inclusion in the multilateral trade regime was a sacrifice of territorial sovereignty, public good regulation, and individual human rights to corporate demands under the guise of (unsubstantiated) utilitarian claims. TRIPS protects corporate interests in intellectual objects and, due to the nature of knowledge, creates the rules for the private provision of public goods. But the exclusive nature of these rights, now internationally required, raises issues of equity and distributive justice traditionally most effectively addressed by the nation state. If as a society we want to encourage cooperation and discourage autarchy, then each nation must be assured by words and action that this is not a zero-sum game whereby gains realized by some members are based on losses by others: 'If no country is to be made *worse* off as a result of cooperation, equity considerations have to be borne in mind'.[80]

We live in one world indivisible yet sated with contradictions. Destitution persists, according to the World Bank, 'even though human conditions have improved more in the past century than in the rest of history – global wealth, global connections, and technological capabilities have never been greater'.[81] In 2000, Sarah Anderson and John Cavanagh reported that of the world's top 100 economies, 51 were corporations and only 49 were countries.[82] The world's Top 200 corporations accounted for over a quarter of global economic activity while employing less than 1 per cent of its workforce. Additionally, they found, the top 200 corporations' combined sales surpassed the combined economies of all countries other than the largest ten, and the combined sales of the top 200 were 18 times the size of the combined annual income of the 1.2 billion people (24 per cent of total world population) that live in 'severe' poverty. Eighty-two (41 per cent) of the Top 200 corporations were American, followed by the Japanese who occupy 41 of the Top 200 spots;[83] it is no wonder

80 Inge Kaul et al., 'How to Improve the Provision of Global Public Goods', Kaul et al., *Providing Global Public Goods*, op, cit., 21–5.

81 The World Bank, *World Development Report 2000/2001: Attacking Poverty* (Toronto: Oxford University Press, 2001) at 3.

82 'Top 200: The Rise of Corporate Global Power', Key findings, <http://www.ips-dc.org/reports/top200text.htm>. The report found the 1999 sales of each of the top five corporations (General Motors, Wal-Mart, Exxon Mobil, Ford Motor and Daimler Chrysler) are bigger than the GDP's of 182 countries. A list of the top 200 is also provided and is based on a comparison of corporate sales and country GDP, <http://www.corporations.org/system/top100.html>.

83 See 'Top 200: The Rise of Global Corporate Power', Institute for Policy Studies, <http://www.ips-dc.org/downloads/Top_200.pdf>. For a list of the 135 countries with less GDP than the smallest of the top ten corporations, see Quigley, 'Catholic Social Thought', op. cit. at 49.

these trading 'nations' advocated for TRIPS; it is they who trade, not their nation-states.

In the life science industry, consolidation of market share means a few transnational corporations having a dangerous amount of control in bioindustrial products related to agriculture, food, and health. The Crucible Group, a multisectoral group with representatives from industry, indigenous peoples, third world farming interests and lawyers, provides some startling statistics. The world's top ten agrochemical corporations account for 91 per cent of the $31 billion worldwide agrochemical market while the top ten global seed companies control an estimated one-quarter to one-third of the $30 billion commercial seed trade. The top five vegetable seed companies control 75 per cent of the global vegetable seed market (69 per cent of the North American maize seed market is controlled by four companies and a single company, by the end of 1998, controlled 71 of the US cotton seed market).[84] Concentrated corporate control exists independent of the biopatenting debate and needs to be addressed by competition policy since such power erodes the vigour and benefit that free and robust competition brings to the market and individual consumer. However, patents provide additional incentives for the consolidation of markets with acquisitions of patent portfolios as a primary basis for mergers and acquisitions that expand exclusive rights to exploit licensing revenue. Corporate restructuring may also keep the larger firm's licensing costs low where patent rights are significantly fragmented amongst a number of (smaller) competitors. As the value of IP assets have surpassed physical assets for corporate valuation, the quest for IP acquisition provoked a restructuring of the industry that created the pressure for greater IP protection.[85]

Patents confer a major revenue generating asset for many of the large transnational life-science corporations. The vast majority of IP holders tend to be corporate owners or assignees[86] animated to 'persons'.[87] If patents are only valuable because of the tropes of property, the corporate *person* is similarly only meaningful because the state fictionally and institutionally recognizes it as a live entity entitled to 'rights'.[88] This is a moral choice with

84 See Crucible II Group, *Seeding Solutions*, op. cit., IDRC < http://web.idrc.ca/en/ev-31619-201-1-DO_TOPIC.html> at 16.

85 Ibid. citing W. Bratic, P. McLane and R. Sterne, 'Business Discovers the Value of Patents', *Managing Intellectual Property*, September 1998.

86 For statistics on the top 300 patent owners granted US patents in 2005 and each year preceding since 1984, see Intellectual Property Owners Association, General Publications, <http://www.ipo.org>/.

87 American patent laws require the application to name real persons as inventors but no similar restriction exists in Canada. See discussion in Chapter 5 on whether IPRs are *human* rights.

88 Elizabeth Wolgast, *Ethics of an Artificial Person: Lost Responsibility in Professions and Organizations* (Berkeley, CA: Stanford University Press, 1991) examines

clear costs. Having personified the corporation, society is unable to restrain its will or contain its (dysfunctional) personality[89] and, apparently, compete with its economies. The corporation amasses property and capital but does not suffer the destitution of which the World Bank speaks. It is impervious to the social costs of its assets and therefore any sense of duty, obligation, or justice that may correspond with its *rights*.[90] The corporation may, occasionally, voluntarily assume a commitment to social consciousness, such as with the United Nations initiated Global Compact of 1999 inviting international businesses to support universal and social principles, but this too is in anticipation of the economic and goodwill gains to be generated for shareholders. The transnational nature of business seems to have served a facilitative function in ushering in international laws for corporate capitalist interests. Not surprisingly, civil groups and NGOs display a growing distrust of corporations[91] and by extension globalization.

The animosity is unqualifiedly further extended to the WTO, which is credited with serving the corporate agenda for a republic deviating from the *virtuous* one from which our global social conscience and twin international institutions (UN and WTO) for governance were born.[92] The ideals of America's Virtuous Republic were grounded in Christian values and the rule of law and served as a 'new rallying cry for global missionaries'.[93] Expanding the international legal system towards greater state inclusion was a critical component of the proselytization of a political (democratic), religious (Christian), and economic (free market) ideology and pushed forward the commitment to Republican values integral to America's search for economic and missionary opportunities abroad. The institutions for global governance today espouse ideals communicated with a certain rhetorical eloquence that

---

'Corporate Persons' and social problems associated with stressing the idea of 'person' as western cultures do and then ascribing that title to corporations, a moral choice with clear costs. See also Jon Chipman Gray, *The Nature and Sources of the Law*, 2nd edn (New York: Macmillan, 1931). Compare this to the views of legal theorist Hans Kelsen, who writes that 'the concept of … person means nothing but the personification of a complex of legal norms. Man … is only the element which constitutes the unity in the plurality of these norms'. Hans Kelsen, *General Theory of Law and State*, trans. by Anders Wedberg (New York: Russell and Russell, 1961) at 95.

89 Joel Bakan, *The Corporation: The Pathological Pursuit of Profit and Power* (New York: Free Press, 2004).

90 See UN, 'What is the Global Compact', <http://www.unglobalcompact.org>.

91 Bakan, *supra* note 89.

92 See Mendes and Mehmet, *supra* note 2.

93 Yves Dezalay and Bryant G. Garth, eds, *Global Prescriptions: The Production, Exportation, and Importation of a New Legal Orthodoxy* (Ann Arbor, MI: University of Michigan, 2005) [*Global Prescriptions*], Introduction to an insightful set of essays on the new legal orthodoxy.

paradoxically recognize a paternal and charitable duty towards others,[94] while simultaneously contributing to the social conditions that manufacture the need to assist 'them' by fostering hierarchies of dependence through coercively distinct imprints of dislocation, dispossession, and prescriptions for legal title concomitant with colonialism.

By granting proprietary rights to life and defining these rights in terms of exclusion, society perpetuates dependency in creating a need to seek permission (licence) to use new technology of vital potential for food and health delivery. This hierarchy of need enables existing power struggles to continue with unfavourable outcomes for some nations. Innovating nations are privileged while imitating nations continue to have their activities rendered trespassory and illegal even within their own borders. The plethora of welfare-promoting possibilities biotechnology has introduced to further individual human rights interests are consequently surrendered to the corporate ideals of mass commodification and ownership of all things tangible and intangible, with a correspondingly anti-competitive 'warehousing' of these assets. While corporations are not the only IP holders, they are the most vocal, persuasive, and predatory. Individual scientists holding letters patent over their inventions rarely have the time or inclination to mobilize a lobby, or to sue; they are too busy with the science. Perhaps if they did, a more balanced patent system would result nationally and internationally allowing for generous research uses and preserving the basic stock of knowledge needed for competitors in developing down-stream applications. Rather, US industry has pushed its government to transplant strong domestic IPRs into international trade law but without the corresponding necessary common law protections hard-won by a small few litigants in courts or the publicly-minded safeguards legislated into our patent system to maintain the necessary balance between private incentives and public interest. Economist Gerald Helleiner writes,

> [l]arge private corporations purchase influence within all so-called democratic societies. As all Geneva trade diplomats know, their influence over ostensibly international negotiations is also considerable; witness the role of the pharmaceutical industry in intellectual property debates … The international activities of business lobbies are subject to no limits or registration requirements or regulation. The bulk of their activity is <u>un</u>transparent to the public. If the public sees them at all, it is only 'through a glass darkly'.[95]

94 See generally Theodore J. Lowi, 'The Welfare State, The New Regulation, and the Rule of Law' in Allan Hutchinson and Patrick Monahan, eds, *The Rule of Law: Ideal or Ideology* (Toronto: Carswell, 1987), 17–58

95 G. Helleiner, 'Markets, Politics and Globalization: Can the Global Economy be Civilized?' UNCTAD 1-th Raul Prebisch Lecture, Palais des Nations, Geneva, 11 December 2000, <http://www.utoronto.ca/cis/working_papers/2000-1.pdf> at 6.

Governments are very responsive to corporate pressure and, Heilleiner adds,

> if money impacts heavily upon both domestic and international political processes, there should be no illusions as to where the bulk of the power in decision making relating to the global economy is likely to continue to rest – that is, with those countries, firms and organizations that are economically (and, it must be added, militarily) the strongest.[96]

As early as the 1960's and 1970s, it was becoming apparent that representation of the needs and interests of the 'weaker' nations would not be met by the growing trend to harmonize patent law. The 'Andean Group' led a group of DCs through a review and reassessment of IP under the Paris Convention (Paris) and its implications for domestic development within their economies. As a result of this review, many DCs with existing patent legislation reportedly introduced measures or amendments geared at weakening domestic IPRs in an effort to make them more balanced and compatible with development goals.[97] This trend coupled with the appreciating value of knowledge-based technologies in an information economy resulted in a counter-push by well resourced industries in developed countries (particularly pharmaceutical and entertainment) for the IP-trade linkage in the 8th Round and the incorporation of key Paris provisions into TRIPS. Developed countries of origin (mainly the US, EU, Japan and Switzerland) claimed that exports to countries with more lax IP standards were reduced as consumers could acquire similar created imitation goods, such as generic drugs, locally produced in their domestic markets, cheaper without licence. But, as Trebilcock and Howse have written, '[t]he strength of the general claim for policy equivalence in this area is not straightforward'.[98] As stand-alone measures, TRIPS is likely to reduce welfare in DCs by undermining their comparative advantage in imitation, imposing costs on knowledge diffusion, and limiting access to technology. That welfare loss is a net gain subsidy of increase in first-world corporate profits. As domestic IPRs were strengthened by the administrative system and judiciary in the United States, IP industries began to politicize themselves further by aggressively lobbying their politicians for a trade-based approach to IP.[99]

As an initial response, these industries were successful in securing the IP-trade linkage in Section 301 of the US Trade and Tariff Act 1974, which

96 Ibid.

97 Drahos, 'Standard Setting', *supra* note 14, on the extent DCs influence standard-setting outcomes.

98 Trebilcock and Howse, 'Regulatory Diversity', *supra* note 47 at 18.

99 See Peter Drahos and Ruth Mayne, *Global Intellectual Property Rights: Knowledge, Access and Development* (New York: Palgrave, 2002); Sell, *Private Power*, *supra* note 14.

allowed the US to pressure IP infringing states with the threat of unilateral sanctions.[100] Under Section 301, the American president could take appropriate and feasible action to enforce US rights under trade agreements or to eliminate any act, policy or practice unreasonable, unjustifiable or discriminatory that burdens or restricts US commerce. The USTR can be petitioned by American industries or firms, trade associations and individual corporations, such as MGL, to investigate actions of foreign governments (i.e. Canada) that are believed to harm US commerce. If the USTR investigates, it will consult with the foreign government in an attempt to resolve the problem, failing which, within a year, the USTR recommends such appropriate action to the President as unilateral trade sanctions and/or the removal of GSP privileges. Threats to withdraw voluntary concessions are persuasive in eliciting desired compliance with protection of American held IP. As Helleiner notes, '[c]orporate influence over US and other major powers' political decision-making can obviously carry profound spillover effects for rule-making in the global economy'.[101] Two watch lists (one a priority list) are kept under Section 301 with out-of-cycle reviews on countries which pose a threat to the interests of US IP holders. Canada was on the 2005 watch list and was one of the five countries reviewed out-of-cycle. Canada may be subject to unilateral action if it continues to disregard US pressures for greater protections and enforcement of IPRs.[102]

100 In addition, section 337 of the US Trade Act provides inbound (at the border) protection by providing for unilateral retaliatory trade action against foreign products imported into the USA based on violation of IPRs of US producers under American law in the making of those foreign products prior to import. Section 337 does not require that injury to the American producer be established. A finding in violation of Section 337 will result in exclusion of the product from the United States unless the parties enter voluntarily into a licensing agreement and, in this manner, has been considered a form of extra-territorial enforcement of domestic IPRs or a sanctioning of unfair competitive practices. But the remedy provided is an import ban rather than a countervailing duty and the impact is disparate in the sense that local competitors, taking an unfair advantage in domestic trade or trade for export by violating another American's IPRs, will not be affected by this section. Special 301 of the Trade Act 1974 (s. 1303 of the Omnibus Trade and Competitiveness Act of 1988) allows for *outbound* protection by providing for unilateral trade sanctions to be taken against countries identified pursuant to it as engaging in unfair trade practices due to a more relaxed IPP regime. The provision is designed to enhance the ability of the US to negotiate improvements to foreign IP regimes. Usually, any level of IP protection that falls below that provided by US law is likely to be considered unfair. But, according to Trebilcock and Howse, such sanctions are likely to be against the GATT MFN principle. See Trebilcock and Howse, *supra* note 51 at 407 and ch. 16.

101 Helleiner, *supra* note 95 at 6.

102 See USTR's Special 301 Report 2006, <http://www.ustr.gov/Document_Library/Section_Index.html>.

Susan Sell provides a detailed history of the political economy of internationalized IPRs and concludes that effectively 12 corporate executives of US-based MNCs mobilized to form the Intellectual Property Committee (IIPC)[103] which was in turn successful in using states as agents for the convergence of IP standards and their enforceability with the inclusion of TRIPS in the Uruguay Round. Part of the appeal was the naive belief that locating IP within the multilateral trade system would insulate against unilateral action, such as those available under US law,[104] by making them GATT inconsistent for failure to deploy the multilateral regime's mechanisms. This promise helped secure the DCs' and LDCs' commitment to TRIPS; multilateralism held the prospect of more *just* outcomes than unilateralism because 'a standardized rule system can be a bulwark against bullying by the strong'. 'But', Helleiner qualifies, 'if the rules are constructed by the strong to protect their own interests or if there is imbalance in capacity to implement them or both, such a system may be worse than useless to the relatively weak ... Evidently, process is everything in these matters'.[105]

As it turns out, Helleiner is correct. Drahos and Braithwaite observe that the threat of use of unilateralism seems to have *increased* rather than decreased with TRIPS:

> Perhaps the most stunning achievement of the 301 system has been its continued growth and use in the period after the creation of the WTO. There were intimations that the creation of a WTO dispute resolution system would see the US ease off on aggressive unilateralism. But if anything, 301 has acquired a more machine-like efficiency in the post-TRIPS period ... The Clinton Administration ... strengthened 301 by introducing immediate action plans for foreign countries on intellectual property rights as well as out-of-cycle 301 reviews, pushing developing countries into accelerating their implementation of TRIPS[106]

With time, it has become apparent that the national representatives of the DCs and LDCs negotiating the Round may not have fully appreciated

103 Sell, *Private Power*, *supra* note 14, chs 4 and 5 on the domestic origins of a trade-based approach to intellectual property and ch. 5 on the Intellectual Property Committee effectively translating domestic changes into international law; see also Drahos and Braithwaite, *supra* note 2, ch. 6.

104 For example, the United States used the threat of GSP withdrawal and priority watch lists under Special 301 to attack Thailand for inadequate pharmaceutical patent protection it offered American firms in 1991. See Anthony D'Amato and Doris Long, eds, *International Intellectual Property Law* (Boston: Kluwer Law International, 1997) at 67-70.

105 Helleiner, *supra* note 95 at 12.

106 Drahos and Braithwaite, *supra* note 2 at 106–7.

the implications of their agreement. It is now commonly criticized that '[i] nternational corporate power has effectively curbed national sovereignty in the field of intellectual property policy'[107] and given the South a 'raw deal' with TRIPS.[108] Critics assert that institutional factors at the WTO methodologically conspire to maintain positional relations, reinforcing while exacerbating the vast North/South inequities that remain the focal point for correction in the Doha 'Development' Round. Others see the possibilities of TRIPS despite its political history. TRIPS is a significantly more facilitative instrument than it is credited for. Jerome Reichman has examined the potential use of TRIPS to support the positive adoption of desired domestic policies where the agreement is ambiguous, and popularized Jayashree Watal's idea that TRIPS provides ample 'wiggle room' for members to formulate domestic regulatory preferences.[109] The extent to which countries may adopt diverse measures to respond to the need for balanced biotech regulation, the first branch of the bifurcated approach to state agency prescribed in this text, requires a review of TRIPS.

Trade Dimensions of IP Regulation: TRIPS Overview

TRIPS introduces universal minimum standards for IP protection and under the auspices of the WTO provides a formalized multilateral dispute-

107 See Vaver, *IPL*, op. cit. at 3; see also Vaver, 'Uniformity', *supra* note 27 at 5.

108 Martin Khor, 'How the South is Getting a Raw Deal at the WTO' in Anderson, op. cit., at 7.

109 See Jayashree Watal, 'Trips and the 1999 WTO Millennium Round: Some Reflections on Future Issues Related to IPRs in the WTO and the Way Forward for Developing Countries' (2000) 3 J. *World Intell. Prop*. 3, 7–24. See generally, Reichman 'Fair Followers', op. cit., at 27–51, fn 51 and 72 and accompanying text. For a discussion of the pressure and demands for higher levels of protection pressing for a strict interpretation of existing norms and the 'wiggle room' see J.H. Reichman, 'The TRIPS Agreement Comes of Age: Conflict or Cooperation with the Developing Countries' (Summer 2000) 32.3 *Case W. Res. J. Int'l L.* at 454–70 [Reichman, 'Conflict or Cooperation']. Reichman contends having government officials in their economic capacity at the bargaining table would '… ensure that any deals struck between foreign rights holders and local firms would also redound to the public interest, without, however, compromising the larger political organs' ability to maintain official positions on intellectual property law and policy in the relevant international forums' (at 467). See also J.H. Reichman and David Lange, 'Bargaining Around the TRIPS Agreement: The Case for Ongoing Public-Private Initiatives to Facilitate Worldwide Intellectual Property Transactions' (Fall 1998) 9:1 Duke J. of Comp. and Int'l L 11, discussing diplomatic ways to use strategic initiatives that rely on the built-in TRIPS capabilities for implementing obligations. They suggest that agents negotiating within a cooperative framework characterized by goodwill can achieve better success in resolving conflicts through agreement on a transactional basis than by using the conflict system of dispute resolution.

settlement process. TRIPS (Article 2) incorporates some of the key obligations from the Paris Convention for the Protection of Industrial Property (Paris Convention).[110] The Paris Convention was the primary international instrument providing substantive standards in governing the protection of patents as industrial property prior to TRIPS and provided for complaints to be made to the International Court of Justice (ICJ). In theory,[111] a state could even retaliate under the Paris Convention for another's violation but practically speaking that rarely, if ever, happened because states would voluntarily submit to the jurisdiction of the ICJ and their decisions lacked coercive enforcement mechanisms. What was really germane about the Paris Convention was the National Treatment Principle (NTP), which was also a core pillar for the GATT, and priority rights; TRIPS preserves both these features.

The NTP in Article 3 of TRIPS requires that '[e]ach Member shall accord to the nationals of other Members treatment *no less favourable* than that it accords to its own nationals with regard to the protection of intellectual property'. This is in contrast to the Paris Convention which only required the 'same' treatment, the *same protection,* and *same legal remedy* for nationals of other contracting states in the Union (Article 2(1)). Effectively, under TRIPS foreigners can be treated *better* with more protection. 'For the Purposes of Articles 3 and 4 of this Agreement, Protection shall include matters affecting the availability, acquisitions, scope, maintenance and enforcement of intellectual property rights as well as those matters affecting the use of intellectual property rights specifically addressed in this Agreement.' TRIPS adds the Most Favoured Nation principle (MFN, Article 4), the second main pillar from the GATT which was absent from the Paris Convention. The MFN essentially provides that any advantage, favour, privilege or immunity granted by a member to the nationals of any other country shall be accorded

110 Established in 1883 with the agreement of 14 countries and now with 169 signatories of 20 March 1883 last revised at Stockholm on 14 July 1967. Of these 169 contracting parties (countries only), by development status breakdown statistically as follows: countries in transition, 16 per cent, developing countries 44 per cent, industrialized countries 19 per cent, least developed countries 21 per cent. See WIPO, Treaties Statistics, <http://www.wipo.int/treaties/en/statistics>.

111 The United Nations Charter (UNC) provides for the establishment of the International Court of Justice to adjudicate on matters relating to the achievement and maintenance of the purposes and principles relating to the UN including human rights which could be carried out on either an adjudicatory or an advisory capacity. Each member pledges to comply with the decision of the court in any case to which it is a party. But the Statute of the Court annexed to the Charter made the compulsory jurisdiction of the Court voluntary to member states. It is worth mentioning that social and economic rights cannot be adjudicated before the ICJ and rely entirely on domestic means of enforcement (UNC, United Nations, <http://www.un.org/aboutun/charter/>). For these reasons, I propose in Chapter 7 that the WTO, with its sophisticated dispute settlement mechanism, serve as steward of trade *and* other values.

immediately and unconditionally to the nationals of all other members. Subject to limited exceptions, there are no playing favourites amongst the foreigners although as a group they may be favoured over nationals.

By adopting key Paris Convention provisions, TRIPS indicates a preference by the negotiators to update and adapt existing rules rather than create a new system calibrated to different needs and stages of development of members as they face the challenges of new technology.[112] In the case of patents and utility models, the applicant is given 12 months as a priority right over other applications filed by competitors in other countries during the same period for the same invention. Priority is based on a regular first application filed in one of the Contracting States. This allows applicants to consider in which countries they want to patent while assessing varied requirements for due diligence, administrative compliance, application processes and processing and legal fees. Other Paris Convention rules imported into TRIPS include, for example, that patents granted for the same invention in different states continue to be independent of each other. The sovereignty of nations is maintained in that the mere grant of a patent in one jurisdiction will not oblige other Contracting States to grant the patent, and a patent cannot be refused, annulled or terminated in one state on the basis that it has been so in another Contracting State (Paris Article 4bis). Canada could not, for example, revoke MGL's BRCA1 and BRCA2 patents merely because France has done so. The Paris Convention also provides that regulatory restrictions on the sale of a patented product or a product derived from patented processes may not be used to invalidate a patent (Article 4quater). This provision, also incorporated into TRIPS by Article 2, enables governments to use other regulatory mechanisms to monitor and regulate market *accessibility* and the *commercialization* of a patented invention and could inform public policy exceptions and exemptions. Human embryonic stem (hES) cell research offers a suitable example. Stem cell research is critical for future preventative and therapeutic medicine but the private monopoly right, a much contested patent over hES, belongs to the University of Wisconsin's Alumni Research Foundation (WARF).[113] While regulation alone could not invalidate the patent per se, parliament, if it saw fit, could impose restrictions or requirements as precondition to the *commercialization* of that patent by adopting an ethics review approval as suggested by the Nuffield Council on Bioethics, or a regulatory price mechanism similar to the Patented Medicine Prices Review Board as suggested

112 See Gervais, *supra* note 40 at 68–70, 'The Logic of TRIPS'.

113 See Carl Gulbrandsen, 'Stem Cell Patent Holder's View of the California Challenge' Wisconsin 16 November 2004, <http://wistechnology.com/article.php?id=1352>. See also 'Foundation's stem cell patents impede search, scientists say', *San Diego Union Tribune*, 30 July 2006.

by Austin and Amani for gene patents.[114] Article 5 of the Paris Convention provides for conditional use of compulsory licenses to eschew patent abuses and allows forfeiture after all else, including the compulsory licence, fail to avert the ongoing abuse by a patentee. TRIPS Articles 31 and 40 also expressly address compulsory licences and abuse of patents.

Part II, Section 5 (Articles 27–34) of TRIPS sets out minimum national standards for patent protection within member states; Article 27 articulates patentable subject matter:

> … patents *shall* be available for any invention, whether products or processes, in all fields of technology, provided that they are *new, involve an inventive step and are capable of industrial application* … [and] shall be available and patent rights enjoyable *without discrimination* as to the place of invention, the *field of technology* and whether products are imported or locally produced.

Canada meets this obligation in that it provides patent protection for any invention (product or process) in all fields of technology so long as the novelty, utility, and non-obviousness requirements are satisfied. TRIPS does not establish any substantive ceilings or floors for the content of the domestic patentability requirements. Tightening the application of these requirements administratively and judicially is a TRIPS-compliant response to modifying biopatent policy and can be reinforced through validity reviews, opposition proceedings, and infringement determinations. Article 27 prohibits discrimination, on the availability of a patent or rights enjoye,d based on the *field* of technology. It does not, however, foreclose raising the requirement bar or the degree of knowledge attributed to the PHOSITA, the necessary disclosure to be made, or the scope of what may be considered prior art. These are viable means of providing acceptable legal nuances to help calibrate the standards of patentability in accordance with the technology. The standards for the PHOSITA are already permissibly variable across both technologies and jurisdictions. While a state cannot refuse to confer patent rights to all biotech inventions as a class, more optimal domestic policy outcomes may be achieved from regulatory reform that focuses on adopting more stringent standards for patentability, disclosure and commercialization. TRIPS obligations are only universal minimums. Article 27 states that a patent 'shall be available'; it does not state that a patent 'shall be *granted*', or that if it meets the established criteria it cannot be refused based on additional criteria like an ethics review, or that even if granted the patent may not be subjected to non-infringing uses permitted within domestic law.

Article 28 sets out the rights to be conferred: generally the exclusive right to prevent third parties without consent from making, using, offering for sale,

114 The Nuffield Council on Bioethics Report, op. cit.; Austin and Amani, 'Patents on Genes, op. cit.

selling or importing for these purposes that product, and Article 33 requires the minimum term of protection to be 20 years regardless of the field of technology; this too is based on Western standards such as s. 44 of the CPA. A number of safeguards are, however, directly incorporated into the text of Article 27 and contribute to the 'wiggle room' states have to protect their public's interest. Article 27.2 provides:

> Members may exclude from patentability inventions, the prevention within their territory of the commercial exploitation of which is necessary to protect *ordre public* or *morality*, including to *protect human, animal or plant life or health or to avoid serious prejudice to the environment*, provided that such exclusion is not made merely because the exploitation is prohibited by domestic law.[115]

The wording demonstrates the intent to locate patent rights amidst a hierarchy of pressing domestic regulatory objectives for members, allowing for a general exclusion from patentability *ex ante* inventions to protect human health amongst other things. Reasonably, this provision would extend to enable regulatory diversity regarding biopatenting more broadly speaking because of the words '*ordre public* or morality'. However, any contested measure must be justifiable by objective standards and not merely as a result of domestic prohibitive laws. What this means is, as with the Paris Convention, governments may make regulatory restrictions on commercialization, by requiring for example labelling of all GM products,[116] or creating some form of regulatory approval similar to that required for drugs, or may prohibit exploitation of an invention, but such restrictions cannot alone render the invention *unpatentable*. Daniel Gervais explains: 'any exclusion under the Article would need to be justified ... [T]he fact that the exploitation of the invention is prohibited under national law is, at best, one of the arguments necessary to meet the condition prescribed'.[117] In other words, the domestic prohibition on commercial exploitation *is evidence* of the public policy or moral objective that is being protected and not the *basis* on which 'Members may exclude from patentability'. The exemption may in fact be narrowed through future judicial interpretation as national courts increasingly demonstrate their inability to address broader public policy issues and refuse to engage in moral claims in

---

115 See also NAFTA 1709(2) for the parallel provision with almost identical wording.

116 See e.g. Commission of the European Communities, 'Proposal for a Regulation of the European Parliament and of the Council concerning traceability and labeling of genetically modified organisms and traceability of food and feed products produced from genetically modified organisms and amending Directive 2001/18/EC, <http://ec.europa.eu/food/fs/biotech/biotech09_en.pdf>.

117 Gervais, *supra* note 40 at 223.

the context of domestic litigation. It may also be further narrowed by the subsequent interpretation of the WTO decision-making bodies, especially of the words '*necessary to protect*' which may be construed conservatively to mean *least restrictive means* as in prior GATT XX decisions.[118] Ultimately, the domestic measures must be justifiable by objective standards and not constitute a means of arbitrary discrimination or disguised restriction on trade.

A national standard setting instrument could reaffirm the intention to incorporate more expressly the Article 27.2 exclusion in patent *practice*. This can be reflected in domestic industry manuals such as the MOPOP[119] and could include an express provision for *morality* or *ordre public*, articulate a position on gene patents, and establish the threshold for the requirements of patentability. But, Article 27.2 cannot be replied upon *ex post facto*, as the limited exclusions are from granting patents over inventions that would otherwise meet the patentability criteria and not as an independent ground to *defend* the infringement of patent rights once conferred, such as those alleged against the Canadian government for unauthorized use of MGL's BRCA patents. This is even if the infringement, as in the Canadian context, is to meet the same public policy objectives (human health) the Article permits.[120]

Finally, TRIPS Article 27.3 allows a specific content based exclusion for domestic protection over life forms. Members may exclude patents '... for diagnostic, therapeutic and surgical methods for the treatment of humans or animals' and '... for plants and animals other than microorganisms, and essentially biological processes for the production of plants or animals other than non-biological and microbiological processes'. This provision is broad and vague but has significant potential benefit in supporting deferential treatment for domestic regulatory measures where human health is concerned. It is at the core of debates on the patentability of life as well as the patentability of inventions that have a direct effect on health policy. Genetic tests can be excluded from patentability even if they are characterized as 'diagnostic' or 'therapeutic' under Article 27.3(a). There has been some debate whether predictive genetic testing is in fact *diagnostic* as it only correlates with the *propensity* to develop a medical condition and not the diagnosis of one *per se*. The discovery of the BRCA1 gene in chromosome 17, for example, means that certain women can be tested to determine with 85 per cent *likelihood* whether they will develop breast or ovarian cancer sometime in the future. But genetic diagnostic methods are much like other diagnostic test in providing

118 GATT Article XX and the interpretation of 'necessary' in reference to Article XX(b) which uses the same language are discussed in detail in chapter 7. Full text of GATT 1947, <http://www.wto.org/english/docs_e/legal_e/gatt47_01_e.htm>.

119 See Chapter 3.

120 See Chapter 7 for a detailed account of how this provision should be interpreted and 'adjudicated' by the WTO panel or Appellate Body in the case of a TRIPS dispute.

a risk assessment. Canada's MOPOP seems to allow diagnostic methods as patentable but Canada is in the position that if it were to amend its laws to exclude the patentability of diagnostic methods, it would remain TRIPS compliant under Article 27.3(a), even though on the wording of the provision it too cannot be relied on to justify any violation of valid patent once granted. Existing gene patents can be revoked, however; TRIPS Article 32 only requires that '[a]n opportunity for judicial review of any decision to revoke or forfeit a patent shall be available'.

Members can exclude plants or animals, but not non-biological microbiological processes for their production, as patentable subject matter under Article 27.3(b). The only condition is that if patent protection is not extended to plants, some form of protection (not necessary for animals) be provided, if not under patent then by an effective *sui generis* system. Canada is already in compliance with this requirement. The SCC in *Harvard* established that animals and plants are not legally patentable and Canada already provides *sui generis* protection for plants, meeting its obligation through plant breeders' legislation. Microorganisms, non-biological processes, and microbiological processes must be afforded patent protection under Article 27.3(b) of TRIPS. There are several problems that become apparent here. First, some critics have noted, genetically modified organisms and DNA sequences may potentially be patented as microorganisms or microbiological processes. None of these terms are defined in TRIPS and the method by which a gene sequence may be introduced into its host is often by way of a 'vector' or similar microbiological means recapturing it within the scope of the universal minimum standard for patentability. A firm then could obtain *de facto* patent rights to a GMO life form by patenting the vector and the process of its introduction even if the life form itself was not patentable. After *Schmeiser*, in Canada at least, this is unfortunately not merely conjecture since patent protection now extends to products in which the patented invention, be it a gene, vector, or some other microorganism is embedded, creating dire consequences for the practical operability of the TRIPS exclusion and the regulatory discretion that TRIPS strives to maintain in this contested area. Worse still, the realization of universal human rights that Canada has committed to may be effectively undermined by judicial agency and the misapprehension of international obligations. Judges are not immune from political and industrial pressures for convergence with the patent law of more protectionist regimes. Such a tendency may also undermine the discretion afforded under Article 6 of TRIPS which leaves regulatory scope over 'exhaustion' of IPRs within sovereign discretion. This section would otherwise allow governments to decide their own rules on *exhaustion*. The first sale of patented life, such as seeds, could exhaust internationally all the rights attached to the product, making the patents unenforceable on the basis of infringement arising from the progeny.[121]

To summarize, regulatory governments have a number of available responses to address the issue of whether life and its genetic building blocks should be patentable within domestic law and these responses would help ensure the necessary balance is maintained between competing interests in patented inventions. In addition to tightening the patentability standards, national patent laws may be amended to exclude patentability of all life or specific objects (such as DNA sequences, or human/animal chimera), redefine the statutory definition of 'invention', specifically exclude products of nature and impose an ethics review or other regulatory requirements before commercialization is allowed. It is worth mentioning that the courts can also continue the tradition of safeguarding competing stakeholder interests by taking a more conservative approach to private property rights in claims construction and in patent construction and by creating common law exclusions, exemptions, or defences in relation to diagnostic and biopatenting, just as courts and legislatures have done in Canada and the United States for methods of medical treatment – in Canada, to judicially hold that such inventions are not patentable and in the United States, to amend patent laws to allow non-infringing uses. Incidentally, this latter approach of 'non-infringing uses' may be a good way to meet TRIPS obligations while retaining significant latitude for regulatory diversity but may constitute poor patent policy and contribute to confusion and therefore litigation. Amendments or judicial review could provide these specific exclusions from patentability, exemptions to infringement, or even recognize relevant statutory defences and remain TRIPS compliant. If as a matter of industrial policy it is decided that Canada does want to continue to allow patents on genes for example, legislated special provisions like those governing the Patented Medicines Prices Review Board[122] could be drafted. These would respond to the concern over prohibitive prices that biotech patents create for the delivery of health by providing that patented genetic tests and health products receive similar treatment as patented medicines.

The Article 27.2 and Article 27.3 exclusions of TRIPS help ease any concerns regarding the adoption of health and human-rights friendly patent policies and further the instrumentalism of domestic patent regimes. Some additional regulatory discretion over cultural diversity in the biopatenting debate may be found in the broader exclusions in Article 27.2 which relate to morality and public policy. Governments may choose to enact provisions

121 For an insightful discussion of the exhaustion debate in relation to agricultural biotechnology in the context of a recent American case, see Chris Holman's blog 'Quanta and Agricultural Biotechnology', 16 June 2008, <http://holmansbiotechipblog.blogspot.com/2008/06/quanta-and-agricultural-biotechnology.html>.

122 See ss. 83–85 of the CPA, op. cit.

domestically to reflect these exclusions. Perhaps the greatest deference to state sovereignty, however, comes in the provisions of general application. Article 7 articulates the general objective of a strengthened trade-related IPR regime and Article 8 establishes the principles that are to guide members in the formulation of their domestic IPRs regimes. Article 8.1 and 8.2 provide further grounds for co-coordinating patent and health policy:

> 1. Members may... adopt measures necessary to protect public health and nutrition, and to promote the public interest in sectors of vital importance to their socio-economic and technical development, provided that such measures are consistent with the provisions of this Agreement.
> 2. Appropriate measures, provided that they are consistent with the provisions of this Agreement, may be needed to prevent the abuse of intellectual property rights by right holders or the resort to practices which unreasonably restrain trade or adversely affect the international transfer of technology.

Article 8.1 expressly acknowledges the need to safeguard health and food priorities while Article 8.2 provides ample leverage in custom designing domestic measures by broadly defining what constitutes 'abuse' and 'unreasonably'. This discretion is supported by the general objective for the protection and enforcement of IPRs. Article 7 confirms that the objective of providing patent protection and enforcement is to promote technological innovation, transfer and dissemination 'to the *mutual advantage* of producers and *users* of technological knowledge and in a manner conducive to *social and economic welfare, and to the balance of rights and obligations*'.[123] It is important to note that Article 7 adopts the utilitarian aspirations of patent rights but does not restrict industrial policy exclusively to this regulatory tool. This is a sensible approach given the strong evidence that scientific progress and technological advancement would continue (as it had in pre-patent periods) to different degrees depending on the innovation sector and particularly with such adapted, incentivizing, regulatory oversights in place as public education, prizes, procurement, and tax incentives. The nation state will necessarily take an inward look at the balance of such policies; any true objection to the underlying technology requires a patent policy response *and* regulation external to the patent regime. The US Supreme Court in *Chakrabarty* observed,

> [t]he grant or denial of patents on microorganisms is not likely to put an end to genetic research or its attendant risks. The large amount of research that has already occurred when no researcher had sure knowledge that patent protection would be available suggests that legislative or judicial fiat as to patentability will not deter the scientific mind from probing into

123 Emphasis added.

> the unknown any more than Canute could command the tides. Whether respondents' claims are patentable may determine whether research efforts are accelerated by the hope of reward or slowed by want of incentives, but that is all.[124]

TRIPS is less onerous than it may seem at first glance and appropriately recognizes that the degree and nature of patent incentives offered by any government remains discretionary. TRIPS Article 1.1 states expressly that 'members shall be free to determine the appropriate method of implementing the provisions of this Agreement within their own legal system and practice'. The statutory basis for the creation and scope of patent rights, their historical origins and their public policy objectives all lead to the conclusion that these are functional public policy tools institutionally justified on the basis that state sanctioned interference ultimately serves the public interest by seeking to optimize *balanced* regulation of various stakeholders' interests in knowledge as a public good. If there are problems in the patent system, they are due to conceptual misunderstandings of intellectual objects, the abandonment of the role of the state and its institutional actors in regulating access to these objects, the proclivities of a patent friendly and not technically trained judiciary, and the resulting failures of design and operation compounded by strong industrial lobbies and predatory practices. The public, in whose name IPRs are heralded, more often than not has been neglected in industrial policy development and patent office practice. Biopatent law has developed incrementally and primarily through jurisprudence. But, this is *not* due to regulatory limitations imposed under TRIPS.

Recital 4 of the TRIPS preamble recognizes the 'underlying public policy objectives of national systems for the protection of intellectual property, including developmental and technological objectives [.]' Domestic patent laws often provide *more* property rights than internationally required under TRIPS. To protect the public policy objectives from being trumped by protected private interests, citizens should demand appropriate state action by their regulatory governments to remedy any imbalances and failures of current IP systems and to stem the tide of judicial extension of IPRs to sensitive biological and hereditary material – formerly precluded and properly excluded by doctrinal standards – as patentable subject matter. Dealing with the issue at this level would be efficient as it would reduce search and examination where a claimed 'invention' would otherwise be insufficiently inventive for the PHOSITA to survive the non-obviousness criteria. This is also effective in addressing the need to keep scientific discovery 'infrastructure' within the public domain, free from incremental encroachment.[125] For now, the hegemony of property within

124 *Chakrabarty*, op. cit.

125 Amici Curiea Brief, <http://www.wcl.american.edu/pijip/go/research-and-advocacy/ip-policy-and-law-reform>.

Western ideological discourse has led to the assumption that 'more is better' and a 'what's mine is not yours'[126] attitude that permeates administrative and judicial thinking. These are not amoral choices, even though they are apparently void of ethics:

> Inevitably … the law is drawn into the debate, and this is especially so when there is a conflict of individual interest … It is pointless to attempt to disengage the moral from the legal dispute – when we talk about legal rules, we are inevitably drawn into a discussion of moral rules.[127]

While the jurisdiction of courts and their impact is always remedial, governments are forward looking and, suggests Jillian Clare Cohen, can anticipate situations and write laws to meet them. It may take constant pressure to get governments to act, however, because '[p]olitical decision makers do not want to have to make tough decisions, so they pass the buck to the court system'.[128] There is a danger to this inactivity, as these *de facto* domestic interpretations may also come to infuse the interpretation of contested measures under the hard-bargained-for Article 27.2 and Article 27.3 exemptions under TRIPS, just as the pressure for convergence towards more protection has reciprocally resulted in gradual extension of patents to DNA and life. In Canada, the judiciary has all but vacated the possibility of optimizing available 'wiggle room' under TRIPS. Parliament should legislate the definition of life and its patentability to overcome some of the indeterminacies and inefficiencies created by the Supreme Court. All states should be concerned over any level of IP protection that effectively allows innovation policy to impede the future of (basic) research or to compromise the delivery of health with the risks and exigencies of litigation.[129] While state agency is essential to ensure that these risks are not left unchecked, it is also imperative in the context of a state's broader international obligations. It seems then, whether nationally or internationally, proper interpretation of legal texts is key. Unfortunately, while common law derogations continue to threaten the public's interest, the corporate capitalist imperative has achieved near hegemony and despite impositions on sovereignty, human rights, and the public interest, new regional and bilateral agreements undertaken by agents of our governments – the 'technocratic trade elites' – threaten far worse implications than TRIPS' multilateralism.

---

126 David Vaver, 'What's mine is not yours: Commercial appropriation of personality under the Privacy Acts of British Columbia, Manitoba and Saskatchewan' (1981) 15 *U.B.C.L. Rev.* 241.

127 Mason and McCall Smith, *Law and Medical Ethics*, op. cit. at 4.

128 Quoted in Laidlaw, 'Reckoning with Technology Law in Court', op. cit.

129 Amani and Coombe, 'The Human Genome Diversity Project', op. cit.

TRIPS, Plus or Minus: Deviating From Prescribed Standards

A growing number of regional negotiations are modelled on TRIPS. These collateral arrangements dealing with American perceived deficiencies in TRIPS create 'a very complex legal environment' potentially at odds with the TRIPS MFN principle in Article 4 and giving rise to potential *de facto* MFN discrimination, according to Frederick Abbott.[130] The United States has actively pursued a number of bilateral trade treaties that strive to mirror the substantive protections prescribed by TRIPS and to exert political pressure for the adoption of 'TRIPS Plus' protections which would further reduce critical policy space for signatory countries.[131] Not surprisingly, 'US PhRMA stands strongly behind these efforts'.[132] 'TRIPS Plus' refers to the requirement of more stringent substantive IPRs protection than mandated by TRIPS.[133] For example, it may include a commitment that a negotiating party provide patentability for plants or animals; or that there be accession to UPOV[134] (which provides stricter and stronger protection for plant varieties than TRIPS requires); or to the Budapest Treaty on the Deposit of Microorganisms for the Purpose of Patent Protection[135] (not mentioned in TRIPS); or to conform with the EPC (that allows patenting of transgenic plants and animals).[136] Having signed on to these TRIPS Plus agreements, countries can no longer take advantage of the flexibilities inherent in TRIPS or of the extended exemptions

130 Frederick Abbott, 'Towards a New Era of Objective Assessment in the Field of TRIPS and Variable Geometry for the Preservation of Multilateralism' (2005) 8.1 J. of Int'l Econ. L. 77 at 97 [Abbott, 'New Era']. See e.g. *Free Trade Area of the Americas* (FTAA), <http://www.ftaa-alca.org>.

131 A full list of US bilateral initiatives and their texts are available at the Office of the United States Trade Representative at <http://www.ustr.gov>. These include The US-Australia FTA (1 January 2005); the US-Chile FTA (1 January 2004), draft US-Central America Free Trade Agreement , and US-Singapore (15 January 2003).

132 Frederick Abbott, 'The WTO Medicines Decision: World Pharmaceutical Trade and the Protection of Public Health' 99 AJIL 317 at 349 [Abbott, 'WTO Medicines'].

133 For a detailed and insightful discussion of criteria for TRIPS Plus status and a helpful table of the bilateral and regional agreements, where developed states have secured these for biological material in DCs, see 'TRIPS-plus' through the Back door: How Bilateral treaties impose much stronger rules for IPRs on life than the WTO' July 2001, <http://www.grain.org/briefings/?id=6>.

134 *Supra* Chapter 3, note 72.

135 *Supra* Chapter 3, note 77, <http://www.dsmz.de/patents/bptreaty.htm>.

136 For a list of existing or under negotiation agreements requiring TRIPS Plus protection, see: 'Bilateral Agreements imposing TRIPS Plus Intellectual Property Rights on Biodiversity in Developing Countries' August 2005, <http://www.grain.org/rights_files/bilats-TRIPSplus-0805.pdf>. See also Susan Sell, 'Industry Strategies for Intellectual Property and Trade: The Quest for TRIPS and Post-TRIPS Strategies' 10.1 Cardozo J. Int'l and Comp. L. 79.

offered under the transitional periods. It will also limit a party's ability to invoke these provisions as TRIPS based defences at the WTO.

Frederick Abbott explains the trend towards bilateral and regional arrangements as an attempt by the United States to *shift* to more restrictive and substantively protective regimes 'as a result of the somewhat more balanced approach to TRIPS now achieved at the WTO ..., calling into question the relevance of the TRIPS Council deliberations'.[137] Abbott recommends that the new IP agreements be subject to objective prior impact assessment, that the WTO members give greater recognition to the fact that IP rules have significantly different public welfare implications depending on a country's level of development and their field of application, and with the gradual ending of transition periods, that the TRIPS Council assessment of its rules and provision for plus or minus adjustments be restructured along industrial subject matter and development lines to ensure positive outcomes are supported empirically. In light of the important human development implications that TRIPS and TRIPS-Plus regimes have for public health, technology and biological resources, the UNDP endorses greater domestic regulatory diversity be preserved unimpeded by international obligations. One option the UNDP recommends is reverting to a 'TRIPS-minus model that significantly reduces the length of protection and scope of coverage and increases national decision-making authority on standards and coverage of protection while maintaining a minimalist agenda at the international level'.[138]

In the meantime, the *harmonization* offered by TRIPS through the imposition of universal minimum standards for IP protection which members can implement as they see fit (Article 1) and the TRIPS Plus arrangements made in parallel are not enough to quell the protectionists' quest. New efforts are underway at the World Intellectual Property Organization (WIPO) for the complete standardization of patent law *vis-à-vis* a Substantive Patent Law Treaty (SPLT) that would pave the way for a singular global patent. To this end, WIPO in 2001 adopted a 'patent agenda', endorsed by its Director-General,

> aimed at the further harmonization of patent law worldwide. The initiative, called the 'WIPO Patent Agenda' was envisaged as a process that would lead to the preparation of a blueprint for the future development and harmonization of the patent system. The Patent Agenda initiative has

137 Frederick Abbott, "Towards a New Era of Objective Assessment in the Field of TRIPS and Variable Geometry for the Preservation of Multilateralism" (2005) 8.1 J. of Int'l Econ. L. 77 at 97 [Abbott, 'New Era'].

138 UNDP, *Making Global Trade Work for People* (New York: Earthscan Publications Ltd., 2003), <http://www.undp.org/dpa/publications/globaltrade.pdf> [UNDP, *Making Trade Work*] at 221.

> therefore placed the further harmonization of patent law as a top priority in WIPO's activities. The main activities under the Patent Agenda relate to the efforts to promote the ratification of the Patent Law Treaty (PLT); the reform of the Patent Cooperation Treaty (PCT); and, the ongoing negotiations on a Substantive Patent Law Treaty (SPLT). These separate, but interlinked, activities are ultimately oriented to set up an international legal framework for a global patent *with the result that the limited policy space left in the hands of national governments under the World Trade Organization's (WTO) Trade-Related Aspects of Intellectual Property Rights (TRIPS) will be further eroded.*[139]

To the extent that new restrictions on important public policy space are ushered in through WIPO, WIPO's agenda will effectively facilitate a *corporate* agenda by narrowing the scope for state agency and is not, therefore, inconsequential if its standard setting role contributes to a trade-enforced system designed to compromise justice for charity.

Sacrificing Justice for Charity

Because of distributional inequities and barriers to access, social welfare and justice are often usurped by the privatization of public goods and then rationalized on the economic basis of efficiencies. The imposition of strict IP regimes on DCs is justified by an inherently flawed theory which posits that IP producing companies (mainly drug, life science and biotech companies) are significantly harmed and disadvantaged in their R&D by the lax or absent IP regimes of developing countries and that this in turn has a negative impact on people in that community. Alan Sykes, for example, defends TRIPS for the welfare gains it provides to DCs and LDCs on the basis that strong patent protection in these countries protects R&D interests necessary to provide external firms incentive to undertake research important to these specific countries (i.e. malaria vaccines/drugs) with the assurance that the costs of the benefits would not simply be externalized through lax IP regimes.[140] One might counter that such a Faustian bargain would not be necessary if

139 Carlos M. Correa and Sisule F. Musungu, 'The WIPO Patent Agenda: The Risk for Developing Countries', <http://www.southcentre.org/publications/wipopatent/toc.htm>, from the executive summary [emphasis added]. The authors stress the importance of representation for the developing countries in the norm setting process at the WIPO and the need to foster an appreciation of the importance of IP as a tool for development policy and its prospects.

140 Alan Sykes, 'TRIPs, Pharmaceuticals, Developing Countries, and the Doha 'Solution' (February 2002) U. Chicago Law and Economics, Olin Working paper No. 140, criticizing the Doha Public Health Amendment.

greater attention were directed at ending poverty-linked diseases through the alleviation of poverty; it is well established that malaria is largely preventable with bed nets and environmental controls and the best approach to AIDS is prevention through education and condom use. Jagdish Bhagwati expounds on the underlying fallacy of the welfare gains position, arguing that while poor countries have need, they do not have effective *demand*:

> There is little money to be made to recover normal profits on your invented drugs, if you think of poor markets. To see why the drug companies nonetheless see IPP [IP protection] in the poor countries as a money-spinner, it is necessary to distinguish between two types of diseases: those, such as malaria which are primarily in the poor countries, and those, such as AIDS which afflict rich and poor countries alike .... [In the latter], the drug companies make money in the rich country markets; IPP there is clearly something they value. But then they see piffling effective demand for those drugs in poor countries. So their strategy is to sell there at very low marginal costs and then charging the little that these poor markets will bear.[141]

Why, if poor countries are scarce able to afford the medications and other patented products of foreign firms (at least those that also have developed markets) would IPRs matter in this impoverished context?[142] Although price discrimination can prove profitable for foreign firms, these typically large MNCs, would like to *maximize* their return in poor markets rather than price discriminate by pricing at marginal cost. Their motivations for securing strong IPRs in poor countries are jaundiced and subtly compromise justice for charity. They reveal the perversity of incentives created by a system of *ex post* rewards rather than the system intended to spur R&D through *ex ante* incentives. Bhagwati contends:

> The way [to raise *that* return] is to increase effective demand by using aid moneys addressed to health programs, so that the excess of what they will charge over their marginal cost is increased, raising profitability in the poor country markets. Medical economists have known for years that medical groups, for instance, favour insurance schemes that improve the patients' ability to pay (such as Blue Cross and Blue Shield insurance programs in

141 Jagdish Bhagwati, 'Pricing Medicines to Benefit Poor Countries' *UNDP Cooperation South 2002: Creativity, Innovation and Intellectual Property Rights* (2002) at 38 [Bhagwati, 'Pricing Medicines'].

142 The value will differ according to the price sensitivity: 'Empirical evidence shows that in developed countries industry recoups fifteen to twenty percent of its R&D costs through patents whereas in a country like India, the figure for a domestic inventor is 0.5 to two per cent.' Shiva, *Protect or Plunder*, op. cit. at 5.

> the US), but oppose insurance schemes like the National Health Service of England, which instead reduce the returns to doctors.[143]

Effectively, merchants create the *need* for missionaries – often their own governments, international monetary and health agencies and NGOs. Through state agency towards greater intellectual *property* rights, together they perpetuate colonialist vestiges of dependency, and entrench, in addition to a financial indebtedness, a moral debt deep into the psyche and pocket books of developing countries thus defying any legitimate efforts at promoting economic self-sufficiency and sustainable development by these peoples for and by themselves. Strong internationally mandated IPRs protection is short-sighted in contributing to the artificial growth of foreign debt for developing and least developed nations and increases the need for debt relief and foreign aid to help pay for the patent protected technology.[144] It creates, according to Ostergard, a bit of a quandary:

> the primary concern for developing countries is to adapt technology that will help maintain the physical well-being of their people. By putting up barriers, the developed countries delay the creation of markets that could support entry of technologically advanced IP, thus cutting short the potential profits that could be obtained if the developing countries could sustain themselves.[145]

Not only do the LDCs and DCs lose out, but so do developed countries. Firms, once assured of IP protection in other jurisdictions, may in fact move their capital there to take advantage of lesser regulatory compliance costs and cheap labour. While TRIPS should be applauded for retaining a fair bit of scope for regulatory diversity, a less legally-invasive, market liberalization-seeking TRIPS would have focused on increasing border measures to prevent counterfeits and not have linked substantive standards creating monopolies, which TRIPS did for the first time, with trade law. TRIPS is almost always discussed in terms of the state obligations it creates for WTO members but a state is also conferred *rights* by TRIPS and these are important for maintaining critical policy space. Within this instrument, some of the state rights recognized could have been made more explicit but are, nevertheless, rights. For example, the scope for TRIPS compulsory licensing, now extended to generic exports to countries lacking internal manufacturing capacity under limited conditions, should have been clearly linked to other fields of patenting,

143 Jagdish Bhagwati, 'Pricing Medicines to Benefit Poor Countries', *UNDP Cooperation South 2002: Creativity, Innovation and Intellectual Property Rights* (2002) at 38 [Bhagwati, 'Pricing Medicines'] at 39.

144 See generally Ostergard Jr, 'Intellectual Property', op. cit.

145 Ibid. at 177.

as should the availability of defences directed towards protecting the public interest in IP regulation. The domestic implementation of high prescribed standards of IP protection based on the executive branch of government's undertakings in multilateral negotiations is implicitly instrumental to an existing economic pattern that simultaneously compromises attendant civil and political liberties and creates potential conflict with inter-related social and economic rights. Even if market citizenship (as shareholders) is presented as the modern manifestation of global citizenship, it is fallacious to assume that all can participate fully in a market-driven society in a manner that would preserve the value of equality of that citizenship or its universality; with the immense income disparity in the world, market participation is neither feasible nor imperative for peoples concerned with the satisfaction of more immediate needs like clean water, food, medicine and other essentials for their physical and spiritual survival. In response to the announcement by developers of Vitamin A rice about their deal with agricultural biotech giant Zeneca to license and distribute the crop, Day-cha Sirpat of the Alternative Agriculture Network in Thailand asserted: 'The poor, they don't need vitamin A rice. They need vitamin 'L', that's Land. And they need vitamin 'M, that's Money. Malnutrition is a problem of poverty, not technology'.[146] Genetically engineered vitamin A rice ('golden rice') was

> heavily promoted as a solution to hunger and malnutrition. Yet these promotional campaigns are clouding the real issues of poverty and control over resources, and serving to fast-track acceptance of genetically engineered crops in developing countries … 'The main reason we are against GE rice', says BIOTHAI [Thai Network on Biodiversity and Community Rights] 'is the issue of control. Small farmers in the Third World can't achieve security when transnational corporations control these technologies and give away GE seeds like others give out food aid. It doesn't work, it doesn't help the farmers'.[147]

To help refocus the direction of IPRs' objectives, practices, and standard setting modalities in an effort to create a better balance than previously achieved – and against the backdrop of the WTO 'Development Round' and the 'Millennium Development Goals' – WIPO's General Assembly adopted a resolution for a 'Development Agenda' in 2004 which was formally passed in October 2007 when the General Assembly adopted the 45 recommendations offered for its implementation. The recommendations included the creation of the Committee on Development and IP (CPID) to oversee it. The Development Agenda was formed to reflect the integral need to consider development in

---

146 See Indo-Asian News Service, 'GE-Rice Good for PR, Not the Poor' Masipag, *Third World Network* 7 June 2000, <http://www.twnside.org>

147 Ibid.

regulating IP with the realization that IPRs are cross-cutting with other vital regulatory sectors. This is an important step towards a normative principle for IPRs that requires such rights to be matched to local conditions and founded on a substantive equality principle that strives to measure welfare-generating outcomes of IPRs, and not only in terms of economic growth but their distributional effects.[148] While reference to 'Development' in the agenda name has since been removed at the behest of the United States,[149] important commitments have been made towards technical assistance, assessment, evaluations and impact studies of IPRs, substantive norm setting activities that account for public policy differences and with due regard to flexibilities in international agreements, and to generally

> approach intellectual property enforcement in the context of broader societal interests and especially development-oriented concerns, with a view that 'the protection and enforcement of intellectual property rights should contribute to the promotion of technological innovation and to the transfer and dissemination of technology, to the mutual advantage of producers and users of technological knowledge and in a manner conducive to social and economic welfare, and to a balance of rights and obligations', in accordance with Article 7 of the TRIPS Agreement.[150]

This reflects an institutional response to a global reality: while *peoples* are the intended beneficiaries of various governance regimes, it has been sovereignty's[151] new delegates – the so-called elite *mercatocracy*[152] – that have effectively facilitated, expedited, and navigated the process of translation (of the nation state into its colonial forms) towards the full obliteration of difference; because a single homogenous market with one language, set of

148 See Margaret Chon (2006) 'Intellectual Property and the Development Divide' 27 *Cardozo Law Review* 2821–912.

149 William New, 'WIPO Development Committee Ends First Year On Mostly Agreeable Note' 12 July 2008, o<http://www.ip-watch.org/weblog/index.php?p=1147>.

150 Recommendation 45, Report of the Provisional Committee on Proposals Related to a WIPO Development Agenda (PCDA), A/43/13/Rev., 17 September 2007, <http://www.ip-watch.org/files/WIPO%20Dev%20Agenda%20report%20doc%20a_43_13_rev.doc>.

151 For a discussion of what is meant by the term, see Dennis Lloyd, *The Idea of Law: A Repressive Evil or Social Necessity?* (Harmondsworth: Penguin Books, 1983): '[i]n the sphere of international relations, state sovereignty meant that each state was entirely free to regulate its relations with other states' (at 174).

152 '[T[ransnational corporations are identified as "the central organizers", the "engines of growth", the "key agents of the new world economy", and the "dominant private institutions of the world economy" ... [and] most importantly, are crucial participants in the creation and enforcement of merchant laws'. See Cutler, *Transnational Merchant, supra* note 15 at 21, footnotes omitted.

regulatory compliance requirements, and standard price, is much easier to corner and requires little political accountability.[153] If the corporate imperative has gained international authority without the accountability of a public participatory process or a politic of ethics, it is because some states have used their authority to facilitate this agenda. Claire Cutler comments on sociologist Boaventura de Sous Santos' central thesis that 'transnationalization of the legal field' is a 'constitutive element of globalization':[154]

> [t]he modern law merchant is a form of transnationalized law embodying both the globalization of local law, as Anglo-American corporate laws are adopted throughout the world, and the localization of global law, as states are subjected to increasing discipline from legal regimes developed by international, transnational, and global organizations. Moreover, forms of transnationalized legal relations are discontinuous and uneven. In some cases, merchant laws operate dialectically creating deterritorialized transactions and agreements, but then reterritorializing them to facilitate enforcement.[155]

TRIPS exemplifies this point as WTO members commit to internationally prescribed protective patent standards that are universal and transcend domestic authority baselines offered yet rely on domestic enforcement that in many instances exceeds them. For political scientist Susan Sell, this is part of the inexorable march of *global capital/Western culture:*

> One hardly needs agency to account for the fact that the economically most powerful transnational actors acted in concert with the economically and

153 Cutler explains: 'Derived from the medieval reference to the law merchant, *lex mercatoria*, the mercatocracy is comprised of transnational merchants, private international lawyers and other professionals and their associations, government officials, and representatives of international organizations … [and] operates globally and locally to develop new merchant laws governing international commerce … the mercatocracy blurs the distinction between public and private commercial actors, activities, and law. It exercises near hegemonic influence through its material links to transnational capital and through its monopoly of expert knowledge, thought, and institutional structures. Indeed, the mercatocracy and its law are deeply implicated in the ordering of state-society relations because they operate to recast "public" concerns as 'private' and thus are not subject to democratic methods of scrutiny ….' Ibid. at 4–5, 20–22.

154 Boaventura de Sousa Santos, *Toward a New Common Sense: Law. Science and Politics in the Paradigmatic Transition* (New York: Routledge, 1995) at 252, cited in Cutler, op. cit. at 19–21.

155 Cutler, *supra* note 15 at 20 [footnotes omitted].

> politically most powerful states to devise global rules to benefit them all (and at the expense of most others).[156]

Governments of the day were neglectful of domestic economic costs and moral and political obligations owed to their respective 'publics' when subscribing to TRIPS. Indeed, current patent practice in tandem with the push for TRIPS *plus* protection under bilateral agreements and within national courts may effectively impair access to agricultural and pharmaceutical products, vaccinations and methods of medical or diagnostic treatment, slow progress in the sciences and public health genomics, and privatize the common genetic and cultural heritage of humankind far worse than the commitments made in TRIPS. The human rights of all *peoples* are laid open for compromise, but most significantly for the most disadvantaged people of all nations, and disproportionately for those located in developing and LDCs which lack the relevant institutional capacity and a substantial innovation sector for IP development. How this plays out will depend on how the provisions of TRIPS are interpreted in disputes and competing international obligations reconciled. In order to have secured consensus to conclude the Round, TRIPS retains enough ambiguity to allow for regulatory discretion in patenting life. In determining how to utilize the discretion preserved by TRIPS, states must remain sensitive to the normative authority and legal priority demanded of human rights obligations under UN instruments.

156 'In the material account, the process was eliminating obstacles to international commerce under the economic might and ideological orthodoxy of the transnational capitalist class. Global IP rules were just the latest triumph, neither the first nor the last.' Sell, *Private Power, supra* note 14 at 3.

Chapter 5
TRIP'ing over Human Rights? The Legitimacy Crisis at the WTO

A law is valuable not because it is law, but because there is right in it.

Henry Ward Beecher

Introduction

With the review of the global governance regime of the WTO and the TRIPS' textual flexibilities, it is apparent that without further legislative activity, the common law has arguably graver consequences for national regulatory responses than does TRIPS. This must be adequately addressed along with a critical review of the patent administration system for compliance with a state's obligation to its citizens on fundamental human rights (HR). Until recently, HR rarely entered the fray or were considered as part of the broader costs of national and international patent laws. Where critical patent review occurred, it was traditionally intent on improving the incentives for achieving more optimal industrial and economic policy. Biotech's far reaching promises, however, have had the ironic effect of drawing attention to the limitations on delivery of such promises that biopatent disequilibria foster while highlighting the individual HR costs of a utilitarian based proprietary regime. The HR dimension of patent law and policy can therefore no longer be ignored. Governments must respond by taking appropriate action and may need to be reminded of their HR obligations to do so. Now trade-related, patent policy need not take priority over health policy or other policy arenas affecting individual rights. This chapter reviews the nexus of trade-related IPRs with HR and thereby seeks to complete the review of the institutional and historical evolution of the trade and HR regimes, and to further demonstrate the complementarity of their reciprocal objectives in global governance.

The International Human Rights System

The League of Nations was created by the 1919 Treaty of Versailles in order to promote international cooperation, improve global welfare, and to achieve peace and security through *jus gentium* (the law of nations) which governed relations between states. Still, little affected how states treated their own citizens. The Great War made apparent the atrocities that could result at the hands of one's own government. With the failure to prevent World War II, the League ceased its activities in 1946 but was succeeded by the United Nations, which formed the institutional framework for overseeing the promotion, protection, and attainment of human rights for all as a means of achieving peace and security. The GATT 1947 shared this objective in so far as human rights would invariably be advanced by the goal of 'peace and prosperity through trade'.[1] The political liberalism of Anglo-American constitutional law formed the basis of HR instruments and, what former UN High Commissioner for Human Rights Mary Robinson referred to in 1998 as a 'culture of human rights', inclusive of values integral to all world religions.[2]

Just as economic interdependence was encouraged through multilateral trade, parallel pains were taken to eradicate unauthorized use of force and to protect individual liberties against the tyranny of the state.[3] With mounting pressure for the US to assist in the British war against Nazi Germany, an important meeting between President Roosevelt and Winston Churchill occurred at Placentia Bay, Newfoundland in 1941 and by January 1942, 26 nations, later expanded to 46, endorsed the Declaration of the United Nations.[4] The Atlantic Charter was an ideologically principled plan of action which preceded the GATT and Bretton Woods system, and 'linked the imperative for a new global security institution and respect for human rights with improved labor standards, economic advancement, and social security'.[5] The United Nations Charter (UNC) prefaced the United Nations that was formed under

1 See Mendes and Mehmet, *Global Governance, Economy and Law*, op. cit. at 9. See generally Debra Steger, 'Peace and Prosperity through Trade', <http://www.rsc.ca/files/publications/transactions/2001/>.

2 'Today the Universal Declaration of Human Rights stands as … one of the great documents in world history … the authors sought to reflect in their work the differing cultural traditions in the world. The result is a distillation of many of the values inherent in the world's major legal systems and religious beliefs including the Buddhist, Christian, Hindu, Islamic and Jewish traditions.' Mary Robinson, Forum 98, <http://www.unhchr.ch/huricane/huricane.nsf/0/5624DB25C40FBE268025666E005E709C?opendocument>.

3 Carlos Manuel Vazquez, 'Trade Sanctions and Human Rights: Past, Present, and Future' (2004) 6(4) *J. Int'l Econ. L.* 797. The UNC Art. 2.4 provides 'All Members shall refrain in their international relations from the threat or use of force against the territorial integrity or political independence of any state, or in any other manner inconsistent with the Purposes of the United Nations'.

4 <http://www.ibiblio.org/pha/policy/1942/420101a.html>. Mendes and Mehmet, *supra* note 1 provide a detailed account drawn from the work of the prominent historian, Paul Gordon Lauren.

5 Mendes and Mehmet, *supra* note 1 at 2.

it at a more inclusive conference in 1945 in San Francisco. Article 1 of the UNC provides that the role of the UN is to maintain international peace and security (Art. 1.1), develop friendly relations among nations based on respect for the principle of equal rights and self-determination of peoples (Art. 1.2), promote the achievement of international cooperation in solving international problems of an economic, social, cultural, or humanitarian character, and encourage respect for human rights and for fundamental freedoms (Art. 1.3).[6] Post-war, the commitment from the start was to *cooperate* for mutual benefit and with a liberal commitment to the idea of 'rights' that now function as 'the legitimate representation of a global moral economy'.[7]

Under the UNC, the Economic and Social Council was born as the principal organ to oversee a broad mandate pertaining to international economic, social, cultural, educational, health and related matters. The 1948 Universal Declaration of Human Rights (UNDHR)[8] became the first international instrument to recognize 'a common standard of achievement for all peoples and all nations'[9] and made no distinction or priority between civil and political rights (CPRs) and economic, cultural, and social rights (ECSRs). Since then, the UNDHR has served as the model for codifying human rights protection in multilateral conventions including the *International Covenant on Civil and Political Rights* (ICCPR)[10] and the *International Covenant on Economic, Social and Cultural Rights* (ICESCR).[11] Most states are signatory to at least one of these covenants. Canada has ratified both and the United States of America is one of only a handful of states to have ratified only the ICCPR.

6 The UN is to be a centre for the harmonization of state actions for the attainment of these 'common ends' (Art. 1.4). *Charter of the United Nations*, 26 June 1945, arts 62–72, Can. T.S. 1945 No. 7 amended 1965, 1968 and 1973. Full text <http://www.un.org/aboutun/charter/>.

7 See Hastrup, *Accommodating Diversity*, op. cit. at 15. For a critical view, see Pierre-Marie Dupuy, 'Some Reflections on Contemporary International Law and the Appeal to Universal Values' (2005) 16.1 *European Journal of International Law* 131–7.

8 *Universal Declaration of Human Rights*, GA Res. 217(III), UN GAOR, 3d Sess., Supp. No. 13, UN Doc. A/810 (1948) [UNDHR], full text at <http://www.unhchr.ch/udhr/lang/eng.htm>.

9 Hurst Hannum, 'The UDHR in National and International Law' (1998) 3:2 Health and Hum. Rts 145.

10 *International Covenant on Civil and Political Rights*, 19 December 1966, 999 U.N.TS. 171, Can. T.S. 1976 No. 47, 6 I.L.M. 368 (entered into force 23 March 1976, accession by Canada 19 May 1976), full text at <http://www1.umn.edu/humanrts/instree/b3ccpr.htm> [ICCPR].

11 *International Covenant on Economic, Social and Cultural Rights*, GA Res. 2200A(XXI), UN GAOR, Supp. No. 16, UN Doc. A/6316 (1966), 993 U.N.T.S. 3 (entered into force 3 January 1976), full text at <http://www.hrweb.org/legal/escr.html> [ICESCR].

While HR recognize the equal entitlements of individuals as their subject, they do not assure equality amongst the rights recognized. The preamble to the UDHR provides 'equal … rights of all members of the human family'. Yet, some rights allow for *progressive realization,* limits, restrictions, and optional protocols or reservations, while others, such as the CPRs proclaimed under the ICCPR, are more absolute and immediate. The fact that an internal hierarchy of rights may exist does not in any way diminish their solidarity or indivisibility.[12] CPRs are free and only require functioning political and infrastructural institutions for their realization. They are negative rights that exist independent of state action imposing obligations of non-interference: the right to liberty requires state forbearance and restraint. ESCR, conversely, are considered costly rights that impose positive duties on the state to fulfill welfare providing obligations to individuals: the right to education, food and health require state action for their delivery. These rights to (public) goods are captured by social democracies and welfare states, such that 'in essence … the classical liberty rights are concerned with processes, whereas welfare rights are concerned with outcomes'.[13] ESCR may be qualified by the availability

12 See discussion of Vienna Declaration and Program of Action, *infra* in text. Whether these rights are naturally derived inherent rights is contested in the literature. See e.g. J.P. Gardner, ed., *Human Rights as General Norms and A State's Right to Opt Out: Reservations and Objections to Human Rights Conventions* (London: British Institute of International and Comparative Law, 1997) [HR General]. In the introduction to this text, Rosalyn Higgins writes, '[a]t the heart of it is the balance to be struck between the legitimate role of States to protect their sovereign interests [not to suppose that sovereign interests are necessarily inimical to the promotion of human rights] and the legitimate role of the treaty bodies to promote the effective guarantee of human rights'. See also Markus Schmidt, 'Reservations to United Nations Human Rights Treaties – the Case of the Two Covenants' at 20–35; Catherine Redgwell, 'The Law of Reservations in respect of Multilateral Conventions' at 2–30; Graham Hand, 'Policy Aspects of Reservations and Objections to Human Rights Conventions' at 117–20 in HR General. The Human Rights Committee, General Comment 24(52), on issues relating to reservations made upon ratification or accession to the Covenant or the Optional Protocols or in relation to declarations under Article 41 of the Covenant, Un Doc. CCPR/C/21/Rev.1/Add.6 (1994) states under para. 1 that as of 1 November 1994, 46 of the 127 States parties to the ICCPR 'had, between them, entered 150 reservations of varying significance to their acceptance of the obligations of the Covenant … [T]he Committee, in performance of its duties … must know whether a State is bound by a particular obligation or to what extent'. John Rawls identifies a priority set of HR, *The Law of Peoples*, op. cit. at 80, that have generated consensus and 'set a necessary, though not sufficient standard for the decency of domestic political and social institutions'.

13 David Kelley, *A Life of One's Own: Individual Rights and the Welfare State* (1998) in Henry Steiner and Philip Alston, eds, *International Human Rights in Context: Law, Politics, Morals*, 2nd edn (New York: Oxford University Press, 2000) at 258.

of regulatory resources and thus the ICESCR allows for *progressive* implementation.[14]

Given the ideological and political commitment to human rights, that there are two instruments is an artifact of the political economy at the time of their first recognition and does not detract from the normative interrelation of *all* human rights. As Eide and Rosas have argued, 'fundamental needs should not be at the mercy of changing governmental policies and programmes, but should be defined as entitlements'.[15] Yet separate legal instruments as 'twin covenants' allowed states to assume only obligations which they were willing to undertake, recognizing that some HR protection would be better than none and alleviating in part criticisms that internationally mandated HR protections were themselves a form of cultural imperialism.[16] Today, the challenge with ESCR is giving them *content* and thereby practical legal relevance to make ESCR justiciable: how much food or health satisfy this right? The 1948 UNDHR was adopted without a dissenting vote but its enforceability and legal status remains contentious because '[a] Declaration of the General Assembly is not, by definition, legally binding though it has strong moral force'.[17] Still, this particular Resolution of the UN General Assembly is arguably more significant. Mark Freeman and Gibran Van Ert confirm that declarations, observations, guidelines, and certain international instruments are part of international soft law and thus not binding as part of current international law (*lex lata*) but suggest that they may indicate what international law may or should become (*lex fernda*).[18] However, if the norms in the soft law are

14 Article 2.1 of the ICESCR, *supra* note 11 provides: 'Each State Party to the present Covenant undertakes to take steps, individually and through international assistance and co-operation, especially economic and technical, *to the maximum of its available resources*, with a view to achieving progressively the full realization of the rights recognized in the present Covenant by all appropriate means, including particularly the adoption of legislative measures'. See also related General Comment 3, 5th Sess., *The Nature of States Parties Obligations (Art. 2, par. 1 of the Covenant)* UN Doc. E/1991/23 (1990) [General Comment 3].

15 See Asbjørn Eide, Catarina Krause and Allan Rosas, eds, *Economic, Social and Cultural Rights: A Textbook* (Saskatoon: Kluwer Academic Publishers, 1995) at 18.

16 Draft International Covenant on Human Rights and Measures of Implementation: Future work of the Commission of Human Rights, General Assembly Res. 421 (V), UN GAOR, 5th Sess. (1950). In 1950, the General Assembly adopted a resolution emphasizing interdependence of all HR and called on the UN Commission on Human Rights to adopt a single convention to bind ratifying states but later reversed this.

17 Rhona K.M Smith, *Textbook on International Human Rights* (New York: Oxford University Press, 2003) at 39.

18 Mark Freeman and Gibran Van Ert, *International Human Rights Law, Essentials of Canadian Law* (Toronto: Irwin Law, 2004) at 68.

followed in state practice out of a sense of legal obligation (*opinio juris*) such that states do not generally feel free to disregard the norm, then it may be elevated to customary international law and is thereby binding on all states, including non-signatory third parties, as a source of international law. The UNDHR is said to have achieved this status and is considered declaratory of binding customary law.[19] Similarly, many of the rights articulated in the UN Charter are so widely accepted that they now form part of the corpus of general principles of law although they have yet to mature fully into customary law. 'The matter is particularly important' writes Martti Koskenniemi,

> in regard to norms intended to safeguard basic human rights ... If the only states bound to respect such rights and freedoms are the states that have formally become parties to the relevant instruments – and even then only within the scope of their often compromised wordings and multiple reservations – then many important political values would seem to lack adequate protection ... Some norms seem so basic, so important, that it is more than slightly artificial to argue that states are legally bound to comply with them simply because there exists an agreement between them to that effect ...[20]

Most major UN HR treaties or their related protocols provide state undertakings to accept the oversight of specialized international supervisory committees (such as the HR Committee which monitors compliance with the ICCPR) known as treaty bodies in order to ensure compliance with treaty obligations. The exception is the Committee on Economic, Social and Cultural rights which was established by ECOSOC under Resolution 1985/17 rather than by a treaty and monitors compliance with the ICESCR.[21] Under the ICCPR, there is some provision for use of an inter-state complaint mechanism (under Art. 41, optional declaration) but none have been filed to date, and for individual complaint mechanisms (under ICCPR-OP1). Neither the inter-state or individual complaints/petitions procedures are available

19 Ibid. See also Robert Howse and Makau Matua 'Protecting Human Rights in a Global Economy: Challenges for the World Trade Organization' (2000) Rights and Democracy, <http://www.dd-rd.ca/english/commdoc/publications/globalization/wtoRightsGlob.html>.

20 Martti Koskenniemi, 'The Pull of the Mainstrain' (1990) 88 Mich. L. Rev. 1946, in Steiner and Alston, *supra* note 13 at 78.

21 Art. 16 of the ICESCR, *supra* note 11, requires states to submit periodic reports to the UN Secretary General. Each body is comprised of state elected expert members and is responsible for four different procedures: (1) the review of periodic reports submitted by state parties; (2) investigation of systemic violations; (3) review of petitions filed by one state against another; and (4) and review of petition made by individuals against states parties. See Freeman and Van Ert, *supra* note 18 at 385.

internationally under the ICESCR.[22] Unlike customary international law, which is automatically binding, and just as the common law needs no further domestic statutory incorporation, other international treaty norms including the ESCR require domestic implementation (even by Canadian courts for example) in order to be justiciable; it is domestic law that provides for a legal remedy. Simply put, while civil and political rights can be enforced in international fora, one cannot enforce social and economic rights such as the right to health in an international forum. For these rights, the ability to have an international human rights claim judicially determined and enforced relies on domestic mechanisms such as constitutional law and HR legislation.

Considered 'birth rights', in some contexts HR have reached optimal integration into state laws by virtue of being *constitutionalized*. Ideally, constitutional protection for ECSR would be as express as CPR and require *positive* baselines to be delivered by public authorities along with the negative freedoms and liberties. South Africa's 1996 Constitution, however, remains virtually alone in this approach. The South African Constitution expressly includes the right to education, an environment that is not harmful to human health or well-being, access to health care services, food, water, social security, and housing.[23] Where ESCR have not been constitutionalized, one approach is to incorporate ESCR protection within domestic constitutional rights protecting civil and political liberties. The right to health (and food), for example, may be subsumed within a right to life and security of the person,[24] reaffirming the indivisibility and interdependence of all HR.

---

22 Historically, only states had standing in international fora. By the twentieth century, non-state actors and individuals acquired some limited international legal personality for legal relations, to enjoy rights and assume prescribed obligations. See e.g. Hugh M. Kindred et al., *International Law Chiefly as Interpreted and Applied in Canada*, 6th edn (Toronto: Emond Montgomery, 2000), c. 2 at 11–91.

23 Constitution of the Republic of South Africa 1996, No. 108 of 1996. For a comparative chart on protection of ECSR, see HR and Constitutional Rights, <http://www.hrcr.org/chart/index.html>.

24 See e.g. Bita Amani, 'Patents, the *Charter,* and a Healthy Dose of Rights in Wrongs: *The Poison is the Elixir for Life, Liberty, and Security of the Person*' (2007) 57 U.N.B.L.J. 163–214 exploring public authority liability for gene patenting and whether the right to health has acquired constitutional status under the *Canadian Charter of Rights and Freedoms*, Part I of the *Constitution Act, 1982*, being Schedule B to the *Canada Act 1982* (U.K.), 1982, c. 11 as a result of *Jacques Chaoulli and George Zeliotis v. Quebec (Attorney General)* [2005] 1 S.C.R. 731, 2005 SCC 35. The paramountcy of CPR is fundamental to any liberal democracy. For a discussion of the challenges to subsuming ECR in Canada, see *Gosselin v. Quebec (Attorney General)* [2002] 4 S.C.R. 429, 2002 SCC 84 [*Gosselin* (SCC)], where the SCC rejected Charter challenges to a Quebec law excluding citizens less than 30 years old from receiving social security benefits primarily due to factors characterizing them as programmatic and available at the discretion of government (paras 277–83).

Canada's government has increasingly been criticized by the UN human rights bodies for failure to ensure the realization of social and economic rights for all Canadians: 'in 1993 and 1998, the United Nations Committee on Economic Social and Cultural Rights – which monitors Canada's compliance with its social and economic rights obligations – criticized Canada for its poor record of upholding these rights.'[25] Although Canada continues to rank well overall, it did not fair much better in its May 2006 review. In its interim report, the Review Committee criticized Canada for:

> [t]he lack of legal redress available to individuals when governments fail to implement the Covenant [ESCR], resulting from the insufficient coverage in domestic legislation of economic, social and cultural rights, as spelled out in the Covenant; the lack of effective enforcement mechanisms for these rights; the practice of governments to urge upon their courts an interpretation of the Canadian Charter on Right and Freedoms [sic] denying protection of Covenant rights, and the inadequate availability of civil legal aid, particularly for economic, social, and cultural rights.[26]

Citizens cannot directly invoke human rights duties of their state in domestic legal forums, nor can they complain in international fora regarding their ESCR. While the challenge is to encourage governments to give these rights some content, some governments are simply ambivalent to do so even though arguably, '... human rights most urgently need asserting and defending, both theoretically and practically, where they are most denied'.[27] Without a mechanism for individuals to hold their governments accountable, the rights pursuant to the ICESCR are practically speaking no more than mere statements of aspiration despite the assertion in General Comment 3 that: '[a]ny suggestion that the provisions indicated are inherently non-self-executing would seem to be difficult to sustain'.[28]

In Canada, just as there is a presumption that all domestic law will conform with the Constitution,[29] there is a presumption that Canadian law conforms

25 The Centre for Equality Rights in Accommodation, <http://www.equalityrights.org/cera/>.

26 An unedited version of the UNCESR Review (1–19 May 2006), future E/C.12/CAN/CO/5 is available at <http://www.ohchr.org/english/bodies/cescr/docs/E.C.12.CAN.CO.5.pdf>, criticizing Canada for not doing enough to address social and economic rights such as food shortages and poverty.

27 See David Beetham, 'What Future for Economic and Social Rights' (1995) 43 *Political Studies* 41 in Steiner and Alston, *supra* note 13 at 255, suggesting common ground for HR advocates and proponents of development economics '... on a minimum core of rights such as ... the right to food of an adequate nutritional value, to clothing, to shelter, to basic (or primary) health care, clean water and sanitation ...' (at 256).

28 *Supra* note 13 at para. 5.

29 Hogg, *Constitutional Law of Canada,* op. cit. at 286.

with international treaty obligations such that even in the absence of domestic implementing legislation, 'while a litigant may not be able to place direct reliance on a right guaranteed by a Canadian treaty obligation, she is permitted to raise that right and rely on it for the purposes of interpreting a domestic provision'.[30] This approach presumably extends to patent law. In England, a more express approach is taken. Domestic implementing legislation is used to interpret government measures compatibly with international HR obligations. K.D. Ewing, writing on the UK Human Rights Act of 1998 contends that the *duty* to *construe* legislation to comply with Convention rights is imposed on all courts and if domestic legislation is incompatible with Convention rights, a higher court is empowered under implementing legislation to declare it as such. Ewing argues, the UK Act imposes an obligation on public authorities to comply with Convention rights and this is directly enforceable in domestic courts. Summarizing the view of the Lord Chancellor, Ewing asserts that the duty of the courts to act compatibly with the Convention extends beyond cases involving public authorities to cases between private citizens; 'Convention rights may be relied upon in litigation between private parties, but cannot themselves be the basis of a cause of action'.[31] Rosemary Coombe puts it differently:

> The States' obligations to protect involve duties to prevent abuse of rights by third parties, including non-State actors, whereas obligations to fulfill involve active duties to take appropriate measures that create a framework for creating accountability.[32]

From these discussions we may conclude that human rights and their corresponding duties could be raised directly as defence to domestic actions for patent infringement against the government, if for example Ontario was sued by MGL over BRCAnalysis, but also in proceedings between private parties. The profound potential of such legal arguments being invoked in domestic courts has yet to be realized in creating government accountability and greater human rights compliance. Canada has allowed patents to issue on human genes and evidently these interfere with the realization of one's right to health under ICESCR. Constitutionalizing health rights may allow individuals to sue their governments for compensation or to have violating legislation struck down. But in the absence of this highest level of express protection, the ability to invoke

30 Freeman and Van Ert, *supra* note 18 at 350–51.

31 See K.D. Ewing, 'The Human Rights Act and Parliamentary Democracy' (1999) 62 Mod. L. Rev. 79 in Steiner and Alston, *supra* note 13 at 1010.

32 See Rosemary Coombe, 'Intellectual Property, Human Rights and Sovereignty: New Dilemmas in International Law Posed by the Recognition of Indigenous Knowledge and the Conservation of Biodiversity' (1998) 6 *Ind. J. Global Legal Stud.*, 59 [Coombe, 'IP, HR and Sovereignty'] at 68.

HR entitlements directly in domestic legal proceedings and to raise these as a defence may provide a sufficiently suitable alternative. It is on this principled model of direct invocation that I advocate later in this text, for a similar approach to human rights compatibility in dealing with trade complaints later in this text. Such an interpretation is consistent with Article 28 of the UNDHR confirming that: '[e]veryone is entitled to a social and international order in which the rights and freedoms set forth in this Declaration can be fully realized.'[33]

Article 28 creates a duty on governments to ensure that the 'social' and 'international' order allow for the articulated rights (including ESCR as per Arts 25–27, UNDHR). This means that international legal obligations must be reconciled coherently. Some critics of TRIPS argue that if human rights, indivisible and interdependent as they are, have achieved the status of customary law, the entire TRIPS Agreement may be a nullity. Rather than pursue this line of inquiry, a more conciliatory approach to resolving potential human rights conflicts would be to interpret the permission creating norms in Article 27 of TRIPS and the Article 8 Principles in a manner consistent with the Article 7 Objectives and the diverse regulatory policies adopted and measures taken to fulfil HR obligations. In this approach, 'such rights can increasingly be justiciable and amenable to individual petition and complaint'.[34] We should be mindful, as Freeman and Van Ert are quick to point out: 'the relative ease or difficulty of implementation of any right or category of right has no bearing on the moral, legal, or political value of that right. A right is not accorded greater or lesser priority based on the degree of intervention or restraint required'.[35]

The case-by-case and defensive nature of invoking international human rights obligations in domestic litigation, however, makes it less than ideal for affecting patent policy changes as well as protecting the underlying human right(s) on a population basis. Of course, there are good reasons for governments to prioritize industrial and trade policy over health policy; to date, the risk of liability in domestic and international law as well as the simple cost of defending a patent infringement action was sufficient incentive, especially since there are no similar costs associated for disregarding the individual's HR. The provision of constitutional (*Charter*) remedies and compensation for human rights violations would provide some countervailing cost incentives for policy change, address enforceability concerns expressed by the Review Committee, and is consistent with Canada's obligations under Article 6 of the UN Declaration on the Right to Development, which provides: 'All human rights and fundamental freedoms are indivisible and interdependent; equal attention and urgent consideration should be given to the implementation, promotion and protection of civil, political, economic, social and cultural

33 UNDHR, *supra* note 8.
34 Beetham, *supra* note 27 at 256.
35 Freeman and van Ert, *supra* note 18 at 33.

rights.'[36] Also, the Preamble to the ICESCR mirrors that of the ICCPR, and both assert that 'the ideal of free human beings enjoying freedom from fear and want can only be achieved if *conditions are created whereby everyone may enjoy his economic, social and cultural rights as well as his civil and political rights*'.[37]

The Vienna Declaration and Program of Action 1993 (Vienna Declaration)[38] obviates former distinctions, reiterating the duty of the state in Article 5:

> *All human rights are universal, indivisible and interdependent and interrelated.* The international community must treat human rights globally in a fair and equal manner, on the same footing, and with the same emphasis. While the significance of national and regional particularities and various historical, cultural and religious backgrounds must be borne in mind, it is the *duty of States, regardless of political, economic and cultural systems, to promote and protect all human rights and fundamental freedoms.*[39]

Protections for ECSR are integral to the operations of the welfare state. Civil and political rights impose obligations through the state on all other human beings for their respect. Welfare rights impose obligations on the state for the attainment of basic necessities of life but so far as implementation is through government programmes, the obligation is also distributed to all taxpayers.[40] David Kelley writes that to protect liberty rights of individuals, governments are under a duty to prevent incursions by other individuals whereas the implementation of welfare rights requires a much more activist government in terms of large-scale transfer programmes, social safety nets, and other distributive means. However, in order to retain a conceptual operational understanding that coherently reconciles *all* HR, we can recast welfare rights, albeit reducing in the short term the entitlements they confer, to at least *negative* rights that oblige states to *refrain from interference* in their realization in the tradition of classic liberalism's 'freedom from'. Rather than determine what is the appropriate amount of food, or welfare, or health that must be delivered to fulfil its obligation, ECSR as negative rights would require

---

36 1986 GA Res. 41/128.

37 ICESCR, *supra* note 11, emphasis added.

38 *Vienna Declaration And Program of Action*, World Conference on Human Rights, Vienna, 14–25 June 1993, UN Doc. GA/CONF.157/23 (1993). Report of the Conference, GA/Conf. 157/24, 12 July 1993.

39 Available at <http://www.unhchr.ch>. See also Article 6 of the *Declaration on the Right to Development*, GA Res. 41/128, annex, 41 UN GAOR, Supp. No. 53 at 186, UN Doc. A//41/53 (1986) which provides that states should cooperate towards these ends, and should 'eliminate obstacles to development resulting from failure to observe civil and political rights, as well as economic social and cultural rights'.

40 Kelley, *supra* note 13 at 258.

at a *minimum* that the state avoid implementing strict IPRs (what one might characterize as *private takings* legislation) that would unduly interfere with the individual's self-realization and fulfilment of her rights. From this perspective, but for state created IPRs, individuals would be fully *free* to use the knowledge as a public good; the right to do so may in fact itself be a human right.

In short, the universal minimums of TRIPS should not be implemented in a manner that violates the universal minimums of human rights obligations. Current IPRs as institutional instruments of the state create private monopolies on genetic resources that directly affect plant varieties, food varieties, nutraceuticals and human health and thereby *violate* a state's *negative duty* owed (a duty of forbearance) not to interfere with the ICESCR rights. The universal declaration of these rights creates legal entitlements making them 'less amenable to trade offs in the face of competing claims and interests'.[41] Because it is with justice that human rights are concerned, ESCR as negative rights removes the definitional limitations impeding governmental delivery and paves the way for accommodating cultural, economic, and developmental differences of a state in relation to its own peoples, making the culture of human rights more functional and fulfilled.[42] It is the first step towards the *progressive* realization of rights and remains morally imperative for distributional justice.[43] These quasi-constitutional values successfully exported

41 Freeman and Van Ert, *supra* note 18 at 27. The Committee on Economic, Social and Cultural rights, charged as the authoritative interpreter of the ICESCR, has 'provided specific guidance on how to implement the general and potentially conflicting responsibilities of State Parties'. See David Weissbrodt and Kell Schoff, 'Human Rights Approach to Intellectual Property Protection: The Genesis and Application of Sub-Commission Resolution 2000/7' (2003) 5 Minn. Intell. Prop. Rev. 1, <http://mipr.umn.edu/archive/v5n1/Weissbrodt.pdf> at 4. They add, 'The Committee has declared that States Parties have a 'minimum core obligation to ensure the satisfaction of, at the very least, minimum essential levels of each of the rights'. In particular, the Committee 'emphasize[d] that any intellectual property regime that makes it more difficult for a State party to comply with its core obligations in relation to health, food [or] education … is inconsistent with the legally binding obligations of the State party'. Citing General Comment No. 3, *supra* note 13, para. 10, and *Substantive Issues Arising in the Implementation of the International Covenant on Economic, Social and Cultural Rights*, Economic and Social Council, Committee on Economic, Social and Cultural Rights, 27th Sess., Agenda item 3, para. 12, 17–18, U.N. Doc. E.C. 12/2001/15 (2001).

42 This would include the presence of cultural and group rights that may be indispensable to the realization of individual rights. No person is completely a separate atomistic entity from their social context. According to Michael Ignatieff, '[i]ndividual rights without collective rights may be difficult to exercise, but collective rights without individual ones end up in tyranny' at 89–90, 'Human Rights as Politics and Idolatry' in A. Gutman (ed) *Human Rights as Politics and Idolatry* (Princeton, NJ: Princeton University Press, 2001).

43 Kristen Hastrup writes, that 'few would openly question the basic assumption that they [HR] can further justice between people. In that sense, the universal

into the UN system's protection of human rights formed a complementary institutional mechanism to the GATT 1947 and together sought to ensure that the utilitarian principles of improving aggregate social welfare through trade liberalization, for securing of territorial boundaries during times of war, and creating incentives for new welfare improving innovations, were not purchased with a compromise of individual liberties.

Even in a welfare state or the kind of deliberative democracy envisaged in classic civic republicanism, individual liberties have, at least notionally, always triumphed. The 'public interest' to be meaningful requires protection of the interest of the individuals that comprise it including their ECSR. Universal IPRs standards denigrate libertarian ideals that seek to safeguard individual interests from utilitarian claims by conferring not just a right to *use* but to *exclude*. Consequently, IPRs raise an interesting paradox in that they espouse the individualism of liberal rights rhetoric, impinge on a differing set of individuated (human) rights, and yet are most commonly theoretically accepted for their utilitarian functionality. The problem is that despite the conceptual strangeness of IPRs, because of theoretical rights-based rhetoric used to justify their governmentality, there are those who assert that IPRs are themselves *human rights.* Next we will explore this contention at its nexus with *other* human rights under the ICESCR. We do this for two important reasons: first, it will eliminate the discussion of patenting life as a dialogue of *competing human rights* and thereby avoid the need to establish a just and equitable means of prioritizing amongst what are indivisible HR. Second, it will provide us with a greater understanding of the kind of intellectual right that *is* provided under HR instruments; one which is consistent with a public domain or alternatively, a liberally defined commons inclusive of humanity.

Human Rights Dimensions of IP

Proponents of IP may argue that IPRs are human rights pursuant to Article 15(1)(c) of the ICESCR which recognizes the right of everyone 'to benefit from the protection of the moral and material interests resulting from any scientific, literary or artistic production of which he is author'. Article 27.2 of the UNDHR was its obvious predecessor.[44] There are two important restrictions, however, in recognizing IP as a HR. First, as a natural *human* right, fundamental to their enjoyment is that the holder of such rights be

declarations of human rights have come to represent a 'common good', something we should actively strive to realize on a global scale. They have taken root in the collective imagination of the natural order'. *Supra* note 19 at 9.

44 'Everyone has the right to the protection of the moral and material interests resulting from any scientific, literary or artistic production of which he is the author'. UNDHR, *supra* note 8.

'human'. While self-identification with a particular group or collective may mean that some HR may be based on a collective identity (such as the right to one's religion, to practice one's culture), the individual *human* is an essential feature for these rights and it is clear that the requisite is not simply of legal personhood. Article 1 of the UNDHR asserts that '[a]ll human beings are *born* free and equal in *dignity* and *rights*'.[45] Regardless of any legal fictions putatively impregnating domestic understandings of the corporation as 'person', the essential matter remains that *that* person is *not* human nor of 'woman *born*'.[46] This may help explain the preference in the US system for having a real person named as the inventor on the application even if the rights are assigned to a corporation. Even where IP can be *held* and owned by corporations, only humans could claim it is a *human right* – individually or as a member of a cultural community (on the basis of a cultural right). Article 15 ICESCR provides further insight into the content of IP as a *human* right:

> *Article 15*
> 1. The States Parties to the present Covenant recognize the right of everyone:
> (a) To take part in *cultural life*;
> (b) To *enjoy the benefits* of scientific progress and its applications;
> (c) To benefit from the protection of the moral and material interests resulting from any scientific, literary or artistic production of which he is the author.
> 2. The steps to be taken by the States Parties to the present Covenant to achieve the full realization of this right shall include those necessary for the conservation, the *development and the diffusion* of science and culture.
> 3. The States Parties to the present Covenant undertake to respect the freedom indispensable for scientific research and *creative activity*.
> 4. The States Parties to the present Covenant recognize the benefits to be derived from the encouragement and development of international contacts and co-operation in the scientific and cultural fields.

This brings me to a second point: taken in the context of the other provisions in Article 15, Article 15(1)(c) does *not* confer a right of *exclusion* or 'property' or the related market value based on the labour of 'authorship' – all

45 The presence of both words suggests that the two have different meanings but are complementary concepts: respect for rights promotes respect for dignity and vice versa.

46 'All human life on the planet is born of woman' Adrienne Rich, 'Introduction from *Of Woman Born*' in Adrea O'Reilly, ed., *Maternal Theory: Essential Readings (*Toronto,: Demeter Press: 2007) at 6.

matters of social policy[47] –but a right of *participation* and *access* consistent with the ideals of open sharing in science (supported by Article 15.3 and 15.4). It also confers the right to take part in cultural life and creative activity (Art. 15.1(b) and 15.3) and concomitant with the idea of a collective commons in which we may or may not locate (communal) property. To demonstrate, let us imagine that the concept of IPRs was captured with different rhetoric that did not centrally rely on exclusion or the idea of 'property'. For example, let us call this creature 'information/innovation/creative contribution and enjoyment rights' (ICCERs). Termed this way as a HR, it confers on everyone the right (a claim) to be able to *participate* in contributing to the information and creative commons as a personal right rather than a *proprietary* claim to the products of that creative contribution once made, as is suggested under the rhetoric of intellectual 'property rights'. This is not only a sound interpretation of the international HR instruments, but is consistent with the institutionalist approach to patent legislation also adopted by the UN Secretary General's Report, which recognized that such legislation,

> [h]as never been based solely on the concept of the patent as the confirmation of an inherent rather than the creation of a statutory property right. Such a concept would have left no room for such restraints on the patent grants as its fixed duration, its exclusion for inventions in certain fields … and the forfeiture or compulsory licensing of patents for failure to work them.[48]

---

47 See Edwin C. Hettinger, 'Justifying Intellectual Property' (1989) 18 Phil. and Pub. Aff. 31. It is the exclusive nature of the rights conferred that has lead to the view that these instruments are fundamentally in conflict with other human rights. See e.g. Intellectual Property Rights and Human Rights, Res. 2000/7, UN. ESCOR Comm'n on Hum. Rts, Sub.-Comm'n on the Promotion and Protection of Human Rights, 52nd Sess., UN Doc. E/CN.4/sub.2/Res/2000/7 providing: 'actual or potential conflicts exist between the implementation of the TRIPS and the realization of economic, social and cultural rights'. See also Weissbrodt and Schoff, *supra* note 41, discussing the significant influencing role this resolution had leading into Doha and the adoption of the declaration on public health at the ministerial meeting.

48 See *The Role of Patents in the Transfer of Technology to Developing Countries*, GA Res. 1935 (XVIII), UN GAOR, 18th Sess. (1963) 28, *Report of the Secretary-General, United Nations* (New York: Martinus Nijhoff, 1964) at 9. This report recognized that the Secretary General's task, as per a GA resolution, to examine the national patent systems of a number of different jurisdictions could not be completed in time for this GA due to the breadth of geographic and substantive coverage, <http://daccessdds.un.org/doc/RESOLUTION/GEN/NR0/186/10/IMG/NR018610.pdf?OpenElement>. In 'The Progressive Development of the Law of International Trade: Report of the Secretary General of the United Nations, 1966', it was reported that '[s]ince 1961 the United Nations General Assembly has had before it the problem of the role of industrial property legislation in facilitating the transfer of patented and unpatented technological and managerial know-how to developing countries.

Consequently, while IPRs are appropriately refuted as human rights, our reconception of the international provisions is marked by a notable absence of *exclusive* rights while allowing for right of access, a human right of use, creativity (including arguably auto-creativity and procreativity), and enjoyment, which may or may not be communal or cultural. Under Article 27.1 of the UNDHR, 'everyone has the right freely to participate in the cultural life of the community, to enjoy the arts and to share in scientific advancement and its benefits'. This is because intellectual products are social products located within a cultural context related, as the scientific paradigm reveals, to the mountain of preceding intellectual contributions as well as innovation supporting institutions and infrastructure (including those for public health and education).[49] In fact, the right to participate is integral for some individuals' (i.e. scientists, academic researchers, etc.) right to work provided under Article 23 of the UNDHR.[50] Moreover, Article 12(b) of the Universal Declaration on the Human Genome and Human Rights (HGHRD) provides that

> [f]reedom of research, which is necessary for the progress of knowledge, is part of freedom of thought. The applications of research, including applications in biology, genetics, and medicine, concerning the human genome, shall seek to offer relief from suffering and improve the health of individuals and humankind as a whole.[51]

HGHRD was unanimously adopted together with a resolution for its implementation by UNESCO.[52] The real threat, according to Drahos and Braithwaite, to human (natural) rights, cultural rights, and the scientific

---

At its sixteenth session, the General Assembly adopted resolution 1713 (XVI) on the role of patents in the transfer of technology to underdeveloped countries, in which it requested the Secretary-General to study the issues involved, including specifically, the effects of patents on the economy of developing countries, patent legislation in selected developed and developing countries, and the characteristics of the patent legislation of developing countries in the light of economic development objectives'. In addition, 'the study emphasized, *inter alia*, that *properly adapted* patent legislation was essential if the patent system was to be beneficial to economic development and advancement of industry in developing countries'.

49 See e.g. Brian Martin, *Information Liberation: Challenging the Corruptions of Information Power* (London: Freedom Press, 1998), <http://www.uow.edu.au/arts/sts/bmartin/pubs/98il/ilall.pdf>.

50 UNDHR, *supra* note 8.

51 *The Universal Declaration on the Human Genome and Human Rights,* ESC Res. 27 C/5.15, UN ESCOR, 29th Sess. (11 November 1997), online: <http://portal.unesco.org/en/ev.php-URL_ID=13177andURL_DO=DO_TOPICandURL_SECTION=201.html> [HGHRD]>.

52 (Article 1.1). UNESCO strives to 'promote the free flow of ideas by word and image' (Article 1.2). UNESCO *Manual of the General Conference*, texts adopted by the

information commons posed by the property rights conception of IPRs is to *liberty*:

> [T]he greater danger of intellectual property lies in the threat to liberty. When a group of scientists stop working on a protein molecule because there are too many intellectual property rights that surround the use of the molecule, a basic freedom, the freedom to research, has been interfered with. The liberty cost of intellectual property rights may seem remote because most of us do not carry out research on proteins. But we all have an interest in seeing public research programmes into diseases and health research being carried out. We want ... public researchers to continue working on the genes for breast and ovarian cancer and helping to develop cheaper, more effective clinical tests ... Companies are entitled to protect their treatment for disease but not, through use of their patents, to deter or prevent others from access to genes which are linked to the origins of disease.[53]

The reciprocal relationship between knowledge, liberty, and culture has long been appreciated. Founding fathers of modern political thought, such as Thomas Jefferson and Adam Smith, stressed the importance of education, knowledge and freedom of information, now underscored by Brian Martin:

> culture – which makes ideas possible – is built not just on intellectual contributions but also on practical and material contributions, including the rearing of families and construction of buildings. Intellectual property is theft, sometimes in part from an individual creator, but always from society as a whole.[54]

This perspective is implicitly affirmed by the UN Educational Scientific and Cultural Organization. Under its Constitution, UNESCO provides that 'the wide diffusion of culture, and the education of humanity for justice and liberty and peace are indispensable to the *dignity* of man and constitute a sacred *duty* which all the nations must fulfil in a spirit of mutual assistance and concern'.[55] The Constitution provides:

General Conference at its 31st session (2002) at 8. Constitution of UNESCO, online <http://www.icomos.org/unesco/unesco_constitution.html>.

53 Drahos and Braithwaite, *Information Feudalism*, op. cit. at 3. See discussion of Wisconsin's stem cell patent, *supra* note 506.

54 Martin, *supra* note 49, ch. 3 'Against Intellectual Property', at 55.

55 (Article 1.1). UNESCO strives to 'promote the free flow of ideas by word and image' (Article 1.2). UNESCO *Manual of the General Conference*, texts adopted by the General Conference at its 31st session (2002) at 8. Constitution of UNESCO, <http://www.icomos.org/unesco/unesco_constitution.html>.

> the States Parties to this Constitution, believing in full and equal opportunities for education for all, in the unrestricted pursuit of objective truth, and in the *free* exchange of ideas and knowledge, are agreed and determined to develop and to increase the means of communication between their peoples and to employ these means for the purposes of mutual understanding and a truer and more perfect knowledge of each other's lives; [i]n consequence whereof they do hereby create the United Nations Educational, Scientific and Cultural Organization for the purpose of advancing, through the educational and scientific and cultural relations of the peoples of the world, the objectives of international peace and of the common welfare of mankind for which the United Nations Organization was established and which its Charter proclaims.[56]

Articles 13–19 of the HGHRD establishing the conditions for the exercise of scientific activity places the responsibility of regulating human genomic research and its ethical and social implications on the public and private science policy-makers (Article 13). As well, it argues the implementation of 'measures to foster the intellectual and material conditions favourable to freedom in the conduct of research on the human genome and to consider the ethical, legal, social and economic implications of such research, on the basis of the principles set out in this Declaration'.[57] Article 15 clearly acknowledges the relationship of key commitments:

> States should take appropriate steps to provide the framework for the free exercise of research on the human genome with due regard for the principles set out in this Declaration in order to safeguard respect for human rights, fundamental freedoms and human *dignity* and to *protect public health.* They should seek to ensure that research results are not used for non-peaceful purposes.[58]

It is the responsibility of the state to ensure that these declarations are realized. According to Article 18 of the HGHRD, states

> should make *every effort,* with due and appropriate regard for the principles set out in this Declaration, to continue fostering the international dissemination of scientific knowledge concerning the human genome, human diversity and genetic research and, in that regard, to foster scientific and cultural co-operation, particularly between industrialized and developing countries.

56 Ibid. emphasis added.

57 Article 14, HGHRD, *supra* note 51.

58 Ibid.

Rhetoric is crucial, but how does one get beyond rhetoric that effectively converts the meaning of discretionary crown privileges (related to the introduction of new industry and labour through imports) to exclusionary 'IPRs' and ultimately to liberty-restricting property, legally defined as a right in content, tenuously abstracted – or perhaps more accurately *extracted* in the genomic context – from the 'intellectual'?

Emphasizing the *propertization*, rather than the *creative contribution* of the individual has led to what Drahos and Braithwaite label 'information feudalism', modernity's new distributional inequality transforming the relations of production through the search for new property acquisitions (like land holdings) ever present as biogopolies:

> all of the prior institutional projects of world history, conceived narrowly for our purposes as projects to redistribute property unequally, have important surviving features today. They are never fully supplanted. So too with information feudalism. It will certainly not supplant industrial and financial capitalism, or the persistent residues of colonialism, or the king's power to tax centrally, or serfdom and slavery, or patriarchy ... [Information feudalism] contributes another layer of inequality.[59]

If patents do continue to issue in the field of genomics, a very generous research exemption should be upheld in domestic legislation to signal the recognition of the 'information contribution and enjoyment right' provided as a human right in Article 15 of the ICESCR. This imposes no significant additional cost on regulatory governments for their realization; it only requires state forbearance much like the right to liberty. Moreover, the very rights that biotechnology promises to advance, discussed with greater detail next, may be endangered by modern patent law and practice which, for distributional reasons, threaten to expand the 'genetic divide' in the interest of developed countries and their TNCs. The warning in the Bogève Declaration should not be disregarded: 'any new technology introduced into a society which is not, by its nature, a 'just' society will exacerbate the gap between rich and poor'.[60] Biotechnology does this by affecting a range of human rights of all peoples but disparately so for those in poor and developing countries who continue to suffer disproportionate burdens of disease.

---

59 Drahos and Braithwaite, *supra* note 53 at 199.

60 ETC group <http://www.etcgroup.org/key_defs.asp>. Bogève Declaration (1987) 'The Socioeconomic Impact of New Biotechnologies on Basic Health and Agriculture in the Third World' resulting from a meeting in France of 28 civil society organizations from 19 countries, 7–12 March 1987.

The Right to Health, Food, Benefit from Scientific Progress and Farmer's Rights

The regulation of industry, innovation, commerce, and the economy, may create tensions and potential unintended effects in other fields, such as individual and population health. In addition to income, other non-income measures of poverty, such as health, are affected by stronger IPR protection and, at once, aggravated by it. In Botswana and Zimbabwe, where one in four adults is infected with HIV/AIDS, AIDS orphans compound the poverty problem, in turn threatening civil and political security and stability and revealing the integrated relationship of social and economic with political and civil rights. Human health is polycentric and relates as much to social and political factors as it does to the insular solutions offered by technology and medical science. The prevalent conception of health in industrialized societies has traditionally been a fragmented and *medicalized* one. Its social determinants (such as poverty, dislocation, and unemployment) are ignored; this approach is progressively changing. A House of Commons Standing Committee on Foreign Affairs and International Trade Report of June 2003 called on Canada's federal government to respond to the crisis in sub-Saharan Africa by tripling its current contributions to the Global Fund to Fight AIDS, Tuberculosis and Malaria and to make access to medicines in poor countries a priority in its negotiating position within the multilateral trade organization. The Committee recognized that the AIDS programmes 'need to reflect the fact that in many African countries, HIV/AIDS is inextricably linked with other issues, including food shortages and famine, armed conflict and political violence, and problems of governance'.

Until recently, in most developed countries, the conceptual relationship with food was also limited (to consumption or comfort).[61] In other parts of the world, foods have a long history as preventative or alternative medicines and peoples have embraced a more holistic approach to health. In fact, some of the notable patent controversies galvanizing the 'no patents on life' campaign relate to the appropriation of traditional indigenous knowledge of plant genetic resources by American researchers and corporations asserting patent rights over plant resources such as turmeric or the neem tree. The cases dealt with the medicinal use of food and the desire to appropriate the knowledge of these uses to the potential exclusion of the very communities from which they

---

61 The past decade has witnessed a paradigm shift however, see the population health programme of the Canadian Institute for Advanced Research (CIAR) and related publications considering social environmental factors affecting health care. See also R.G. Evans, M.L. Barer and T.R. Marmor, *Why are Some People Healthy and Others Not? The Determinants of Health of Populations* (New York: Aldine de Gruyter, 1994) and Monica Townson, *Health and Wealth: How Social and Economic Factors are Affecting our Well Being* (Ottawa: Canadian Center for Policy Alternatives, 1999).

were taken.[62] The World Health Organization (WHO), a specialized agency of the UN established in 1948 for the attainment of the *highest available standard* of health for *all* people, has helped to change insular perceptions of health. WHO recognizes human health is integral to the 'happiness, harmonious relations and security of all peoples' and therefore defines health holistically in its Constitution as 'a state of complete physical, mental and social well-being and not merely the absence of disease or infirmity'.[63] WHO also stresses the importance of state responsibility for ensuring the protection of health and draws attention to the need for state cooperation because health is a global public good creating positive spillover effects for the rest of the world. But what underlies WHO's commitments is the goal to protect the essential right of each individual to the full realization of the *highest available standard* of health without discrimination based on the ability to pay ('economic or social condition').

While the right to food is separately protected as a human right, food rights are also an integral part of health rights properly understood. Article 25 of the 1948 UNDHR reflects the confluence of health and food rights by recognizing these as integral to the 'standard of living' and provides that '[e]veryone has the right to a standard of living adequate for the health and well being of himself and of his family, including food, clothing, housing and medical care and necessary social services …'. Implicit in this definition is the idea that a minimum standard is essential for a life of dignity. The ICESCR, on the other hand, gives the right to food and the right to health distinct recognition and adds a positive obligation on states to cooperate to ensure that these human rights are met through full access to the gains in scientific knowledge.

Article 11 of the ICESCR recognizes the

> right of everyone to an adequate standard of living for himself and his family, including adequate food … and to the continuous improvement of living conditions … [64][and] recognizing the fundamental rights of everyone to be free from hunger [State Parties] *shall* take, individually and through *international co-operation*, the *measures* … needed … [t]o *improve* methods of production, conservation and distribution of food by *making full use*

62 See e.g. Emily Marden, 'The Neem Tree Patent: International Conflict over the commodification of Life' (Spring 1999) 22 *Boston College Environmental Affairs Law Review* 279 and Dutfield, *IPR*, op. cit.

63 See World Health Organization Constitution, Preamble, <http://www.who.int/about/en/>.

64 '[P]arties will take appropriate steps to ensure the realization of this right, recognizing to this effect the essential importance of international co-operation based on free consent'. ICESCR, *supra* note 11.

> of technical and scientific knowledge, by disseminating knowledge of the principles of nutrition ….[65]

Industry has also caught on. There is a growing market for R&D investment in 'functional foods' and 'nutraceuticals' which enhance the nutritional and therapeutic value of foods (typically plants and grains) or provide some resistance-bearing qualities for cultivation (as in the case of GM maize). This convergence of food and medicine obviates traditional distinctions while simultaneously encouraging massive mergers of food processors, agrobiotech firms and drug companies on the basis of mutual interests in food, biotechnology, agrochemicals and pharmaceuticals. 'Biogopolies' emerge with potentially dire results:

> Within those [Northern countries], the introduction also of patent protection for genetic resources has resulted in the concentration of agricultural biotechnologies in a few northern corporations ('biogopolies'). The effect of this concentration in restricting the number of seed producers, may be to restrict the number of seed varieties available to farmers, as those corporations focus upon the marketing of blockbuster seeds. Thus, it is suggested that in planting elite varieties, which have market acceptance, farmers, particularly in traditional areas, neglect the land-races and farmers' varieties which have acted as reservoirs of genetic diversity. [66]

Changes in the agricultural production industry also affect human rights in more implicit ways. HR advocates contend that agrarian communities, their lifestyles, right to work, culture and way of life are being destroyed by globalization and the hegemony of property as dislocated farmers migrate to the nearest urban centres and are replaced by large corporate agribusiness in the rural landscape. The mass migration is said to exacerbate conditions of poverty and unemployment, particularly in some third world communities while evidence suggests terminator technology may be depriving farmers of their traditional right to preserve and replant seeds each year from the prior year's crop. This technology is genetically engineered into the plant and is a practical way of enforcing IPRs.[67] But the patents legally restrain the farmer's traditional rights to save seeds for future harvests and transform their role from cultivators to accidental infringers, even unintentional 'consumers' of seeds or GM plants containing patented product. Because of these

65 Ibid.

66 Blakeney, 'Stimulating Agricultural Innovation', op. cit. See also Drahos and Braithwaite, *supra* note 53, ch. 13.

67 Jeremy P. Oczek, 'In the Aftermath of the "Terminator" Technology Controversy: Intellectual Property Protections for Genetically Engineered Seeds and the Right to Save and Replant Seed' 41 B.C.L 627.

concerns, the very active Canadian based non-governmental organization Rural Advancement Foundation International (RAFI), now the ETC Group (Action Group on Erosion, Technology, and Concentration), mobilized on behalf of civil society with a mass mailing to 140 countries' senior officials responsible for agriculture, environment, and patent offices imploring their states to assert national sovereignty over their seed supply and to ban seed sterilization technology outright.[68]

Where farmers do stay on the land, they are increasingly reduced to labourers toiling the land – losing control over their means of production and becoming ever dependent on companies to supply seeds, fertilizers, and chemicals, most of which require heavy licensing fee payments. Some HR advocates assert that the proper forum to address farmers' rights would be under the right to food. A 1999 Economic and Social Council (ECOSOC) study on the Right to Food submitted to the Commission of Human Rights urged that farmer's rights be promoted as part of the right to food since the future of our food supply may depend on it.[69] NGOs and farmers agree, asserting that 'food security cannot be achieved without taking full account of those who produce food'.[70] But the role of food security extends beyond the achievement of individualized health as the Committee on ESCRs recognized long ago. The Committee acknowledged the delicate relationship that is shared between food, poverty, and dignity, affirming that the

> right to adequate food is indivisibly linked to the *inherent dignity of the human person* and is indispensable for the fulfillment of other human rights enshrined in the International Bill of Human Rights. It is inseparable from social justice, requiring the adoption of appropriate economic, environmental and social policies, at both the national and international levels, oriented to the eradication of poverty and the fulfillment of all human rights for all.[71]

---

68 See 'Call for "Seed Sovereignty Ban" on Terminator Patents', 29 May 1999, The ETC Group <http://www.etcgroup.org/documents/news_callfor.pdf>.

69 See Crucible Group II, *Seeding Solutions*, op. cit. at 76.

70 Cia Campesna 'The Right to Produce and Access to Land', <http://www.voiceoftheturtle.org/library/1996%20Declaration%20of%20Food%20Sovereignty.pdf>.

71 See General Comment 12, *The Right to Adequate Food*, 12 May 1999, E/C.12/1999/5. The Committee observes that 'a disturbing gap still exists between the standards set in Article 11 … and the situation prevailing in many parts of the world. More than 840 million people throughout the world, most of them in developing countries, are chronically hungry; millions of people are suffering from famine as the result of natural disasters, the increasing incidence of civil strife and wars in some regions and the use of food as a political weapon'. It should be noted that general comments by the Committee on ESCRs are 'authoritative interpretations of the relevant international human rights law …'. See Peter Prove and Miloon Kothari, 'Human Rights Bodies Gear Up on TRIPs' 1 *Bridges* Comment, 13–14, <http://www.IPRsonline.org/

For thousands of years farmers have been cultivating seeds, breeding new varieties, saving and exchanging them to advance the best crop-producing seeds for food. Farmers have rallied to protest the erosion of their customary human rights[72] by MNCs which drive for greater privatization under plant variety protection legislation, *sui generis* regimes or patents with strict (criminal) enforcement provisions.[73] Farmers and their growing group of supporters criticize governments for not doing enough to protect biodiversity, seed sovereignty, and food sovereignty when faced with trade pressures:

> The Indian Government, bowing to World Bank and W.T.O. pressures, is pushing through laws such as patents on plants and plant variety legislation that allow private monopoly over seed. Multinational corporations have already taken out patents on basmati and other rice varieties, corn, cotton, mustard, and other agricultural crops. Corporations like W.R. Grace have taken out patents on neem, and others have patents on turmeric, ginger, pepper, jeera, karela, jamun, brinjal, and other gifts of biodiversity.[74]

At the same time, giving seed patents to first world corporations ignores the work of farmers as historical creators, locating authorship in corporate investment by differentiating the 'other' creative claims of farmers and rural communities in preserving traditional knowledge and biodiversity by cultivating new, improved, and field-adapted plant varieties. Ignored too is the work of *cultures* in developing crops specific to geographic and climatic conditions, culturally appropriate for consumption.[75] Part of the concern is that the corporate impetus towards profit will steer traditional plant variety production (through natural cross-breeding) towards inter-species genetic modification resulting in GMO products that may eventually flood the entire market (through gene flow and contamination of 'pure' crops or through selective breeding out of traditional varieties) and thereby eliminate consumer ability to choose 'organic' or 'GM free'. This may even be a deliberate business strategy. Don Westfall, Vice President of the biotech consulting company Promar International, offers: 'the hope of the industry is that over time the

---

ictsd/docs/ProveKothariBridgesYear4N6JulyAugust2000.pdf#search=%22peter%20prove%20miloon%22>.

72 Martin Khor, 'A Worldwide Fight against Biopiracy and Patents on Lfe' *Third World Resurgence* No. 63 November 1995, 9–11. Special issues on patenting life: No. 57 (May 1995) and No. 84 (August 1997).

73 Farmer protests predate TRIPS, See 'Indian Farmers Rally Against GATTt, Bio-Patents', 4 October 1993, <http://www.sunsonline.org/trade/areas/intellec/10040093.htm>. More recently, Navdanya, Bija Swaraj, <http://www.navdanya.org/earthdcracy/seed/bija_swaraj.htm>.

74 Ibid.

75 Vandana Shiva, 'Seeds of Discontent' (1997), <http://www.organicconsumers.org/vandana.html.

market is so flooded [with GM food products] that there is nothing you can do about it. You just sort of surrender.'[76] The issue of biodiversity and access to it becomes an important issue that is compounded then by perceptions of risk creation, assessment, management and tolerance over the unknown potential health implications of transgenic foods and environmental contamination of farmer's fields.[77] A contemporaneous debate has been launched on the use of the precautionary principle for GMOs against the reductionist claims of science:[78]

> On Friday 24, June [2006], EU members states inflicted a severe defeat to the Commission by voting down eight proposals aimed at forcing the lifting of national bans on the growing of certain genetically modified organisms (GMOs) … [I]t is the first time that the qualified majority necessary for governments to turn down a Commission's proposal is gathered. As many as 22 out of 25 member states voted against the lifting of national bans … These vote results mark a historic turning point in the debate on genetically modified food and crops in Europe … For Greenpeace, the clear majorities against the bans show that it is time for the EU executive to listen to the 80%

76 See Stuart Laidlaw, 'Starlink fallout could cost billions' *Toronto Star*, 9 January 2001 discussing the ramifications of failure to get drug approval for human consumption of Starlink corn engineered to repel pests, made by drug and agriculture giant Avantis. The GM corn could cause allergic reactions and traces of it were found in over 300 different brands of taco shells that were recalled. See Julia Vorman, 'Recall Climbs to Over 300 Different Items', <http://www.organicconsumers.org/ge/starlink300.cfm>.

77 See Amicus Curiae Brief, WT/DS/291, 292, 293, EC – Measures Affecting the Approval and Marketing of Biotech Products, <http://www.ciel.org/Publications/ECBiotech_AmicusBrief_2June04.pdf>.

78 The debate transcends national politics into the world of international trade. International NGO Greenpeace reports that already several countries (Poland, Germany, Austria, Greece, Luxembourg, France, and Hungry) have bans on GMOs and that recent events show resistance towards removing such bans under pressure from the United States, Canada, and Argentina under threat of WTO trade sanctions. See Amicus Curiae Brief, ibid. The US complaint was against national safeguard clauses and against the European Commission's failure to have them lifted. The Panel's interim ruling (published February 2006) expressed views against the national bans but accepted that if these bans provided a risk assessment in conformity with the SPS Agreement, then they would be justifiable. See e.g. Pennapa Hongthong, 'Thai and Asian Farmer's Rally Against Frankenfood', <http://www.organicconsumers.org/ge/asianmarch.cfm>. The final Panel Report (circulated September 29, 2006) has yet to be adopted by the DSB. The report concludes that scientific studies offered by some banning states, 'in no case did they provide an assessment of the risks to human health and/or environment meeting the requirements of the *SPS Agreement*' at 1069, para. 8.10, available online with a detailed history, at <http://www.wto.org/english/tratop_e/dispu_e/cases_e/ds291_e.htm>.

> of the public who are opposed to GMOs … [and] learn from this lesson and stop hiding behind technocratic procedures and confidential expertise to try and force GMOs into European fields.[79]

One news report headlined: 'New EU States Block Monsanto – Twice.'[80] Helke Ferrie welcomes these events as overdue: 'the nation state is returning: people in rich and poor countries demand that governments act in the public interest'.[81] While imposing additional regulatory requirements outside of the patent system for commercialization of new technology is both Paris and TRIPS-consistent to deal with the public's interest in *risk*, it may be complained against for adding a further layer of compliance complexity for farmers[82] without addressing the IPRs issues. In developed states, farmers tend to be heavily subsidized, which means the public there pays through tax dollars and also in direct costs, passed on to the consumer, of such business expenses as the seed licences imposed by firms such as Monsanto. The World Development Report 2008: Agriculture for Development,[83] draws attention to the need for greater inclusion of civil society groups – in particular farmers' organizations – in setting the agricultural agenda but little effort has been made to retain their stakeholder interests in patent law and policy.

DCs and LDCs are the greatest losers because they are typically not IP creators and their agricultural industry will have greater difficulty competing against subsidized counterparts in foreign markets, particularly if they continuously have to pay added rents under licences for the produce they grow each year but lack the domestic governmental resources that farmers in the developed world enjoy to help offset these transaction costs. The imposition of

79 See Greenpeace, 'Background Briefing: National Bans on GMOs', 25 July 2006, Greenpeace, <http://www.greenpeace.eu/downloads/gmo/NationalBans0507.pdf>.

80 See Biotech Mailout, 'New EU States Block Monsanto – Twice', July 2004, Friends of the Earth Europe, <http://www.foeeurope.org/publications/2004/Biotech_July04.pdf>.

81 Helke Ferrie, 'Thought for Food: Biotech under Siege as European Nations Ban GM Food', 29 July 2006, Straight Goods, <http://list.web.net/archives/food-news/2006-August/000063.html>.

82 Randy Haillier, President of the Ontario Landowner's Association, argues that farmers are disempowered by overregulation: '[f]armers also sold out their future and freedom by accepting government regulations that dictate every use and activity on the farm. Farmers unknowingly surrendered themselves to a mild form of slavery, and have become slaves in the truest sense of the word. They are slaves to their farm groups and to the corporate monopolies that control them, and are caged by the regulations of safety and security. Slaves are never self-reliant, free, nor prosperous'. 'Sowing Socialist Seeds and Expecting a Harvest of Free Enterprise', 16 July 2006, <http://www.quebecoislibre.org/06/060716-5.html>.

83 Link under the 'Data and Research' heading, World Bank, <http://econ.worldbank.org>.

uncalibrated seed IPRs perpetuates cycles of poverty by imposing corporate imperatives on individuals in a manner that compromises human rights. And, the 'poverty-health nexus is so strong that poor health keeps poor [sic] in absolute poverty and poverty pushes them in poor health status'.[84] Braithwaite and Drahos opine that, 'had TRIPS been framed as a public health issue, the anxiety of mass publics in the US and other Western states might have become a factor in destabilizing the consensus that US business elites had built around TRIPS'.[85] Brian Martin laments, if only the 'scale of such protests could be combined with other actions that undermine the legitimacy of intellectual property, the entire system could be challenged'.[86]

The concern over distributive justice is integrated into HR instruments which focus on improving aggregate welfare through the protection of individual rights. The right to share in scientific progress, as articulated in Article 27.1 of the UDHR and Article 15 of the ICESCR, is a means of ensuring this end. Article 15.2 imposes an obligation on states to take steps to achieve the full realization of this right, including 'those necessary for the conservation, the development and the diffusion of science and culture'. Moreover, Article 27.3 uses the language of 'freedom' and 'liberties' similar to the ICCPR and obligates states to 'undertake to respect the freedom indispensable for scientific research and *creative activity*'.[87] Patent laws are *prima facie* intrusions for the period of the patent term on this conception of scientific freedom and the ability of farmers to engage in creative activity by saving, exchanging, planting, and cross-breeding seeds without first securing a licence. Through the creation of artificial monopolies, they impede the direction of future research not just in medicine but also in food supplies and not just in the laboratory but also in the field where over 2000 years of traditional science has taken place. As corporations and publicly funded researchers become more intimate bedfellows through the privatization of science, basic research will be diverted to areas that maximize commercial viability. The 'the lure of profits threatens to diminish scientific integrity in the biosciences' writes Hope Shand.[88] Herder and Gold contend that '[e]ven if we are willing to accept that IP rights (especially patents) provide a necessary incentive for biotechnology companies to engage in the R&D process, it is unmistakably clear that they are

84 N.K. Nayak, 'Imperatives of Global Health Law for the 21st Century' in R.K. Nayak, chief ed., *Global Health Law: Proceedings, Selected Papers and Recommendations Presented at the Inter-disciplinary International Conference on Global Health Law Held December 5–7, 1997* (New Dehli: Indian Law Institute and the World Health Organization, 1998).

85 John Braithwaite and Peter Drahos, *Global Business Regulations* (Cambridge: Cambridge University Press, 2000) at 576.

86 Martin, *supra* note 49 at 53

87 ICESCR, *supra* note 11. See also Beetham, *supra* note 27 at 247.

88 See Shand, 'Gene Giants', op. cit. at 233.

insufficient incentives to address the particular health and industrial needs of the world's poor'.[89] And, Michael Blakeney likewise concludes

> [t]he market dominance of these private corporations also has an important influence upon the sort of biotechnological research which is undertaken. For example, it has been suggested that the dominance of private corporations in agricultural research direct that research towards Northern concerns and away from Southern food priorities.[90]

Many critics worry about potential increases to crop vulnerability and a narrowing genetic basis (genetic erosion) associated with incentives to create phenotype uniformity for yield and processing that fits the precise varietal description required to attract legal protection. Who will safeguard the public's interest? How will peoples have their rights to the *highest attainable standards of physical and mental health*, set out under Article 12(1) of the ICESCR, realized?[91]

For a system based on public health, cost of delivery is a significant determinant of the number of people who can receive the needed medical health intervention; whether governments are willing to undertake obligations for population health and public health genomics programmes; the direction of policy formation; and the ability of a state to meet its international human rights obligations optimally. Patent incentives, in so far as they exist, may be important to promote new drugs, vaccines and, with genomics, prevention-based interventions; in this regard, they can be said to serve a social purpose.[92] According to Jonathan Putnam,

> [p]erhaps more than any other consumer purchase, the acquisition of health-related products and services is governed by norms and expectations that challenge the usual market allocation mechanism. Nowhere is that challenge mounted with greater official backing than in Canada, where universal

89 Herder and Gold, 'Intellectual Property Issues in Biotechnology', op. cit. at 11.

90 Blakeney, *supra* note 66, referring to J.P. Alston, G. Pardey, and J. Rosenboom. 'Financing Agricultural Research: International Investment Patterns and Policy Perspectives' (1998) 26 *World Development* 1045.

91 The provision reads: *The States Parties to the present Covenant recognize the right of everyone to the enjoyment of the highest attainable standard of physical and mental health.* ICESCR, *supra* note 11. It also obligates States to take steps necessary for *the full realization of this right* [which] *shall include* … the improvement of all aspects of environmental … hygiene … [t]he prevention, treatment and control of epidemic, endemic, occupational and other diseases … [and] the creation of conditions which would assure to *all medical services and medical attention* in the event of sickness'. Ibid., emphasis added.

92 The WHO adopts this position in its joint report with the WTO Secretariat, *WTO Agreements and Public Health*, op. cit..

> health is a basic right ... That is why intelligent policy making towards the development of new drugs rests on a kind of economic jujitsu: if one is to reach a public end, one must deflect private means in that direction.[93]

But as the foregoing chapters reveal, there is ample evidence that a proper *balance* has not been struck, and the system is not operating efficiently, particularly in light of practices that seek to increase *ex post* rewards. The jujitsu is jinxed. Ancillary costs include barriers to the delivery of public health, effective cross-policy coordination, efficient allocation of resources, and the ability of a state to respond to the priorities of its constituents, meaning that despite biotech's promises as avowed by industry stakeholders to secure their investments, patents mean that, contrary to Articles 12 and 15.1(b) of ICESCR, a society may not be able to afford the technological gains efficiently and certainly not *universally* or at the *highest attainable standard* for each person.

The biotech industry generated 1.1 billion in sales in 1997, half of which came from health care products according to a March 2001 report prepared for the Canadian Biotech Advisory Committee (CBAC).[94] Statistics Canada reports that Canada's biotechnology sector generated nearly $2 billion in revenues by 1999, $718 million of which was generated from exports. It was expected to exceed $5 billion in revenues in 2002.[95] The health sector continues to be the primary beneficiary of, and therefore is the most affected by biotechnology according to the June 2002 CBAC report: 'more than 90 percent of the advanced biotechnology products on the world market are related to health. It is expected that about three-quarters of global biotechnology demand will continue to be in this area'.[96] The social value of new innovations is undermined by the market value that is tied to their protection under letters patent. Recall the BRCA1 and BRCA2 example. While an estimated 22,200 women will be diagnosed with breast cancer in 2006 and 5,300 die,[97] several provinces continue to be threatened with domestic infringement litigation by MGL. Breast cancer is the most frequently diagnosed and one of the most treatable cancers amongst Canadian women. The Cancer Society posits that

93 Jonathan Putnam argues that patents are consistent with health policy; see 'The Price We Pay for Drug Research' *Innovate* Spring 2004, 26–8, CILP, <http://www.innovationlaw.org>.

94 See the Project Steering Committee on IP and the Patenting of Higher Life Forms, available online, op. cit. Specifically, the CBAC, Biotechnology and Intellectual Property: Patenting Life Forms and Related Issues (Interim Report) (Ottawa: Canadian Biotechnology Advisory Committee, November 2001).

95 *Harvard SCC*, op. cit., para. 16.

96 Patenting of Higher Life Forms and Related Issues (Ottawa: Canadian Biotechnology Advisory Committee, June 2002), CBAC, op. cit. at 2.

97 Canadian Breast Cancer Foundation, 'Media Backgrounder: Canadian Cancer Statistics – Screening: Breast, Colorectal, and Rervical' 11 April 2006.

deaths could be reduced further by one quarter if only 70 per cent of women in this age group had routine clinical breast exams and biannual screening.[98] A population health approach to screening will become important in the future when disease related genes are further identified, and more likely a prospect if the diagnostics and screening become readily available at marginal cost without being subject to the strict royalty demands of patent holders. Some chronic diseases appear with genetic mutation variants disparately affecting not only genders but specific racial populations.[99] As our knowledge of genetics improves, a population health approach to public health genomics may be necessary and the larger the target group, the more significant the social and economic costs of IPRs.

The tensions between patents and HR are not new; they have surrounded law and policy decisions in most jurisdictions involving patented medicines, generic drugs, compulsory licences, the patentability of methods of medical treatment, and the patentability of food. Historically, writes Daniel Kevles, 'urban Americans probably tended, like Europeans, to think of food as a scarce resource and to be reluctant to grant anyone a monopoly right over food products, even for a limited period'.[100] Canada's exclusions have usually occurred without moral conflict. In the early part of the twentieth century, Canada issued patents only for new *processes* for making foods or medicines which were subject to low royalty rates due to compulsory licensing. Food and medicinal *products* were not themselves patentable.[101] A prosperous generic drug industry resulted. However, David Vaver reminds us,

---

98 Genevieve Beauchemin 'More Screening Can Cut Cancer Deaths: Report', CTV report, <http://sympaticomsn.ctv.ca/servlet/ArticleNews/story/CTVNews/20060411/cancer_screening_060411. See also Canadian Breast Cancer Foundation, <http://www.cbcf.org/news/events.html#erin>.

99 For example the Mediterranean variant is the prominent genotype (> 83 per cent) for G6PD deficiency in a newborn screening study. See I. Khneisser et al. 'G6PD Newborn Screening: Incidence of Neonatal Jaundice among Cases of G6PD Deficiency with the Mediterranean Variant', Poster Presentation at *Public Health and Genomics*, op. cit. See also Nadja Kanellopoulou, 'Pharmacogenomics and Racially-Targeted Drugs: A Marriage of Inconvenience?', Poster Presentation discussing population genomic research leading to a 'more refined understanding of ethnic and racial differences in drug response' and the BiDil controversy over the first drug approved by the US Food and Drug Administration to target a specific racial group (the first 'ethnic drug'), for heart disease.

100 Kevles, 'A History of Patenting Life', op. cit. at 4. See also s. 41 of the *Patent Act*, R.S.C. 1985, c. P-4, op. cit. [CPA], now repealed, as discussed by SCC, to explain the former exclusions of food or medicine from patentability.

101 Section 41(1) of the Patent Act R.S.C. 1970, c. P-4; amended R.S.C. 1970, c.10 (2nd Supp.), s. 65, provided that 'in the case of inventions relating to substances prepared or produced by chemical processes and intended for food or medicine, the specification shall not include claims for the substance itself, except when prepared

> [u]nder pressure from the US pharmaceutical drug industry and its proxies in the US government, this system was entirely dismantled between 1987 and 1993. Product patents for medicine and food are now granted, compulsory licensing is gone, and a Patented Medicine Prices Review Board monitors patented medicines for 'excessive' prices. Overall prices are nevertheless higher than when compulsory licensing was in full swing, thus creating one more problem for Canada's embattled health care system.[102]

More recently, the pendulum has begun to swing back. The tireless efforts of NGOs on behalf of the poor in both first- and third-world countries to increase access to patented medicines and generic drugs has been rewarded by amplified international attention to these issues. In Canada, the patent/human-right controversy was (re)-politicized with the review of patent legislation and subsequent statutory amendment to allow the manufacture of generic pharmaceuticals for export to developing countries whereby the hitherto industry-appropriated spaces were reclaimed.[103] What has been omitted from the process is a full scale parliamentary review of the issues surrounding the patenting-of-life debate, a subset of the patent/human-rights debate, and recommendations for a legislated response to maximize related benefits and minimize costs.[104]

In calculating the harm or cost/benefit of granting genetic patents, the welfare loss is not limited to an underutilized technology but includes

---

or produced by the methods or processes of manufacture particularly described and claimed or by their obvious chemical equivalents'. Subsections (3) and (4) dealt with licences in relation to food and medicine respectively. The section has been repealed in its entirety.

102 Vaver, *IPL*, op. cit. at 127. For a food security discussion, see Geoff Tansey, 'Food Security, Biotechnology and Intellectual Property: Unpacking some issues around TRIPS' July 2002, <http://www.geneva.quno.info/>; essential medicines, see Frederick Abbott, 'The TRIPS Agreement, Access to Medicines and the WTO Doha Ministerial Conference' (2001) 5 J. of World Intellectual Property 15, <http://www.geneva.quno.info/> [Abbott, 'Access to Medicines'].

103 See CPA under heading *Use of Patents for International Humanitarian Purposes to Address Public Health Problems* (ss. 21.01–21.20) as implemented S.C. 2004, c. 23, s. 1. This brought into force the *Jean Chrétien Pledge to Africa, Canada* Bill C-9, *An Act to amend the Patent Act and the Food and Drugs Act*, 3rd Sess., 37th Parl., 2004, <http://www.parl.gc.ca/PDF/37/3/parlbus/chambus/house/bills/government/C-9_2.PDF>. See also General Council's statement on the then draft Decision contained in document IP/C/W/405 to implement paragraph 6 of the Doha Declaration on the TRIPS Agreement and Public Health, on which Bill C-9 was based, <http://www.wto.org/english/news_e/news03_e/trips_stat_28aug03_e> (30 August 2003).

104 See the final recommendations of the CBAC in 'Patenting of Higher Life Forms and Related Issues' (Report to the Government of Canada Biotechnology Ministerial Coordinating Committee, June 2002).

the knowledge that a possible measure exists but costs too much to access optimally.[105] If the system of health is a blend of private and public,[106] then the ability for some to pay for a licence and genetic testing *outside* of the public system threatens to widen the *genetic divide* as only the rich will be able to afford the licensing fee. If the tests are funded publicly, then the propriety of higher health delivery costs distributed over the broader tax paying public should also be publicly debated through democratic processes since, as a matter of simple arithmetic, a government can afford proportionately less testing for cancer-related genes and public health genomics if each test costs approximately five times more under licence. Governments may become reluctant to undertake such screening routinely for fear that a standard of care may emerge which they may subsequently be unable to afford, despite vast potential health benefits.[107] Patents also contribute to delays and wait times in testing, further impacting domestic health policy. MGL's exclusive rights to BRCAnalysis empowered it to demand that *all* testing proceed through them or a licenced laboratory. This raises complicated issues of sovereignty. Similar conditions were imposed in France such that samples had to be sent out of the country to MGL laboratories in the United States over which there was no French regulatory oversight or control. Researchers complained they could not continue to test or develop alternative improved methods for testing because of the patents, undermining one of rationales for such incentives and making it difficult to scientifically validate the test. The British Columbia (BC) Cancer Agency in Canada performed some 600 tests but MGL's cease-and-desist letters effectively forced the BC government to discontinue its testing, favouring the rights of the patentee, for approximately two years during which 150 female cancer patients were advised of Ontario's research study and transferred to this province for testing free-of-charge (though still in alleged patent violation). Approximately 30 patients elected to have the testing done directly by MGL.[108] Should a monopoly on the invention confer a monopoly on knowledge or health? TRIPS Article 8 does not require it and actually protects against it.[109]

105 There are other forms of welfare loss as well. See e.g. Trudo Lemmens, ''What About Your Genes?' Ethical, Legal, and Policy Dimensions of Genetics in the Workplace' *Politics and the Life Sciences* 16, 1 (March, 1997), 57–75.

106 *Chaoulli*, *supra* note 24, the SCC held legislation preventing public access to privately insured services as unconstitutional for harming health in violating of s.7 protections of life and security of the person.

107 See Emmanuelle Rial-Sebbag, 'Genetics and State Intervention in France: New Powers' Poster Presentation at *Genomics and Public Health*, op. cit., regarding state failure to prevent harm where the genetic knowledge and technology exist.

108 'Genetic Testing Resumes at B.C. Cancer Agency', 23/02/14, <http://search.phsa.ca/cgi-bin/MsmGo.exe?grab_id=0andpage_id=1377andquery=myriadandhiword=myriad%20> ['Genetic Testing'].

109 Amani, 'Patents and Public Health', op. cit. at 89.

Adopting national or regional health policy based on fear of litigation or a WTO complaint and redirecting cancer screening patients to laboratories *outside* of the province within whose jurisdiction these patients fall is *not* a sound basis for health governance. Although BC resumed testing in 2003, BC Cancer Agency reports that there is now a one year waiting list for counselling with 300 families waiting for testing.[110] The Canadian Cancer Society (CCS) opposes the exclusive rights of gene patent holders if they are used to interfere with an individual's health. In light of the MGL scandal, they recommend that gene patents not be used in a manner restricting the development of knowledge about genes, or health, or limiting the access of Canadians to cancer-related genomic testing.[111] Through all of this, Ontario brazenly continued its testing supported by then provincial, now federal, Health Minister Tony Clement, supported by then Premier Mike Harris, mindful of the limited resources and the duty on his government to respond to the health needs of its constituents.[112] But, because the patent belonged to a foreign firm, it raised the question of whether these actions were legal under domestic law and whether any illegality would invite WTO scrutiny. IP lawyers knew Canada could be sued domestically for patent infringement just as trade law specialists knew that Canada could be held accountable for TRIPS violation if MGL were to successfully lobby its US Trade Representative to bring the complaint.[113] HR scholars and public health practitioners could not ignore the basic governance question: to what extent should patent law, and now trade law, dictate or supersede public health policy and the regulatory preferences of national governments and their constituents? The MGL controversy resonated around the globe. The MGL example demonstrates the difficult and practical issues of law and policy coordination across provincial, national, and international jurisdictions, costs for domestic governments across ministries,

110 'Genetic Testing', *supra* note 108. See Institut Curie, 'Breast and Ovarian Cancer Susceptibility Gene BRCA1: Another Victory for Opponents of Patents Held by Myriad Genetics; European Patent Office Rejects the Essential Points of BRCA1 Gene Patents' (Press Release, 31 January 2005), <http://www.curie.fr/upload/presse/myriadpatents310105.pdf>. In November 2004, Myriad Diagnostics sold its BRCA patents to the University of Utah Research Foundation but continues to hold exclusive licences.

111 <http://www.cancer.ca/ccs/internet/standard/0,3182,3172_31282995__langId-en,00.html>.

112 Canadian Cancer Society and National Cancer Institute of Canada on the Patenting of BRCA1/2 Genes, <http://www.ncic.cancer.ca/vgn/images/portal/cit_86751114/11/12/89486356postionstatements_BRCA1A_en.pdf>.

113 A complainant party to both TRIPS and NAFTA may elect under which to proceed but typically the WTO is preferred because of its formal rules based procedure, appellate review, and the fact that decisions are effective automatically; consensus adoption is unnecessary. See Gary Horlick, *WTO and NAFTA Rules And Dispute Resolution* (London: Cameron May Ltd, 2003).

and to Canadian taxpayers. According to the Annual Report of the Canadian Institute for Health Information (CIFI), health care spending in Canada exceeded $120 billion in 2003.[114] Private enterprise continues to be subsidized by public scientific efforts in basic research and through public education; to pay artificially inflated health care costs based on patents in a publicly funded health system further diverts limited resources from other important government initiatives. Patents on life, from BRCA1/2 genes to the treatment of AIDS,[115] for the genetic sequence for meningitis,[116] SARS or avian flu virus, and over GM vitamin-A rice may affect the full realization of a number of human rights. These range from the most specific individualized rights to health and food, to broader cultural and communal rights, and fundamentally the general right to sovereignty and self-determination so critical for protecting all other rights in a manner that preserves requisite attention to how public policies shape human experiences. Coherent, national, public regulations governing

---

114 Canadian Institute Health Information (CIHI) Annual Report 2003–2004, CIHI, <http://secure.cihi.ca/cihiweb/dispPage.jsp?cw_page=GR_1131_E> as cited in Tony Clement 'Just What the Doctor Ordered?' *Innovate* at 22: 'further analysis showed that health expenditures represented 10% of Canada's GDP in 2003, matching the all time high reached in 1992 and keeping Canada 4th among comparable OECD countries ... [indicating that] in Canada there are too many health care demands chasing too few public resources. It is not, however, a case of too few dollars ... Indeed the CIHI's report makes it clear that public and private investments in health care represent more of Canadian's tax dollars and discretionary income ... The key question becomes whether health expenditures pass the test of being sustainable, accountable and accessible.'

115 Nelfinavir is an AIDS treatment drug patented by the Swiss pharmaceutical giant Roche. Putting the HR to health first, Brazil's then Health Minister, Jose Serra asked the public health laboratory to make a generic version, reducing costs by 40 per cent. The TRIPS complaint launched byt the United States was subsequently abandoned. See 'Brazil to break AIDS Patent', <http://news.bbc.co.uk/hi/english/business/newsid_1505000/1505163.stm>.

116 Human Genome Science (HGS), an American company rival to Celera Genomics, has applied to the European Patent Office to patent one of the bacteria that causes meningitis. It could lead to a demand for payment of royalties on every future treatment if a new meningitis vaccine is found. HGS strives to be the first to own the whole genetic sequence of bacteria. See <http://lists.essential.org/1998/pharm-policy/msg00045.html>. HGS boasts that it holds over 200 genomic patents. Under relaxed utility guidelines in the United States, HGS patented CCR5 even though they did not know of nor disclosed its application at the time. Later, another company discovered its utility as the entryway for the AIDS virus to infect cells but as a result of the prior patent, HGS is entitled to royalties on any drug that targets this protein even though HGS was not involved in inventing or even 'discovering' its industrial application. See Pollack, A. (2000) 'U.S Hopes to Stem Rush Toward Patenting of Genes', *New York Times* 6–28, <http://www.nytimes.com/library/national/science/062800sci-genome-patents.html>.

disease control, vaccination,[117] pre-implantation and newborn and adult screening programmes for genetic susceptibility to hereditary based diseases for population health all require discretion to prioritize these, framed in terms of human rights obligations, over trade obligations. The economics would be prohibitive otherwise, and ethics may require it, but the main determinant is that the formation of policy and direction of research rely on it. The Bellagio Report, 'Genome-based Research and Population Health' provides,

> the human genome is not the only one of relevance to human health. Our genome interacts with those of a myriad other organisms, including plants and microbes, both in maintaining a healthy body function and in many disease processes. An understanding of pathogen genomes, in particular, is elucidating the molecular and cellular processes of infection and of the variation in human susceptibility to infectious diseases. We need, too, to emphasize the importance of human genetic/genomic diversity as a valuable resource for the human species.[118]

WHO's Constitution encourages cooperation to ensure optimal health globally, recognizing the positive and negative externalities of human health. It provides:

> The health of all peoples is fundamental to the attainment of peace and security and is dependent *upon the fullest co-operation of individuals and States.* The achievement of any State in the promotion and protection of health is a value to all. Unequal development in different countries in the promotion of health and control of disease, especially communicable disease, is a common danger.[119]

WHO's cooperation ethic – to maximize health through the agency of the state – is sensible given the fact that individuals are first and foremost citizens of a state while simultaneously citizens of the world. People are as mobile as the infectious diseases that plague them; the 2003 SARS 'epidemic' taught us

---

117 See e.g. Karen Lynn Durrell, 'Intellectual Property Protection for Plant Derived Vaccine Technology: Here they Come are we Ready or Not?' (Winter 2006) 10.3 *Lex Electronica,* <http://www.lex-electronica.org/articles/v10-3/durell.pdf>. Patenting of genetic microorganisms, viruses, and plant derived vaccines will impact population health programmes. See WHO 'Intellectual Property Rights and Vaccines in Developing Countries' 2004, <http://www.who.int/intellectualproperty/events/vaccines_meeting/en/>.

118 Report of an expert workshop held at the Rockefeller Foundation Study and Conference Centre, Bellagio, Italy, 14–20 April 2005, <http://dceg.cancer.gov/files/genomicscourse/bellagio-011807.pdf> at 6.

119 WHO Constitution, *supra* note 63, emphasis added.

this.[120] It will be interesting to see how development of targeted population treatments and national responses will be undertaken the next time there is an epidemic, given that various teams of scientists claim to have patented all or parts of the SARS virus, albeit as a defensive measure, some argue, meant to deny private profiteers from foreclosing liberal access and use.[121]

Not surprisingly, the patent-health relationship is central to a number of important studies in jurisdictions grappling to devise better strategies for health and industrial management.[122] A recent study commissioned by the Department of Health (DOH) in the United Kingdom[123] examined the issue of whether there is a risk in the health sector that biotechnological IPRs may generate disproportionate claims from right-holders creating unacceptable barriers, impeding the delivery of health care, and thwarting the generation of health care products. The health department in commissioning the report recognized the growing importance that IPRs play in the UK health sector and thus adopted an active role in the discussion of IPR issues. The study made a number of recommendations based on the laws of the UK and EU and was specifically geared to the issues faced there. It also made some general procedural recommendations that are inward looking but of potential benefit to other countries as a model. These modalities emphasize the need for consultation between trade and health industries and the possibility of establishing a single (UK) policy on IPRs and healthcare provision (encompassing both internally generated and externally sourced innovation), a framework for partnership between the DOH and commercial providers of IP (such as pharmaceutical companies and universities), and the adoption of a 'robust central policy for "licensing in" designed to moderate

120 SARS stands for Severe Acute Respiratory Syndrome. See Clifford Kraus, 'The SARS Epidemic: Canada; Toronto Officials Seek End to Travel Advisory', *New York Times*, 29 April 2003.

121 See 'Patent Applications for SARS Virus and Genes', 29 May 2003, <http://www.who.int/ethics/topics/sars_patents/en/>. WHO's position on the SARS patent is to monitor its use to ensure that collaboration of health professionals, infectious disease control and the realization of the right to health are not hampered. WHO does not formally oppose defensive patenting.

122 See e.g. the Nuffield Bioethics Report, op. cit., and The European Commission's Report on the *Development and Implications of Patent Law in the Field of Biotechnology and Genetic Engineering* (Brussels: The Commission, 2002), <http://europa.eu.int/eur-lex/en/com/rpt/2002/com2002_0545en01.pdf>.

123 See W.R. Cornish, M. Llewelyn and M. Adcock, *Intellectual Property Rights (IPRs) and Genetics: A Study into the Impact and Management of Intellectual Property Rights within the Healthcare Sector* (Cambridge: Public Health Genetic Unit, July 2003), <http://www.phgu.org.uk/about_phgu/s-ipr1.doc>. The study considers other forms of IPRs that would privatize genetic information including certain analogues to copyright, unregistered design rights in the structure of DNA molecules, and database protection.

excessive demands by licensors by considering, as possible options, the use of compulsory licensing, competition law and Crown use'. What such reports emphasize is that any policies adopted necessarily need to be country specific and correspond to the domestic needs and laws of a given state; this is consistent with WHO's view that '[g]overnments have a responsibility for the health of their peoples which can be fulfilled only by the provision of adequate health and social measures'.[124] Consequently, the state is charged with acting to determine, within each jurisdiction and in accordance with its values, to what extent patent policy imposes inefficient costs, in addition to the regular deadweight loss attributable to a patent monopoly, on the health sector and how these costs can be reduced without major impact on innovation of new medical/health technology and drugs.[125] In May 2008, the 61st World Health Assembly adopted 'Resolution WHA 61.21: Global strategy and plan of action on public health, innovation and intellectual property'[126] to help address the issues of need, access and production of healthcare products with regard to the conditions disproportionately affecting DCs.

124 WHO Constitution, *supra* note 63.

125 Colleen Flood and Greg Hinds argue comprehensive cost-benefit analyses and coordination of cross-sector policies are needed in 'Future of Health Care in Canada: The Who, What and How Much of Publicly Funded Health Care' at 38. In *Innovate* (Spring 2004) 23, they assert that regarding health care expenditure, '[i]n theory, the choices or decisions made should be a function of public values, available resources, and information regarding relative costs and health benefits. For example, if we spend an extra $150,000 on health care, that money will not be available for education. We must ask ourselves, does the health care system benefit from that funding, as compared to the benefit that the education system would have received from the same amount of funding?... Instead, the choices are a function of accidents of history and long held accommodations between governments and the medical profession; inflexible and inadequate regulations and law; and turf protection and lobbying by different stakeholders and interest groups'. The costs of patents should also be considered given the greater institutional and financial resources diverted from health systems administratively, in licence fees, and due to 'bad patents'. Dutfield discusses poor patent quality in natural products (like maca and yellow bean) and its harms in 'Sharing the Benefits of Biodiversity: Is there a Role for the Patent System?. 5(6) *Journal of World Intellectual Property* 899–931 ['Sharing the Benefits'], reprinted in Keith Maskus, ed., *The WTO, Intellectual Property and the Knowledge Economy: Critical Perspectives on the Global Trading System and the WTO* (Brookfield, VT: Edward Elgar Ltd, 2004) at 908–91.

126 See WHA 61:21, <http://www.who.int/gb/ebwha/pdf_files/A61/A61_R21-en.pdf>.

Government Use and Compulsory Licensing

Where a patent in which the government has an interest has been granted, such as for a disease-linked gene or diagnostic testing such as BRCAnalysis, even if a court subsequently upholds validity upon challenge, a viable means for regulatory governments to meet users' interest in the patented invention while meeting their own international human rights and trade obligations is through the statutory provision for government use *vis-à-vis* compulsory licences. In an insightful article, economist Dan Usher argues that the normal distributional implication of a patent is that benefits of the invention are a shared consumer surplus from the newly-invented good but that benefit is reduced by the patent monopoly for its limited term in order to protect the incentive to invent. The result is that consumers are better off than they would be without the invention but worse off than they would be if the good was 'un-monopolized'. Due to monopolistic pricing, the surplus might be enjoyed more by the rich than the poor. However, Usher contends that in some cases, as with indivisible goods, this distributional story is distinctly *wrong*:

> patents direct all benefit to the wealthy when the newly-invented good is what might be called *indivisible* … in this sense of the term when each person wants to buy either one unit of the good or none at all because less than one full unit is useless and any amount over the above unit is of no additional benefit. Indivisible goods are exemplified by medical procedures which you adopt entirely or not at all. You cannot have half a heart transplant. Nor can you have half an AIDS treatment or half a course of antibiotics; you must take the entire treatment if it is to do you any good.[127]

In such cases, Usher contends, 'every penny of benefit of the appearance on the market of a newly-invented good accrues to people whose incomes exceed some critical amount' and some people are excluded altogether. Usher's study has some qualifications but ultimately he uses the findings to rationalize the public provision of medical services and the expropriation of key drugs by the state in order to make these drugs available at their cost of production, which he argues may be appropriate so long as the drug companies are compensated for that use. Compensated taking need not be limited to key drugs, however. Usher's study can be a useful response to dealing with other areas of importance for health provision where the patented goods are indivisible (and demand, as with health care, inelastic due to public funding) including genomics and genomic testing for gene-linked diseases.

---

127 Dan Usher, 'The Distributive Implications of Patents on Indivisible Goods' (April 2004) Queen's Economic Department, Institute for Economic Research Working Paper No. 1018, <http://qed.econ.queensu.ca/pub/papers/abstracts/download/2004/1018.pdf> at 2.

In Canada, if the government wishes to use a patented invention and negotiations for licensing between it and the patentee are unsuccessful, then the government can seek a compulsory licence from the Commissioner of Patents, who has statutory discretion to provide non-exclusive licences when conditions precedent are met.[128] The government must still pay a licensing fee in an amount the Commissioner considers adequate under s. 19(4). This method raises quantification challenges for the Commissioner: what is 'adequate' compensation? Although useful for dealing with specific cases to respond to public health needs *after* the patent has issued, it should not be relied upon often to forge patent-health policies in an ad hoc manner, as it is remedial, offers no long term comprehensive solution to the patent-health tension, and is poor substitute for coherent policy coordination and resource management. But, compulsory licences in limited contexts by a state for national use are a TRIPS-consistent means to make use of patented 'inventions' for public health.

Article 31 of TRIPS provides comprehensive procedural requirements for *other uses* and allows for compulsory licensing and government use as long as the same set of necessary preconditions (negotiation of a licence on reasonable commercial terms have been unsuccessful within a reasonable time of request, etc.) are satisfied, with similar special provision for waiving negotiations in cases of national emergency, other circumstances of extreme urgency, or in the case of *public non-commercial use*.[129] This latter exception can itself offer a broad range of 'wiggle room' for public healthcare systems. Article 31 offers a 'fast track' whereby preconditions for prior negotiations with the patentee may be waived. Otherwise, for ordinary commercial uses, the applicant must first seek a voluntary licence from the patentee on reasonable terms and conditions (Article 31(b)). Licence applications are to be considered on their merits and

128 Section 64 of the CPA, *supra* note 57 allows for compulsory licensing in abuse of patents cases upon application by the Attorney General of Canada or 'any interested persons' to the Patent Commissioner. Clauses 65(2)(c)–(f) define what constitutes abuse under the Act. Section 19 gives the Commissioner statutory discretion to grant government use by way of a non-exclusive compulsory licence over a patented invention in order to supply a domestic market, with the patentee receiving adequate compensation. This provision could have been resorted to in response to the public health concerns surrounding BRCA1/2 testing in a publicly funded health care system so long as the government had first tried and failed to negotiate a licence.

129 This negotiation requirement can be dispensed with under the CPA (s. 19.1(2)) in cases of 'national emergency or extreme urgency or where the use for which the authorization sought is a public non commercial use'. See also TRIPS Article 44 regarding injunctions: '... and provided that the Provisions of Part II specifically addressing use by governments, or by third parties authorized by a government, without the authorization of the right holder are complied with, Members may limit the remedies available against such use to payment of remuneration in accordance with sub-paragraph (h) of Article 31 ....'

the provision does not limit the grounds on which such licences may be issued, so long as the patentee is paid adequate remuneration in the circumstances of the case and the licence is 'non-exclusive'. Once the conditions for invoking this provision to grant the licence are resolved, the licence terminates. This provision assuages concerns over the right to patented food or health inventions in the biotech industry and could potentially support Canada's 'infringement' of a foreign firm's IPRs in relation to food and health.[130] But, unauthorized and uncompensated use by the government, even if for well-intentioned reasons as delivery of public health to Canadians in the BRCA1 example, is effectively considered 'theft' and may not only lead to a domestic infringement suit, but can form the basis of a WTO complaint for undermining property rights. But it is also an ineffective way to deal with the need for comprehensive cross-sector coordination. Article 31 is significantly limited to local use and manufacture since the licence issued should be 'predominantly for the supply of the domestic market' of the member (Article 31(f)) but presumably this means that it could also authorize export of a *non-predominant* portion of production; what it means will have to be determined by WTO jurists. The restriction on exports in Article 31(f) and the conditions necessary for granting a licence under Article 31(b) do not apply when a compulsory licence is issued to remedy anti-competitive practices (Article 31(k)). This leaves some scope for a member to determine what constitutes anti-competitive practices sufficient to warrant a compulsory licence to issue without meeting the conditions set out in the provision. It is foreseeable that a complaint may be raised to challenge the invocation of Article 31(k), and Article 31 may be of little effective assistance to countries lacking the necessary infrastructure and manufacturing capacity to imitate the invention for domestic supply, e.g., to supply genetic tests domestically under a compulsory licence scheme. These countries would need to rely on other markets and parallel imports to meet their demand, with outstanding concern over TRIPS based limitations.

Amending TRIPS: The Patents-Public Health Declaration Arising from Doha

TRIPS Article 71 provides that the Council of TRIPS (Council) Committee as whole shall evaluate TRIPS and its implementation upon expiry of the transitional period set out in Article 65 para. 2. The Council may undertake reviews in light of relevant new developments that might warrant modification or TRIPS amendment. The TRIPS Council is also responsible for monitoring

130 It is worth exploring whether greater government use of patented inventions for the delivery of health can be characterized as *public non-commercial use* for the purposes of s. 19.1 in Canada given our health care system. However, if it were to be used routinely by governments to meet human needs, then R&D in these inventions may be discouraged, paradoxically, in favour of inventions of arguably less social utility.

the operation of TRIPS and members' compliance in domestic law (Article 63.2), affording consultation opportunities, assisting members in the context of dispute settlement procedures if requested, and fulfilling other duties assigned to it by the members (Article 68). Despite the provision for review by the Council, there is no authority to make changes to TRIPS obligations; they can only occur through waiver or amendment by the members but advisory opinions concerning members' interpretations of TRIPS may be provided.

General concerns over the legitimacy of the WTO or the fairness of TRIPS have grown more vocal since their origin a decade ago. The deaths of millions of people, and resulting social, economic, political, and civil instability, in Africa, Brazil, India, Thailand and elsewhere due to the AIDS pandemic rallied civil society and various NGOs against their states in demand for more just distributive outcomes. As the deadliest infectious disease in the world, AIDS claims 8,000 lives each day with over 95 per cent of them in the developing world.[131] WHO suggests that 'patent protection where it does not already exist would result in the average price of drugs rising, with projected increases ranging from 12 to 200 per cent.[132] It soon became apparent that TRIPS-mandated patent rights were severely curtailing the availability of affordable essential medicines and that this was linked to the exclusive rights of patent holders to set the prices on these medicines and preclude others from selling cheaper generics. The public outcry that ensued targeted as much the WTO for prescribing protection as BigPharma for setting the prices.[133] Would these costs be offset by the welfare gains encouraged in terms of foreign direct investment? Not for those bearing them. According to the World Bank, middle income countries may benefit from increased foreign investment but the drug costs are rising due to patent protection spreading throughout the developing world and this has a direct bearing on access to such medicines.

Entering the Doha Round, developing countries, in particular Brazil and India, aligned themselves as a negotiation strategy along common interests and objectives to protect public health and domestic regulatory diversity not to the extent *permissible* under TRIPS but rather in *priority* of it. The language of their draft Ministerial Declaration stated that '*nothing* in TRIPS *shall prevent*

131 See UNAIDS (Joint United Nations Program on HIV/AIDS) 'AIDS Epidemic Update' December 2000. Some DCs used political will and public mobilization to achieve desired health outcomes, see Hannah E. Kettler and Chris Collins, 'Balancing Health Needs and Drug Research Incentives', Cooperation South, UNDP 2002 *Creativity, Innovation and Intellectual Property Rights* 10–37.

132 See 'Access to Essential Drugs May be Undermined by Global Patent Agreement', <http://www.globalpolicy.org/socecon/bwi-wto/wto/2002/1201panos.htm>; see also Panos Report 'Patents Pills and Public Health: Can TRIPS Deliver?'online: <http://www.panos.org.uk/resources/reportdetails.asp?id=1053andnull=1000and>.

133 For a discussion of the role of the access to medicines debate in further derogating the WTO in the public eye see Abbott, 'Access to Medicines', *supra* note 102 and Abbott, 'WTO Medicines', op. cit.

members from taking measures to protect public health'.[134] In contrast, the version backed by Canada and the United States going into the Doha meeting more modestly acknowledged the right to take measures necessary to protect public health *only* to the extent that is TRIPS *compliant*.[135] The latter wanted to keep a narrow focus on pandemics, while the former, a broader gaze on public health. The oft referenced 'North-South' divide going into Doha threatened, according to then WTO Director-General Michael Moore, to be 'the deal breaker at this conference'.[136]

Meanwhile, during the launch of the Doha Round of trade negotiations, developed countries like Canada showed disregard for foreign patent rights. Canada had its MGL controversy but also the violation of patent rights of the foreign firm Bayer in relation to an anticipated national anthrax outbreak emergency post 11 September 2001. These events accentuated the need for a public health exception to patent rules as much for developed countries as for DCs and LDCs concerned with access to essential AIDS and malaria medicines. Cross-policy coordination is needed for efficient patent governance regardless of where in the world you are.[137] Developing countries saw an opportunity and

134 See Draft Ministerial Declaration From Group of Developing Countries, <http://www.wto.org/english/tratop_e/TRIPS_e/mindecdraft_w312_e.htm>. The South-South trend towards greater development joint initiatives and cooperation has continued with six countries (Brazil, China, Nigeria, Russia, Thailand and Ukraine) signing an agreement to increase cooperation in developing and producing generic drugs, with India and South Africa considering participation, at the XV International HIV AIDS Conference. See 'Brazil to Coordinate Six-country Anti-AIDS Network', BBC Monitoring International Reports, 16 July 2004 cited in Frederick Abbott, 'The Cycle of Action and Reaction: Latest Developments and Trends in IP and Health' ICTSD-UNCTAD Dialogue on Ensuring Policy Options for Affordable Access to Essential Medicines, Bellagio 12–16 October 2004, <http://www.IPRsonline.org/unctadictsd/bellagio/docs/Abbott_Bellagio3rev1.pdf>.

135 For a thorough discussion of the how the Ministerial Conference in Doha was influenced by the converging threat of AIDS and terrorism, the Brazilian response to the former and the North American response to the latter vis-à-vis patent rights for medicines, see Bradley Condon, 'The Twin Security Challenges of AIDS and Terrorism: Implications for Flows of Trade, Capital, People, and Knowledge' in Buckley, *Changing Face*, op. cit. at 251–84.

136 Steven Chase, 'Drug Patent Skirmish Threatens WTO Talks', *Globe and Mail*, 9 November 2001, B6.

137 This point is underscored by a dispute between the European Union and Canada on stockpiling and regulatory approval of generic drugs. See e.g. the WTO dispute in *Canada – Generic Medicines* case (Panel Report, *Canada – Patent Protection of Pharmaceutical Products*, WT/DS114/R, Appellate Body Report WT/DS170/AB/R, adopted 12 October 2000) on Canadian policies, allowing generic manufacturers to start testing of patented products three years prior to patent expiry but disallowing stockpiling six months prior.

endeavoured to exploit divisions in the positions of developed countries (the North-North divide):

> The severity of the AIDS crisis in many parts of the developing world, combined with the duplicitous policy of Canada and the United States regarding patent rights, contributed to a North-South divide with respect to this issue. The European Commission and Norway took fairly flexible positions regarding the developing countries' positions. In addition, important sectors of civil society within developed countries shared the same concerns as those of developing countries, as abundantly shown in the international public opinion during the pre-Doha debate.[138]

As a result, not only did the Doha Ministerial Declaration support social and economic rights by locating the trade imperative within the objectives of maximizing welfare through more equitable distribution of benefits, but it also resulted in the very important Declaration on TRIPS and Public Health which reaffirmed within its final text the right of developing countries to take a public health perspective to *interpreting* TRIPS obligations:

> We agree that the TRIPS Agreement *does not and should not prevent* Members from taking measures to protect public health. Accordingly, while reiterating our commitment to the TRIPS agreement, we affirm that the Agreement can and should be interpreted and implemented in a manner supportive of WTO Members' right to protect public health and, in particular to promote access to medicines for all ... In this connection, we reaffirm the right of the WTO Members to use, to the full, the provisions in the TRIPS Agreement, which provide flexibility for this purpose.[139]

The Doha declaration is consistent with human rights-prioritizing recommendations based on UN inquiries into the intersection of TRIPS and HR. In August 2000, leading up to the WTO Ministerial Conference in Doha, Qatar, the UN Sub-Commission on the Promotion and Protection of Human

---

138 Condon, *supra* note 135 at 251–84. Canada's response was 'illegal' as it bypassed proper procedure under national and international law in obtaining a million pills of Cipro, a Canadian generic version manufactured from Apotex, in alleged violation of Bayer's patent rights; the mere *concern* made public health a priority.

139 Doha Declaration on TRIPS Agreement and Public Health, WT/MIN(01)/DEC/2, Adopted 14 November 2001, Ministerial Conference Fourth Session Doha, 9–14 November 2001, <http://docsonline.wto.org/DDFDocuments/t/WT/Min01/DEC2.doc> [TRIPS and Public Health]. Paragraph 6 of the Declaration recognized that some WTO members would have insufficient or no manufacturing capacity in the pharmaceutical sector and accordingly would be unable to make effective use of the compulsory licensing provisions and mandated the TRIPS Council to devise a solution before 2002.

Rights adopted a resolution on 'intellectual property rights and human rights' which reaffirms that the primary obligation of governments under the Vienna Declaration are its human rights responsibilities. The Resolution urged governments to take full account of existing HR obligations when engaged in trade negotiation (naming food and health rights in particular), when reviewing TRIPS in the context of the new Doha Round. In addition, the Resolution recommended for Special Rapporteurs and the Committee on ESCR to consider the HR impact of TRIPS when assessing compliance, and asked the UN Secretary General to provide a report on the IP/HR nexus in 2001. That report strongly urged a human rights approach to trade arguing that not only was it the law, but it made good sense too for welfare maximization, poverty reduction, and development.[140] And, the release of a joint study by the WTO and WHO on the relationship between trade and public health makes it cogently clear that the WTO can no longer operate or evolve in an institutional vacuum.[141] In short, civil society and domestic governments can no longer afford to remain oblivious to the social costs and human rights compromises of the TRIPS' trade-related IPRs regime. IPRS are HR-related.

TRIPS has not been implemented fully in most developing countries and a further extension has been granted for their compliance. When TRIPS took effect on 1 January 1995, developed countries were allowed one year to ensure compliance of their laws and practices with TRIPS; developing countries were given a transition period ending January 2000 (for other than the general principles in Articles 3–5); LDCs were given until 1 January 2006 and possibly longer upon approval to ensure conformity with the agreement.[142] The Doha Declaration granted the LDCs an extension to January 2016 for pharmaceutical patents.[143] Peter Yu observes, '[h]ad the level of intellectual property protection been adjusted to reflect the countries' needs, interests, and

140 See Office of the High Commissioner for Human Rights, 'Human Rights and Trade', 5th WTO Ministerial Conference, Cancun, Mexico, 10–14 September 2003, <http://www.unhchr.ch/html/menu2/trade/cancunfinal.doc>.

141 *WTO Agreements and Public Health*, *supra* note 92. See also reference to TRIPS in a resolution by the World Health Assembly at WHO, Resolution WHA 57.14 on scaling up treatment and care within a coordinated and comprehensive response to HIV/AIDS, Fifty-Seventh World Health Assembly, Agenda item 12.1, 22 May 2004, para 2(6). TRIPS discussions occupy debates in other UN based agencies, such as the Food and Agricultural Organization (FAO), 'The TRIPS Agreement and Agriculture', <http://www.fao.org/documents/show_cdr.asp?url_file=/docrep/003/x6730e/x6730e06.htm>; and the UN HR Commission, 'Intellectual Property and Human Rights', <http://www.unhchr.ch/Huridocda/Huridoca.nsf/0/c462b62cf8a07b13c12569700046704e?Opendocument>.

142 Part VI, op. cit. Summary at <http://www.wto.org/english/tratop_e/TRIPS_e/tripfq_e.htm>.

143 See TRIPS and Public Health, *supra* note 139.

conditions, those transitional provisions in the TRIPS Agreement might not be needed'.[144]

According to a 2003 UNDP report critical of TRIPS and urging governments to take action to replace it,[145] the Public Health Declaration may have a significant effect on the future interpretation and implementation of TRIPS:

> There is little indication, apart from the Doha Declaration, that TRIPS has really been interpreted in the true spirit of balance between rights holders and users. From a legal perspective, the generalist language employed in TRIPS has worked both ways for developing countries; it has allowed for flexible interpretation, but also left the text open to dispute. The latitude in the text requires tremendous specialized legal capacity, which most developing countries lack. Moreover, the experience of Brazil ... [in terms of free access to drugs with extensive health infrastructure supported by national legislation in relation to AIDS treatment and the US challenge that ensued through the request of a WTO dispute panel which it eventually withdrew based on an agreement reached with Brazil] has shown that efforts to use this flexibility provoke strong opposition from the developed world. Finally, the enforcement mechanism – the cross-retaliation mechanism of the dispute settlement process – takes little account of differences in capacity to retaliate. This is costly and harmful for developing countries. Exceptions are limited and specific, and the burden of proof falls on the alleged violator. In practice, this considerably reduces the power of the exceptions.[146]

---

144 Yu, 'Four Common Misconceptions', op. cit. at 135.

145 'Member nations need to begin dialogues to replace TRIPS – and equivalent top-down schemes of substantive IPR harmonization – with alternate intellectual property paradigms that are unrelated to trade sanctions and may include, but are not restricted to: an intellectual property ladder, where more stringent laws apply to countries at higher levels of income and technology use, and countries progress from one level of protection to another with improvements in their Human Development Index/Millennium Development Goals indicators ....' UNDP, *Making Trade Work*, op. cit. at 221.

146 Ibid. at 210 and 222. The IP related portions of the Report are discussed also in Genetic Resources Action International (GRAIN) 'New Report Urges Governments to Replace TRIPS', 20 February 2003, <http://www.grain.org/bio-ipr/?id=188>. The reference to the burden and limits of exceptions relates to the manner in which exceptions have been 'adjudicated' by TRIPS panels or the AB. For example, in *Canada – Generic Medicines* case, *supra* note 137 regarding Canadian provisions which allowed for testing for regulatory compliance and stockpiling of medicines by generic manufacturers, Canada used Article 30 in its defence to argue that such provisions were permissible exceptions that members may provide to patent protection. The panel significantly narrowed Article 30 by stipulating three criteria which Canada had to prove to establish that the exception was 'limited' (the measure did not unreasonably

Some critics did not consider the Declaration a 'win' for developing countries, noting that its significance has often been grossly exaggerated as a major achievement.[147] In fact, while, as *The Economist* aptly put it, 'the declaration is political and not legally binding',[148] what is binding on the parties to a dispute is the TRIPS text on which the Declaration is based. As a pastiche of TRIPS, the Declaration accentuates its built-in flexibilities along a common issue and had the positive impact of leading to an important waiver in 2003. The TRIPS Council and the WTO General Council subsequently decided to convert the temporary waiver of TRIPS' Article 31 (f) and (h), made by the WTO general Council Decision of August 2003 into a permanent amendment to justify compulsory licensing and parallel imports of generic drugs for countries lacking manufacturing capacity for pharmaceuticals. By 2005, the first amendment ever to be made to a 'core WTO agreement' provided substantive change to TRIPS and created permanent conditions allowing production of generic drugs and the effective use of compulsory licences in third party countries for parallel importation, so long as a compulsory licence has also been granted in the importing country.[149] The

conflict with the normal exploitation of the patent, did not unreasonably prejudice legitimate interests of the patentee, and took into account legitimate interest of third parties). For an insightful discussion critiquing the panel's interpretation of 'limited' see Trebilcock and Howse, *International Trade Relations*, op. cit. at 419 where they submit the exception should have been interpreted in light of the purpose of TRIPS set out in Article 7: 'To do justice to this purpose, the panel would, at a minimum, have had to consider the scope implied in the word 'limited' not only from the perspective of how much rights-holders' interests were being curtailed, but also from the perspective of how limited the exception could be while affording sufficient scope to protect public health and the interests of consumers generally'. Dinwoodie and Dreyfuss, *International Intellectual Property Law and the Public Domain of Science*, op. cit., also provide a detailed analysis of Article 30's three part test.

147 Walden Bello, 'Learning from Doha' at <http://www.focusweb.org/publications/2001/learning-from-doha.html> writes that DCs 'were actually losers of the Conference by launching a new round with an ambitious mandate to extend the purview of the WTO to new agreements, like industrial tariffs, and for negotiating new (the 'Singapore issues') non-trade areas such as investment, competition policy, government procurement, and trade facilitation when the over 104 implementation concerns of developing countries leftover from the Uruguay round had yet to be resolved. Little commitment seemed to be made on a quick phase out of agricultural subsidies or textiles quotas'.

148 <http://archives.econ.utah.edu/archives/pen-l/2001m11.3/msg00039.htm>.

149 See Press Release, 'Members OK amendment to make health flexibility permanent', Press 426, 6 December 2005, <http://www.wto.org/English/news_e/pres05_e/pr426_e.htm>. The WTO General Council adopted the Decision on Implementation of Paragraph 6 of the Doha Declaration on the TRIPS Agreement and Public Health (see WT/L/540 (1 September 2003)). For a detailed discussion, see Abbott, 'WTO Medicines', *supra* note 133. Abbott finds that the decision, though 'a helpful piece of a much larger public health puzzle', cannot adequately deal with the economic

flurry of international activity and public protest was finally having the desired impact on accommodating for the disparity of resources affecting members' ability to invoke the government use provisions of Article 31,[150] even though the amendment is limited in scope; it applies only to pharmaceutical patents over essential medicines and does not address lack of manufacturing capacity generally.[151]

The TRIPS preamble recognizes public policy objectives of domestic IPR regimes (para. 5) but simultaneously reaffirms the need for strengthened IP protection through a trade linkage (para 1.) and thereby subjects IP to the underlying normative economic force of the multilateral trade regime. TRIPS subscribes to a 'rights' thesis for IP (para. 4) and thereby reduces them to a single dimension: as exclusively *tradeable* commodities rather than public goods, infrastructure for the delivery of public goods, and the national cultural

---

and political power exerted by the US in its forum shifting strategy to TRIPS plus agreements outside of the WTO. Abbott provides an overview of how the decision is operationalized, the need to give notice to the TRIPS Council of the intention to use the system in whole or in part as an importer, and the determination of whether domestic manufacturing capacity and know-how is insufficient in accordance with criteria set out in the Annex of the Decision (at 327–49). Abbott criticizes Canadian implementation of this decision with legislation to amend the CPA, *Use of Patents for International Humanitarian Purposes to Address Public Health Problems* and the the *Food and Drugs Act, Food and Drugs Act*, R.S.C. 1985, c. F-27 for requiring that a voluntary licence first be sought and for conditioning the licensing procedures on the domestic situation of the exporting state, which is perverse; it 'turns the object and purpose of the Decision on its head. If Canada is going to export ARVs to Africa, this is not because there is a public health crisis in Canada but, rather, because there is a public health crisis in Africa' (at 342). And, Canada's implementing legislation 'does not make any provision for use of the fast track for emergencies and public noncommercial use, but it does provide that thirty days are an adequate period for seeking a voluntary licence' (at 347). And, the Decision does not preclude initiation of dispute settlement for measures contested as inconsistent (i.e. non-violation nullification or impairment complaints).

150 See e.g. International Federation for Human Rights, 'Understanding Global Trade and Human Rights: Report and Resource Guide for National Human Rights NGOs in View of the 2005 WTO Ministerial Conference, Hong Kong, MC6' May 2005, Geneva, <http://www.fidh.org>.

151 Carlos Correa argues that issues affecting the ability to operationalize on the interim waiver actually rendered it of no practical effect. Extensive pre-conditions must be fulfilled (such as differentiating the export product from the patented version in get-up, the responsibility of the generic producer to prove bio-equivalence, and the contestability of the 'capacity' for domestic manufacture by the importing country). The rationale was to contain the economic implications of the waiver and amendments by limiting the solution to enumerated lists of diseases according to Carlos Correa, 'Access to Drugs under TRIPS: A Not so Expeditious Solution' Bridges International Centre for Trade and Sustainable Development, January 2004 at 21.

artifacts that they are.[152] Peter Drahos portrays TRIPS 'as a prime example of 'business sovereignty over regulatory standard setting'. Self-determination is critical to the ability to reset this balance and develop regulatory strategies to optimize the outcomes based on IPRs calibration. TRIPS is an 'instrument of intellectual property law' write Dinwoodie and Dreyfuss, that

> must strike a balance between sufficient levels of protection to stimulate the desired social and commercial activity undertaken by first-comers, and sufficient limits on those rights to ensure the maximum socially useful exploitation of that activity … [L]ike any international agreement [it] must also deal with issues such as sovereignty, diversity, and legitimacy that pervade international relations. It must accordingly allocate power between supranational and national institutions, between national and international laws.[153]

While the power dynamics continue to play out, the state's central role as agent and mediator of its public's interest becomes more imperative. IPRs must remain culturally sensitive while 'pressed into service on behalf of human rights'[154] – a call to duty with proven practical consequences.

152 For a similar analogy within the field of copyright regulation, see Myra J. Tawfik, 'Is the WTO/TRIPS Agreement User-Friendly?: Final Report to the International Trade Treaties Committee of the Canadian Library Association' 30 January 2005, <http://www.cla.ca/resources/tawfik_final_report.pdf> at 11.

153 Dinwoodie and Dreyfuss, *supra* note 146 at 448.

154 Peter Drahos, quoted in Peter Prove and Miloon Kothari, 'Human Rights Bodies Gear Up on TRIPs' 1 Bridges Comment, 13–14, <http://www.IPRsonline.org/ictsd/docs/ProveKothariBridgesYear4N6JulyAugust2000.pdf#search=%22peter%20prove%20miloon%22> at 14.

Chapter 6
The Sovereign Right to Self-Determination: From Agency to Development and Dignity

Introduction

The state's right to self-determination has a moral force: even though states have often stood in colonial opposition to their peoples, human rights need to be implemented in highly stratified contexts and are therefore conditioned on the state's appropriate exercise of its sovereign right. This means that diversity of regulation is constitutive of universal principles and at the same time preserves stratified cultural landscapes.

When the patenting of life first became a public issue, Isidora Acosata, President of the Guayami General Congress, declared:

> I never imagined people would patent plants and animals. It's fundamentally immoral, contrary to the Guayami view of nature, and our place in it. To patent human material ... to take human DNA and patent its products ... that violates the integrity of life itself and our deepest sense of morality.[1]

Debra Harry, for the Indigenous People's Diversity Network similarly objected to the Human Genome Diversity Project (HGDP):

> They have come to take our blood and tissues for their interests, not for ours ... genetics is a violation of our ethics, it attacks our culture's worldview. *We don't view our genes as protein actions ready to be interpreted, for us our genes are sacred.*[2]

---

1 Rural Advancement Foundation International (RAFI), *Indigenous People Protest U.S. Secretary of Commerce Patent Claim in Guyami Indian Cell Line,* Press Release RAFI, 26 October 1993.

2 Debra Harry (1995) 377 *Nature* 372, cited in Baljit K. Dhadda, 'Patenting Human Genetic Information: Is Nothing Sacred?' in Lionel Bentley and Spyros Maniatis, *Intellectual Property and Ethics, Perspectives on Intellectual Property Series* (Carswell: 1998), 89–108, emphasis added.

State signatories to UN human-rights instruments are charged with the duty to protect and promote HR; it may be done by granting patents if the necessary safeguards are there and if doing so is consistent with the cultural values and social mores of a given peoples.[3] Art. 1 of the UNC codifies the supremacy of territorial integrity and political independence of the nation state and provides authority for the state right to give priority to HR; it is not only a moral obligation of states as agents of their people, but also a legal right. The capacity for the state to act as primary causal agent relies on the state's right to self-determination, essentially a sovereign discretion to govern without undue external interference by other states. For states, the right to self-determination is as paramount as the right to liberty for the individual. It may also be the only means of holding states accountable for egregious violations of their duties to their people, since peoples cannot themselves justiciably invoke this right.[4] The right to self-determination provides traction to a states' sovereign status and is cornerstone to all other HR in enabling their provision and protection in domestically contingent and culturally diverse frameworks. In the wake of 9/11, UNESCO's unanimously adopted Declaration on Cultural Diversity[5] (UNDCD) offers some recognition that cultural pluralism is essential to global objectives:

> Previously the notion of sustainable development embraced economic, environmental and social parameters, yet largely ignored those pertaining to cultural issues. A change in strategy is clearly a must, UNESCO believes, if the promotion of cultural diversity is to be given a central … place in the debate.[6]

The preamble of the UNDCD notes that 'culture is at the heart of contemporary debates about identity, social cohesion, and the development of a knowledge-based economy'. Article 1 of the Declaration provides,

---

3 See article 2 of the *International Covenant on Civil and Political Rights*, op. cit. [ICCR] and the *International Covenant on Economic, Social and Cultural Rights*, op. cit. [ICESCR].

4 The Human Rights Committee charged with monitoring state compliance with the ICCPR decided in *Lubicon Lake Band v. Canada* (1990), Comm. No. 167/1984 that ICCPR (Article 27) protects the right of persons to engage *in economic and social activities* that are part of the *culture* of the community to which they belong. The expropriation of 10,000 km^2 of land by the Alberta government for use by oil and gas interests violated Aboriginal group rights to engage in protected economic and cultural activities but the right of 'peoples' to self-determination was considered non-justiciable (para. 32.1).

5 See UNESCO *Universal Declaration on Cultural Diversity*, 31st Sess. of the General Conference of UNESCO, Paris, 2 November 2001 [UNDCD], <http://www.unesco.org/culturelink/review/35/cl35un.html>.

6 UNESCO, News Release, 'Sustainable Development' (3 September 2002).

> Cultural diversity is as necessary for humankind as biodiversity is for nature. In this sense, it is the common heritage of humanity and should be recognized and affirmed for the benefit of present and future generations.[7]

Article 3 recognizes the important link between cultural diversity and development:

> Cultural diversity widens the range of options open to everyone; it is one of the roots of development understood not simply in terms of economic growth, but also as a means to achieve a more satisfactory intellectual, emotional, moral and spiritual existence.

Article 9 further preserves the role of the state and supports the general thesis presented in this text that the matter of appropriately balancing interests in policy making should be left within the sovereign dominion of a regulatory state:

> While ensuring the free circulation of ideas and works, cultural policies must create conditions conducive to the production and dissemination of cultural goods and services … It is for each State, with due regard to its international obligations, to define its cultural policy and to implement it through the means it considers fit, whether by operational support or appropriate regulation.

Article 11 is dedicated to building bridges between civil society and the public/private sectors towards these ends, since 'market forces alone cannot guarantee the preservation and promotion of cultural diversity, which is key to sustainable human development'.[8]

The Relationship between Human Rights and Human Development

Human development (HD) has been defined by the UNDP as 'a process of enhancing human capabilities – to expand choices and opportunities so that each person can lead a life of respect and value', and human rights (HR) as '… claims to social arrangements that … secure the freedom for a life of dignity'.[9] Both share a common vision and purpose: 'to secure the freedom,

---

7 UNDCD, *supra* note 5.

8 'From this perspective, the pre-eminence of public policy, in partnership with the private sector and civil society, must be reaffirmed'. Ibid.

9 UN Development Programme (UNDP), *Human Development Report 2000: Human Development and Human Rights* (New York: Oxford University Press, 2000) [UNDP HR/HD Report 2000] at 2.

well-being and dignity of all people everywhere'.[10] There are other important shared goals: both HR and HD desire to secure 'freedom from want – to enjoy a decent standard of living; [and] freedom to develop and realize one's human potential'.[11] The interdependency of this relationship is expounded by the 2000 UNDP Report on HD that confirms: 'when human development and human rights advance together, they reinforce one another – expanding people's capabilities and protecting their rights and fundamental freedoms.'[12] The report traces the historical parallel paths of the distinctly separate HR and HD agendas and their agents – '[t]he one largely dominated by economists, social scientists and policy makers, the other by political activists, lawyers and philosophers'.[13] But today, there is newfound recognition of their complementarity.

HD agendas are enhanced by HR as the latter tends to draw attention to accountability for respecting, protecting, and fulfilling the HR of all people:

> The tradition of human rights brings legal tools and institutions – laws, the judiciary, and the process of litigation – as means to secure freedoms and human development. Rights also lend moral legitimacy and the principle of social justice to the objectives of human development. *The rights perspective helps shift the priority to the most deprived and excluded,* especially to deprivations because of discrimination.[14]

Moreover, the attainment of HR has much to gain from the perspective of HD because HD initiatives offer a long-term perspective by directing attention to the socio-economic context for realizing or suppressing human rights:

> The concepts and tools of human development provide a systematic assessment of the economic and institutional constraints to the realization of rights – as well as the resources and policies available to overcome them.[15]

'[T]he human rights approach', the report adds, 'links the human development approach to the idea that others have duties to facilitate and enhance human development'.[16] Accordingly, while indivisible and contingent, the rights more clearly link to state duties when tied into a development

10 Ibid., Overview. See also World Bank Report, *Development and Human Rights* 1998, <http://www.worldbank.org/html/extdr/rights/hrtext.pdf>, affirming that HR protection is part of the expansive principle of *sustainable development*.

11 UNDP HR/HD Report 2000, *supra* note 9 at 64.

12 Ibid. at 2.

13 Ibid.

14 Ibid. at 1, emphasis added.

15 Ibid. at 2.

16 Ibid. at 21.

approach. Development is polycentric and requires domestic conditions conducive to its objective of *freedom*,[17] which in turn is integral to a life of *dignity*. While a causal relation between development and freedom is difficult to establish conclusively, a correlation is empirically supported by Michael Trebilcock's findings:

> [I]t is an observable fact that most of the relatively rich countries in the world are democracies, while most of the poorest countries are not. It is also true that most of the relatively rich countries in the world are market economies, and many of the poorest countries are not. Scully (1992) in an extensive empirical analysis of 115 economies over the period of 1960–1980, claims that societies with high levels of political liberty, civil liberty, and economic liberty grow at three times the rate and are two and one-half times as efficient in transforming inputs into national output as societies in which these rights are proscribed.[18]

The various international institutions (such as the UN agencies, the IMF, WB, GATT and WTO) in their collective mandates acknowledge the role of domestic conditions and strive to foster transformative modalities. Whether we call it *development*, *democracy and free markets*, or simply, the desire to provide the means for 'raising the standards of living' by prioritizing HR and respecting cultural diversity and the sovereign right to self-determination, the objective is the same: to serve a *human* (rather than corporate) end.

Former World Bank president James D. Wolfensohn stated that given the dire living conditions of most of the world's population, 'progress is too

---

17 See e.g. Amartya Sen, *Development as Freedom* (New York: Oxford University Press 1999) [Sen, *Development as Freedom*] at 3 defines development as 'a process of expanding the real freedoms that people enjoy. The focus on human freedoms contrasts with narrower views of development, such as identifying development with the growth of gross national produce, or with the rise in personal incomes, or with industrialization, or with technological advance, or with social modernization. Growth of GNP or of individual incomes can, of course, be very important as *means* to expanding the freedoms enjoyed by the members of the society. But freedoms depend also on other determinants, such as social and economic arrangements (for example, facilities for education and health care) as well as political and civil rights (for example, the liberty to participate in public discussion and scrutiny) … Viewing development in terms of expanding substantive freedoms directs attention to the ends that make development important rather than merely to some of the means that, *inter alia,* play a prominent part in the process'.

18 See Michael Trebilcock, 'What Makes Poor Countries Poor?: The Role of Institutional Capital in Economic Development', in E. Buscaglia et al., eds, *The Law and Economics of Development* (Greenwich, CT: JAI Press, 1997) 15 at 20. See also UNDP HR/HD Report 2000, *supra* note 9.

slow'.[19] He asserted the need for a multifaceted and cooperative approach to development:

> Broadly, our sister institution [IMF] has the responsibility for macroeconomic stabilization for our client countries and for surveillance. We have the responsibility for the structural and social aspects of development … these are not isolated roles … [T]he two functions are like breathing in and breathing out. An appropriate macroeconomic framework is essential for our work, but the social, structural, and human agenda, which we share with the regional banks, members of the UN system, and other partners in development, is essential for the IMF which cannot and does not prescribe in a vacuum. Together we must serve the hopes and aspirations of the people in our client countries, or our clients will not achieve their objectives in peace and stability. And together, we must work with and support the work of the World Trade Organization which is so critical to the trading arrangements and future of our client countries.[20]

The multilateral trade regime provides the instruments for economic integration, growth, fiscal stability, and furthers cooperation towards development. Perhaps the failure of members to overcome significant impasses leading to the temporary collapse of the ninth round of negotiations and the current delay is due to the strong commitment that the majority of the WTO's membership has made to secure significant gains in DCs' and LDCs' issues, such as agriculture, textiles, and intellectual property,[21] to further

---

19 'With three billion people still living under $2 a day, with growing inequity between rich and poor, with forests being degraded at the rate of an acre a second, with 130 million children still not in school, with 1.5 billion people still not having access to clean water, and two billion people not having access to sewage we cannot be complacent … Two billion more souls must feed themselves by the year 2025, hampered by wars, with growing inequity, and with distortions of economies and politics as evidenced in crises … With the reduction in Overseas Development Assistance and current instability in the international financial markets, there is much to be concerned about'. See James D. Wolfenshon, *A Proposal For a Comprehensive Development Framework (Discussion Draft)*, January 1999, Worldbank, <http://www.worldbank.org/cdf/cdf-text.htm>.

20 Ibid.

21 But see Jeffrey Schott, 'Unlocking the Benefits of Free Trade' Special Report, *The Economist* (1 November 2003) at 67. Schott believes the promises made to DCs were actually self promoting 'rhetoric' to which some of the failure of the Doha Round, at the Ministerial meeting in Cancun, can be attributed: 'The emphasis on development inspired many [DCs] to justify their demands for new concessions by arguing that they paid too much in the previous trade talks (the Uruguay Round) and got nothing in return – and now it was pay back time. Those countries seemingly believed their own rhetoric and expected Cancun to be a donor conference rather than a trade negotiation.'

their development agendas within the multilateral trading regime. The Doha Declaration clearly supports an emphasis on free trade as a *tool* for economic growth and 'sustainable development':[22]

> International trade can play a major role in the promotion of economic development and the alleviation of poverty. We recognize the need for all people to benefit from the increased opportunities and welfare gains that the multilateral trading system generates ... Recalling the Preamble to the Marrakesh Agreement, we shall continue to make positive efforts designed to ensure that developing countries, and especially the least-developed among them, secure a share in the growth of the world trade commensurate with the needs of their economic development.[23]

A 2003 UNDP Report, *Making Trade Work*, urged: 'the multilateral trade regime will be well governed if it is focused on the achievement of the Millennium Development Goals'.[24] A few years before, economist Helleiner had proposed along the same lines that the WTO become a *development* focused agency connected to

> a vision of humane global governance, with celebration of humanity's diversity, promotion of democratic process, and social and economic justice and human rights for all. Such ... [a] vision will see trade, financial flows and macroeconomic stability as no more than instruments of its objectives rather than as ends in themselves.[25]

Compare with Heather Stewart, 'Mandelson: US Greed caused the Doha Collapse' *The Observer*, 20 July 2006, referencing the EU Trade Commissioner's view that the US was wrong to expect to be recompensed 'dollar for dollar' for every reduction to farm subsidies and that Washington should have made some concessions against its farmers in exchange for gains in manufacturing and service. <http://guardian.co.uk/business/2006/jul/30/usnews.development/print>.

22 Defined by the OECD as 'a development path along which the maximization of human well-being for today's generations does not lead to declines in future well-being'. <http://www.oecd.org/glossary>.

23 '[E]nhanced market access, balanced rules, and well targeted, sustainably financed technical assistance and capacity-building programmes have important roles to play'. Doha Ministerial Declaration, WT/MIN(01)/Dec/1 dated 20 November 2001, para 2 [Doha Declaration] <http://www.wto.org/English/thewto_e/minist_e/min01_e/mindecl_e.htm>.

24 UNDP, *Making Trade Work*, op. cit. See also Dani Rodrik, 'Global Governance of Trade as if Development Really Mattered' Report Submitted to the UNDP 2001, criticizing the substitution of trade liberalization for development ('confounding the ends and means') and the priority of trade policy over 'the critical role of domestic institutional innovations that often depart from prevailing orthodoxy'.

25 Helleiner, 'Markets, Politics and Globalization', op. cit. at 17.

In short, HR and HD have a simultaneously dialectic and mutually reciprocal relationship despite their apparent historical evolutionary insularity. Trade liberalization furthers HR objectives by facilitating development. A HR framework for trade dispute resolution is not inconsistent with the original mandates of the two regimes and would create a coherent international order because diverse instruments and interpretive practices need to be reconciled into a single trajectory; one whose post war evolution speaks to the future realization of a better world measured by improvements to *both* HR *and* HD.

A Global Cooperation Ethic and Renewed Commitments to 'Development'

Multilateral trade law emerged to promote friendly relations through economic interdependence and to facilitate development in poorer countries on the rationale that the world as a whole would be more prosperous (and thus secure) if its nations and peoples were prosperous (and secure). With the rise of the welfare state, social welfare maximization objectives added a responsibility on the individual to act, to some extent, as his 'brother's keeper' (through tax-subsidized state-run welfare programmes) to help ensure just distribution so that no one would be left behind. In accordance with the same political ideals, richer developed countries were commonly recognized to owe a similar duty to assist poorer countries; this sense of responsibility may have significantly impacted the changing face of international law over the last 50 years. The state-to-state interaction that dominated at the inception of the UN has given way to a more comprehensive global legal order transcending territorial boundaries for the 'common good'. In fact, the accompanying intrusion of international trade law on state sovereignty is considered justified because of the recognized duty of wealthier nations to assist poorer countries to become developed. John Ruggie characterizes this as 'embedded liberalism' and it helped garner initial political support (as a projection of dominant national and philosophical trends primarily in the United States at the time) for a multinational trading regime tolerant of domestic regulatory diversity out of Bretton Woods.[26] Trade liberalization 'was embedded within a *political* commitment', writes Howse, that was

> broadly shared among the major players in the trading system of that era, to the progressive, interventionist welfare state; in other words, to a *particular*

26 Nicolaidis and Howse, 'This is my EUtopia', op. cit. at 775 explain this as 'a conception of the complementarity between bargained trade liberalization, on the one hand, and the evolution of the domestic welfare and regulatory state, on the other. This outlook tolerated significant national differences in the manner of delivering the social welfare function [and] … did not need to threaten increasingly liberal co-operative equilibria in the trading regime'.

> political and social vision, including at the same time respect for diverse ways of implementing this vision ....[27]

With time, '[a]t the hands of this trade policy elite, 'embedded liberalism' came to be recast as *economics, and economics became ideology, the ideology of free trade*'[28] such that the original political foundation faded. Trade policies shifted from eliminating tariff barriers and border restrictions to minimizing the trade *impact* of non-tariff barriers including domestic regulatory and policy interventions.[29]

Trade and HR objectives are not inherently inconsistent, however. Economic prosperity facilitates, though it does not ensure for distributive reasons, the realization of all human rights (economic, social, civil, and political) and the improvement of aggregate world social welfare. Increasing global welfare should have the results of increasing human rights which are 'income elastic' – meaning that demands for these will increase with the wealth of a society since HR require infrastructure and credible legal systems which in turn are costly to administer.[30] Nevertheless, even in liberal democracies committed to the greatest good for the greatest number, such utility is always subjected and *subjugated* to the rights of the individual. Therefore, when values compete with trade norms, Pauwelyn asserts,

> [w]hen genuinely pursued, that is, when not abused as disguised restrictions on trade, such goals must trump the instrument of trade, even if they are not set out in the WTO treaty itself. This should be particularly so in case

27 Howse, 'From Politics to Technocracy – and Back Again', op. cit. at 97. See Also Anne-Marie Slaughter, 'Regulating the World: Multilateralism, International Law, and the Projection of the New Deal Regulatory State' in Robert Howse (ed.) *The World Trading System: Critical Perspectives on the World Economy* 50 (1988) reprinted in John Gerard Ruggie, *Multilateralism Matters* (New York: Columbia University Press, 1993) at 125.

28 Howse, 'Technocracy', *supra* note 27 at 99.

29 Howse criticizes the result: 'The very success of the embedded liberalism bargain, along with other phenomena, led to forgetfulness or amnesia concerning the political foundation of the postwar trading regime, its character as a specific and contingent bargain about the interaction between freer trade and the welfare state'. *Ibid,* at 98. See also Dani Rodrik, *Has Globalization Gone Too Far?* (Washington, DC: Institute for International Economics, 1997).

30 Alan O. Sykes, 'International Trade and Human Rights: An Economic Perspective' (2003) John M. Olin Law and Economics Working Paper No. 188, The Chicago Working Paper Series, <http://www.law.uchicago.edu/Lawecon/WkngPprs_176-200/188.aos.human-rights.pdf>, published in Thomas Cottier, ed., *International Trade and Human Rights: Foundations and Conceptual Issues* (Ann Arbor, MI: University of Michigan Press, 2006), 29–40.

> these goals have been defined in other, non-WTO rules of international law as between WTO members that have agreed to those rules.[31]

Steve Charnovitz studies the interstices of global governance and advocates that international trade law should not be isolated from the other dimensions and spheres of international law because, as vital as trade is to economic *and human development*, it is not solitary in its contribution to these ends and may justify restriction in circumstances for the achievement of *other social goals*.[32]

Under the UN Millennium Development Declaration (MDD) a list of Millennium Development Goals (MDGs) were articulated to galvanize world nations to cooperate towards their achievement. The MDD, adopted in 2000 at the largest ever gathering by Heads of state, revitalizes the political commitment to improving the human condition worldwide by making a compact to meet concrete *targets* set for advancing development and reducing poverty by 2015.[33] The MDGs also reinforce national commitments to ECSRs by providing further positive state undertakings in relation to wealth distribution and the realization of welfare rights. These are also being pursued by DCs and LDCs in the Doha Development Round of trade negotiations. To deny these goals would be to deny the significant commitment made in the first half of the twentieth century to the institutions for governing international relations and the normative justification for their existence. The commitment pledged by 191 UN member states is to meet eight stated goals which strive to achieve at a global population level the protection of individual rights articulated in the ICESCR: 1) eradicate extreme poverty and hunger; 2) attain universal primary education; 3) promote gender equality and female empowerment; 4) reduce child mortality; 5) improve maternal health; 6) combat HIV/AIDS, malaria and other diseases; 7) ensure environmental sustainability; and 8) develop a global partnership for development.[34]

Patented seeds and GM plants impact the eradication of poverty and hunger, affecting access to food and farmers' rights; genomic, proteomic and life patents extending to diagnostics and the genetic makeup of viruses and bacteria impede population-based improvements to maternal health, child mortality, and eradication of diseases such as malaria and AIDS, and the widespread economical use of (plant derived) vaccines; and, environmental sustainability and cultural diversity are impaired by the privatization of traditional (cultural) knowledge. Strong uniform standards of IPRs will

---

31 Pauwelyn, *Conflict of Norms,* op. cit. at Preface xi.

32 See Charnovitz, *Trade Law and Global Governance*, op. cit.

33 See *Millennium Development Goals: An Overview setting out universal primary education as the second Millennium Development Goal*, <http://hdr.undp.org/reports/global/2003/pdf/hdr03_overview.pdf>.

34 UN Millennium Development Goals, United Nations, <http://www.un.org/millenniumgoals/#>.

adversely affect the realization of several of the MDGs then, most importantly, the *global partnership for development* which has the potential to subsume all the other substantive goals.

A development partnership set on meeting DCs' 'special needs' would necessarily include a commitment to reducing their debt. It might also, with private sector cooperation, make essential medicines and enriched food available and '[d]evelop further an open trading and financial system that is rule-based, predictable and non-discriminatory, includ[ing] a commitment to good governance, development and poverty reduction, nationally and internationally'.[35] Since development and rights are linked, it would mark an equal commitment to human rights. Discretion over domestic regulation is a manifestation of self determination of protections needed to preserve biodiversity, cultural diversity, and cultural knowledge, simultaneously serving both MDGs and ICESCR obligations. Their interdependence cannot be overemphasized:

> The value of cultural diversity and its relevance to conserving biological resources warrant an effort to address the loss of cultural knowledge ... [C]ultural knowledge can only be conserved by keeping it alive and in use ... Turning public goods into private property is now heavily promoted for conservation purposes. Unfortunately, this is also a high-risk method for societies and cultures that have long been subordinated. Privatization of biological resources could result in greater poverty and exploitation without achieving conservation or equity.[36]

The MDGs transplant to the global community the premise of the welfare state and aim to further the attainment of social welfare and distributive justice simultaneously as part of the economic growth-maximization mandate central to efforts for reducing trade barriers. These goals are consistent with the embedded political liberalism of states that helped usher in the UN and multilateral trade system in a post-war era and reaffirm a moral premise. According to David Kelley, '[t]he welfare state ... rests on an idea. The thinkers and activists who built it insisted that the social provision of goods be treated as a right possessed by all people as citizens, rather than as an act of charity or noblesse oblige, a gift from some to others ...'.[37] The MDGs formally extend the duties corresponding to the ESCR beyond individual states in relation to their citizens to the international community and focus cooperation on the purpose of *development*, a UN goal shared by the Doha Round. But, trade is only one means of attaining this end. Alternative and complementary measures, including domestic ones, must be reflected upon as

35 Ibid.

36 See generally Coombe, 'IP, HR, and Sovereignty', op. cit. at 94.

37 Kelley, *A Life of One's Own*, op. cit. at 257.

> [t]here is no technical reason, for example, why hunger and poverty should not be eradicated. Nor is there any technical reason why remarkable advances in medical science should not greatly improve the health and well-being of people around the world. Economic and political obstacles to realizing these goals are legion … This is a time for questioning – questioning the assumptions underlying current models of development, the concepts used to construct these models, and the methodologies that have been chosen to generate basic social and economic information. It is a time for constructing alternative scenarios.[38]

Economist Jeffery Sachs, Director of the Millennium Project and special advisor to former UN Chief Secretary General Kofi Annan, argues that the goal to halve poverty by 2015 can be met by increasing annual foreign aid to 0.7 per cent of Gross National Product (GNP) starting with an extra $70 billion per year as of 2006.[39] His comprehensive 'clinical economics' approach to solving wealth distribution addresses the web of inter-related problems that perpetuate poverty and hunger. Sachs strongly advocates increasing foreign aid while explaining the inefficiencies surrounding the reality of how the money is spent:

> From the world as a whole, the amount of aid per African per year is really very small, just $30 per sub-Saharan African in 2002. Of that modest amount, almost $5 was actually for consultants from the donor countries, more than $3 was for emergency aid, about $4 went for servicing Africa's debts and 4% was for debt-relief operations. The rest, about $12 went to Africa. Since the 'money down the drain' [that giving makes no difference] argument is heard most frequently in the US, it's worth looking at the same calculations for US aid alone. In 2002, the US gave $3 per sub-Saharan African. Taking out the parts for US consultants and technical cooperation, food and other emergency aid, administrative costs and debt relief, the aid per African came to the grand total of perhaps 6¢.[40]

Sachs' proposal is a practical operationalization of the normative Rawlsian claim to a duty to assist and is confirmed in global agreements like the MDGs, the Monterrey Consensus of 2002, and the renewed commitment to both at the 2005 World Summit,[41] where the United States promised to

38 See the UN Research Institute for Social Development (UNRISD), <http://www.unrisd.org>.

39 Helleiner, *supra* note 25.

40 Jeffrey D. Sachs, *The End of Poverty* (New York: Penguin Press, 2005) and excerpt of his book in *Time*, 14 March 2005, 32 at 44 [citation is to the *Time* article].

41 Official website, <http://www.un.org/summit2005>. For the US agreement to the Monterrey Consensus, see U.S. Department of State official website, <http://www.

give a larger proportion of its annual output (up to 0.7 per cent of GNP) to official development assistance. But relying on aid alone is problematic since signatories have failed to deliver on aid promises and 'failure to follow through has no political fallout domestically, of course, because not one in a million U.S. citizens even knows of statements like the Monterrey Consensus'.[42] Global political leadership is necessary to allay distributional disparities but it cannot be sought from the 'world's one super-power'.[43]

One important criticism of the 'duty to assist' is that it manifests a paternal ideology reinforcing hierarchy and privilege in the world order by perpetuating relationships of dependency by poorer countries rather than fostering *homegrown development*. It displaces and renders peripheral the critically central role of state agency and governmental responsibility for its peoples. In so far as international law restricts diverse regulatory responses, it helps rationalize and entrench a disempowering process. A moral and often legal debt, in addition to the economic one, is created where perhaps on a social-psychological level, the message sent is that 'if you want our help, you must play by our rules', especially when funds are tied to specific (IMF or WB) conditionality. Additional monetary funds, debt relief, or other conditional initiatives to provide fiscal assistance through international financial institutions,[44] are necessary but insufficient for global change.[45] And, some see

---

state.gov/e/eb/rls/fs/53133.htm>. See also UN General Assembly meeting on Financing for Development (Sept. 14) at the 2005 World Summit, at <http://daccessdds.un.org/doc/UNDOC/LTD/N05/511/30/PDF/N0551130.pdf?OpenElement>.

42 Sachs, *supra* note 40 at 44.

43 Helleiner, *supra* note 25 at 8 contends '[t]he world's one super-power … seems to have little interest in multilateral organizations unless they can be used as instruments of its own fairly short-term interests … [O]ne cannot expect leadership towards the kinds of global economic processes and institutions that the world increasingly requires from this source – a country that, despite its wealth, has the weakest aid performance record in the OECD (.1 of GNP of which 30% goes to the Middle East); remains in serious arrears in its financial obligations to the United Nations; is so jealous of its sovereignty that it fails to ratify even some of the most obvious of international conventions relating to the world's most vulnerable (including those relating to the rights of the child; economic, cultural and civil rights; discrimination against women; forced labour; freedom for collective bargaining; and land mines) (UNDP, 2000); and continues to oppose effective international conventions on … the preservation of global diversity'.

44 See e.g. IMF conditionality of resource commitments in Ross Leckow, 'Bringing the Disenfranchised to the Table – Lessons of Conditionality' in *The Measure of International Law: Effectiveness, Fairness and Validity* (New York: Kluwer Law International, 2004) 1–11.

45 William Easterly, *The White Man's Burden: Why the West's Efforts to Aid the Rest Have Done So Much Ill and So Little Good* (New York: The Penguin Press, 2006) argues that foreign aid is actually harmful. For a critical review, see Amartya Sen, 'The

this as yet another means of legitimizing cultural imperialism,[46] by making developing countries obliged beneficiaries for the conferred benevolence that monetary aid, debt forgiveness, and technology transfer provide, whilst external forces are credited with 'civilizing' them. TRIPS demonstrates why the relationship between the national and international is never apolitical.[47] Eric Cheyfitz explains:

> we must recognize a progressive strain in this imperial foreign policy, one that apparently welcomes homogenization, but – and here homogeneity harmonizes itself with hierarchy – only in the terms of the policy maker, who writes a script in which the *other*, in order to be heard, must say: 'for your sake I have become a civilized man … for your sake I will be whatever you will me to be.[48]

The ambition towards homogeneity and harmony with New World (American) standards and the ensuing and enduring international struggle for voice in this process is well rooted in the imperialist foreign policy traditions of both trade and human rights.[49] Concurrent economic and social gains are garnered for the 'donors' in ensuring peace and stability,[50] in creating new markets for the export and consumption of their goods, and fostering healthy new subjects from whom rent seeking is possible at first-world prices. The debt is not one which the developed countries had to endure as part of their maturation where, without TRIPS like restrictions on their regulatory

---

Man Without a Plan', *Foreign Affairs*, March/April 2006, <http:www.foreignaffairs.org/20060301fareviewessay85214/amartya-sen/the-man-without-a-plan.html>.

46 See Charles Gore, 'The Least Developed Countries 2000 Report: Aid, Private Capital Flows and External Debt: The Challenge of Financing Developments in the LDCs' (Geneva: UNCTAD, 2000).

47 For further discussion of the relationship between TRIPS and human rights, see Sol Picciotto, 'The WTO As a Node of Global Governance: Economic Regulation and Human Rights Discourses' Working Paper, Conference on Human Rights and Global Justice, University of Warwick, 29–31March 2006, <http://www2.warwick.ac.uk/fac/soc/law/events/past/2006/rights&justice/participants/papers/picciotto.doc>.

48 Eric Cheyfitz, *The Poetics of Imperialism: Translation and Colonization from the Tempest to Tarzan* (Philadelphia: University of Pennsylvania Press, 1997) at 4–5.

49 International law can be traced to the original formation and ascendancy of the nation state in the sixteenth and seventeenth centuries, with Europe as its birthplace; it 'has been infused with European social and political cultures and values'. Kindred et al., *International Law Chiefly as Interpreted and Applied in Canada*, op. cit. at 1.

50 Writes Buckley, '[i]t is difficult to deny that increasing inequality fuels the bitterness and sense of alienation that underpin much terrorist activity'. Buckley, *Changing Face*, op. cit. at 5. See also D. Bell and M. Renner, 'The War on Terrorism Needs a Marshall Plan', <http://www.net-about-town.com.au/imaginepeace/articles/marshallplan.htm>.

sovereignty, access to knowledge as a public good was determined by *domestic policies* that fostered development. The Uruguay Round's changes deprive the DCs and LDCs of the full benefit of regulatory discretion and developmental opportunities previously enjoyed by states; what remains should be used. Moreover, socially, we should minimize creating the antecedent conditions necessary for assistance by giving with one hand and taking with another through internationally mandated IPRs protection. Must 'trade liberalization', which in the case of IPRs reinforces private monopolies instead of free markets, be purchased by domestic regulatory policies? The matter of balancing patent and health policy is in need of cautious reform to better accord with the public's interest integral to a 'bargain theory' of IPRs and a more efficient use of domestic resources. But the great inefficiencies in the allocation of *global* resources and the awesome (and woefully awful) distributive violence to international humanitarian relief efforts for the benefit of the private at the cost of the 'public' also needs to be addressed because the ostensible system of monetary aid to DCs and LDCs for health, food, and medicinal relief is fiscally inadequate and ethically bankrupt; it provides an additional subsidy for developed countries' innovation-based patented exports by the poorer 'third' world.[51] Inefficiencies in allocation are compounded by directing foreign debt and aid to paying for patented inventions, or enforcement infrastructure, or other such uses of funds. Even if DCs and LDCs were able to shift domestic industries towards innovation, their innovators could hardly afford the costs of protecting their inventions in developed countries:

> [T]he inventors in poor countries find it tough to use patent systems in the rich world … [S]ecuring a patent from America's patent office costs at least $4,000. Defending it in court can cost millions. The commission identifies several ways in which countries could open their domestic IPR system, including discounted fees and subsidized technical assistance.[52]

51 Alan V. Deardorff, 'Should Patent Protection Be Extended to All Developing Countries?' 4 *The World Trading System: Critical Perspectives on the World Economy* 37, argues that global aggregate welfare would be maximized if certain countries were completely exempted from requirements for IPP because the poorer countries' marginal increased rents to the patent holder are unlikely to be significant enough as incentives for further innovation but the losses of being forced out of their comparative advantage in imitation are greater, shift in productive resources to areas of less advantage, resulting in a reduction of global allocative efficiency. See also Keith Maskus, 'Normative Concerns in the International Protection of Intellectual Property Rights' (1990) 13 *World Economy* 387.

52 See 'Patently Problematic' *The Economist* (14 September 2002), <http://www.economist.com/science/displayStory.cfm?story_id=1325219>; Report of the Commission on Intellectual Property Rights: Integrating Intellectual Property Rights and Development Policy, <http://www.iprcommission.org/graphic/documents/final_report.htm>.

It is apparent then that '[i]t is justice, not charity that is wanting in the world'.[53]

To help ensure the WTO remains relevant in serving its purpose and to promote development through justice, international trade rules must be interpreted and implemented in a manner that allows member states to meet their HR obligations and in so doing will retain a politic of ethics. Trade's instrumental operation must necessarily adapt by becoming more inclusive of DCs' and LDCs' interests and offering them technical aid where needed – without compromising the integrity of the *telos* to which their signatories subscribed. The WTO Advisory Centre located in Geneva was created in 2001 to respond to the need of developing countries who felt that the formalism of the post-Uruguay system was a huge limitation to them in their negotiations, compliance levels and use of the Dispute Settlement Mechanism. It is funded by various developed countries who contribute to a fund. An advisory body serves as a resource for national governments in order to ensure DCs and LDCs remain in compliance. The Centre also provides legal expertise, opinions, training, and counsel to ensure that trading partners are also compliant. DCs and LDCs are provided with technical legal support on WTO law and dispute settlement as a response to their concerns over being marginalized by the multilateral trading system.[54]

Locating the Social Dimensions of the GATT/WTO

A member state is the only juridical party with standing to bring a complaint before the WTO but the traders' purported interests protected by members under the multilateral trade regime are supposed to be aligned with national interests and those of individual citizens for whom the state is authorized to act. These are not merely consumers characterized by 'market citizenship' in this age of 'market liberalism'[55] but human beings – or in neoliberal terms,

53 Mary Wollstonecraft, *A Vindication of the Rights of Woman*, <http://womenshistory.about.com/library/etext/bl_vindication000.htm>. Wollstonecraft adds, 'I have sighed when obliged to confess that either Nature has made a great difference between man and man, or that the civilization which has hitherto taken place in the world has been very partial'.

54 See <http://www.ictsd.org/html/weekly/24-07-01/wtoinbrief.htm>. The Uruguay Round established a WTO Committee on Trade and Development to examine how global trade rules affect developing countries whose membership constitutes a majority.

55 Defined by its embrace of 'a neoliberal political philosophy of limited government and strong property rights, coupled with what one might call a cultural exaltation of the market, both as a primary locus for individual growth and expression and as a dominant template for policy design and implementation in those few instances where public action is seen as necessary'. Douglas A. Kysar 'Sustainable Development and Private Global Governance'. (2005) 83 Tex. L. Rev. 1 at 8, <http://

human 'resources' – whose existence and therefore potential for productivity and exploitation is shaped by a more complex (counter) narrative than that provided by any single neoliberal (trade) institution under any singular mandate. For textual analyses, preambular language captures the spirit and intentions of any given agreement and provides interpretative insight. The *Vienna Convention on the Law of Treaties* (VCLT) provides under Article 31.2 that the preamble and annexes as part of a treaty text be included as part of the *context for the purpose of the interpretation of a treaty.*[56] Preambular language is significant for 'setting the foundational constitutional principles of an entire legal text, whether a national constitution, an international treaty, or the charter of an international organization'.[57] GATT 1947's preamble provides that Contracting Parties,

> Recognizing that their relations in the field of trade and economic endeavour should be conducted with a *view to raising standards of living, ensuring full employment and a large and steadily growing volume of real income* and effective demand, developing the *full use of resources* of the world and expanding the production and exchange of goods, Being desirous to contributing to these objectives by entering into *reciprocal and mutually advantageous arrangements* directed to the substantial reduction of tariffs and other barriers to trade and to the elimination of discriminatory treatment in international commerce, Have through their representatives agreed ...[58]

It is clear from the outset that the signatories in 1947 had made a commitment to trade liberalization as a *means* for attaining the goal of enhancing human welfare and development by facilitating what today are considered *social and economic rights* and by restricting monopolies ('full use of resources of the world'). That trade and human rights are linked *in situ* is less an argument about the proper jurisdiction of the multilateral trade system – the dimension commonly focused on in modern linkage debates – and more a testament to their indivisibility for the attainment of certain ultimate goals for the common good of humanity.[59] However, reference to trade for the purpose of *raising*

---

lsr.nellco.org/cgi/viewcontent.cgi?article=1030&context=cornell/lsrp>. See generally Kysar, 'Preferences for Processes: The Process/Product Distinction and the Regulation of Consumer Choice' (2004) 118 *Harv. L. Rev.* 525 discussing 'the effort by regulatory cost-benefit analysts to ground public policies on the values revealed by individuals acting in their roles as market actors'.

56 [VCLT]. Full text at <http://untreaty.un.org/ilc/texts/instruments/english/conventions/1_1_1969.pdf>.

57 Mendes and Mehmet, *Global Governance, Economy and Law*, op. cit. at 74, with reference to the SCC's use of principles found in the preamble to the Constitution Act, 1867 when ruling on important constitutional issues.

58 GATT 1947, op. cit.

59 Ibid. at 75. See generally Sen, *Development as Freedom*, *supra* note 17.

standards of living does more by filling the interstices of trade liberalization with a strong commitment to its social dimensions and thereby puts into perspective the remainder of the text as merely *facilitative*.[60] More expressly, the preamble of the Marrakesh Agreement establishing the WTO states that the parties to the agreement recognize that

> [t]heir relations in the field of trade and economic endeavour should be conducted with a view to *raising standards of living*, ensuring full employment and a large and steadily growing volume of real income and effective demand, and expanding the production and trade in goods and services, while allowing for the optimal use of the world's resources in accordance with the objective of sustainable development ... in a manner consistent with their respective needs and concerns at different levels of economic development ... [;] that there is a need for positive efforts designed to ensure that developing countries, and especially the least developed among them, secure a share in the growth in international trade commensurate with the needs of their economic development.[61]

An addition to GATT made in 1964–1965 also dealt with trade and development and again emphasized the social dimensions of trade by articulating principles and objectives consistent with the preambular ethos of both the WTO Agreement and its predecessor.[62] Article XXXVI, for example, reiterates in clause (a) that the 'basic objective of this Agreement include the *raising of standards of living* and the progressive development of the economies of all contracting parties ...', that (d) 'individual and *joint action is essential* ...' and that (e) ' ... international trade [is] a *means* of achieving economic and social advancement'. The Article reinforces the need to increase market access for the products of DCs as a priority through a commitment to cooperate and assist the less endowed states.[63]

60 See also GATT 1947, Article XVIII.2: 'The contracting parties recognize further that it may be necessary for those contracting parties, in order to implement programmes and policies of economic development designed to raise the general standard of living of their people, to take protective or other measures affecting imports, and that such measures are justified in so far as they facilitate the attainment of the[se] objectives ....'

61 Agreement on Trade-Related Aspects of Intellectual Property Rights, WTO, op. cit.

62 Part IV of the GATT 1947 (Articles XXXVIXXXVII), *supra* note 58.

63 Emphasis added. In fact, Article GATT 1947, Article XXXVI.8 provides: '[t]he developed contracting parties do not expect reciprocity for commitments made by them in trade negotiations to reduce or remove tariffs and other barriers to the trade of less-developed parties.'

The Uruguay Round added 'special and differential treatment extended to developing country members under a number of the WTO agreements'.[64] The provisions recognize that the primary goal of trade liberalization – *to raise standards of living* – according to trade theory and as codified in the preamble, requires some general exceptions from the principles of non-discrimination and reduction of trade barriers on which the international trading regime is based.[65] Most exceptions to members' general obligations provided under GATT 1947 are justifiable on the basis of conformity to the ground rule of *non-discrimination.* The most significant are the exceptions provided for under Article XX[66] which support the conclusion that, in 1947, the goal was that a sovereign realm of discretion free from trade obligations that would be expressly preserved (as a *complementary* means to free trade) to promote welfare and development objectives so long as these domestic measures were not 'arbitrary or unjustifiable discrimination between countries where the same conditions prevail, or a disguised restriction on international trade'. The GATT vision was a good vision of a multilateral regime founded on ideals of mutuality, reciprocity, and non-discrimination; it was expressly committed to its social, HR-promoting, dimensions.

64 See Thomas and Meyer, *The New Rules of Global Trade*, op. cit.at 69 and the 'Decision on Measures in Favour of Least Developed Countries' and the 'Decision on Measures Concerning the Possible Effects of the Reform Programme on Least-Developed and Net Food-Importing Developing Countries' referred to therein.

65 The cornerstone of trade liberalization is the principle of non-discrimination as referred to in the preamble to GATT 1947 and manifested in the doctrines of the Most Favoured Nation Treatment (Article 1 of the GATT) and National Treatment (Article III). Permissible regional trading blocks (Article XXIV) and grandfathering clauses provide certain exceptions to this principle but on the whole, the provision is effective in extending the benefit of negotiated concessions between two members to all other member states usually without the ability to extract further concessions with those members as a precondition to this extension. While there is to be no playing favourites amongst foreign exporters, National Treatment goes further to safeguard against discrimination between domestic and foreign producers.

66 These allow members to take measures 'necessary to protect public morals' (clause a), 'necessary to protect human, animal or plant life or health' (clause b), and measures 'necessary to secure compliance with laws or regulations which are not inconsistent with the provisions of the Agreement, including those relating to … the protection of patents, trademarks and copyrights, and the *prevention of deceptive practices*' (clause d), which could be used to curtail the growing number of questionable patents applied for and the resulting fraud on the public and crown. Article XX has been the subject of a number of contentious disputes.

Sovereignty under TRIPS and over Genetic Resources

TRIPS retains a significant degree of regulatory scope intended by the GATT for governments to meet their public policy objectives in a self-determined way. Members are allowed to implement their TRIPS obligations as they see fit (Article 1) and Article 7 provides that the objectives of protection and enforcement require *mutual advantage* to producers and users 'in a manner conducive to social and economic welfare, and to a balance of rights and obligations'. It will be recalled that Article 8's overarching and guiding Principles respect sovereignty and prioritize it for certain sectors as members may

> adopt measures necessary to protect public health and nutrition, and to promote the public interest in sectors of vital importance to their socio-economic and technological development, provided that such measures are consistent with the *provisions* of this Agreement.

Consistency with the *provisions* in the text is substantially less onerous than broader language used in an earlier draft of Article 8 (the 1990 Brussels draft) to require no *derogation* from TRIPS *obligations*. The latter would have rendered members substantially more vulnerable to non-violation complaints (NVCs)[67] based on domestic regulatory measures taken to protect public health. NVCs, evidenced by their very name, are based on allegations of nullification or impairment and tend to focus on domestic measures – such as a new product subsidy on which a tariff concession has been made – or other actions that are measured against WTO *obligations* rather than *provisions*:

> [S]ome delegations expressed concern about the possible use of the provisions of Article 8 to justify measures which, while not inconsistent with obligations under the Agreement, might have the effect of impairing benefits that other Members could legitimately expect under the Agreement; in other words, that the provision could be used as a *defence* in a non-violation case on the grounds that the taking of the measures envisaged by Article 8 could only have been reasonably expected at the time that the TRIPS negotiations were concluded.[68]

---

67 A member's complaint would allege that the negotiated balance of concessions between members is upset by the application of a measure and would seek monetary compensation as the remedy, regardless of whether the measure is inconsistent with provisions of an agreement. Their inclusion in the GATT 1947 Article XXIII: 1(b) ('nullification and impairment') was modelled on earlier US bilateral trade agreements.

68 Emphasis added. TRIPS Council, Note by the Secretariat, *Non-Violation Complaints and the TRIPS Agreement* 28 January 1999, IP/C/W/124, paras 13–14,

NVCs incorporated into TRIPS by Article 64.1 and the DSU would have intruded further than the TRIPS-substantive obligations by holding different, *technically compliant,* government measures accountable for their (restraining) trade impact if the effect was to deprive members of *expected* trade benefits. Article 64.2 of TRIPS, however, provides temporary relief by banning such disputes for the initial period of five years (1995–1999) which was extended at Doha because of its potential implications.[69] Currently, most members have stated their opposition to the inclusion of NVCs under TRIPS, favouring instead a permanent ban or the continued moratorium the TRIPS Council has recommended pending Ministerial consensus.[70] The potential for such complaints will therefore not be dealt with further here but also because the language of Article 8 makes the consideration of NVCs in the context of ECSRs intersecting with the patenting of life superfluous. Article 8 gives express deference to domestic regulatory diversity directed at certain listed public policy objectives 'provided that such measures are consistent with the *provisions* of the Agreement'. In short, Article 8 appears, at a minimum, to restrict the *kind* of TRIPS-related non-violation complaints that could be brought before the DSU even if consensus is reached on the *status* of non-violation complaints generally under TRIPS. Put differently, even if the moratorium on such complaints were to be lifted, a member may not complain that a domestic 'measure necessary to protect public health and nutrition' or 'to promote the public interest in sectors of vital importance to their socio-

---

<http://www.dfat.gov.au/ip/secretariat_paper.doc>. For country specific submissions, see <http://www.IPRsonline.org/submissions/nonviolation.htm>.

69 See WTO, 'Non-Violation Complaints: Background and the Current Situation', <http://www.wto.org>. In the interim, the TRIPS Council charged under Article 64.3 with examining the 'scope and modalities' of IP NVCs and their relationship with the DSU was further instructed pursuant to para 11.1 of the Doha Declaration on Implementation Related Issues and Concerns to report on its investigation and make its recommendations until which members agreed not to make NVCs. At the Cancun Ministerial Conference in 2003, the Council made four recommendations: (1) ban non-violation complaints in TRIPS completely; (2) allow the complaints to be handled under the WTO's dispute settlement rules; (3) allow non-violation complaints but subject to special 'modalities' (i.e. ways of dealing with them); and (4) extend the moratorium. Failure to reach a consensus on the Draft proposed by the TRIPS Council meant failure to reach any acceptable change according to Article 64.3. While most members favoured the complete ban or the extended moratorium, the US and Switzerland strongly voiced support for TRIPS related non-violation complaints in order to minimize members from 'creative legislative activity' that would alter the 'negotiated balance of benefits' by allowing trading partners to effectively circumvent their TRIPS commitments.

70 Further discussion took place during preparation of the July Package which expressly extended the moratorium until the Sixth Ministerial Conference December 2005 but no substantive agreement was reached.

economic and technological development' can lead to nullification and impairment of an expected benefit *unless* it is inconsistent with the '*provisions of this Agreement*'. That is, a non-violation complaint (based on derogation of *obligations*) will *never* be appropriate in relation to measures captured under Article 8. One necessarily needs a violation complaint if attacking domestic measures embraced by the language of Article 8 because it has internal to it a public-interest defence, in relation to NVCs, due to the inclusion of the word 'provisions'. In the next chapter I argue that this defence logically extends to violation complaints, but for now it is important to appreciate the significance of Article 8 for state agency. As already established, social and economic rights require state action for their realization, in part due to an institutional deficit and instrumental gap in the UN system which makes these *general* laws *only justiciable if domestically implemented* through *particular* laws of the state; the latter necessarily must be culturally nuanced within specific ideological and political frameworks.[71]

According to Article 1.1 of the ICCPR and ICESCR, 'All peoples have the right of self-determination. By virtue of that right they freely determine their political status and freely pursue their *economic, social and cultural development*'. In addition, Art. 1.2 of both agreements adds that '[i]n *no case* may a people be deprived of its own means of subsistence'.[72] *All* human rights then are fundamental to a life of human *dignity*, inalienable and interrelated and dependent on the state for their realization, 'with the net effect that the breach of one will affect the realization of another'.[73] The preamble of the UNDHR declares 'recognition of the *inherent dignity* … of all members of the human family' as 'the foundation of freedom, justice and peace in the world …'. The preamble preserves the primacy of individual human rights and complements the objectives of economic international organizations. The

71 Aristotle, in his seminal work, *Rhetoric*, contends: 'Justice and injustice have been defined in reference to laws … in two ways. Now there are two kinds of laws, particular and general. By *particular* laws I mean those established by each people in reference to themselves, which again are divided into written and unwritten; by *general* laws I mean those based upon nature.' Discussed in Hastrup, 'Accommodating Diversity', op. cit. at 3.

72 See also Article 2.7 of the United Nations Charter (UNC), op. cit., which confirms the sovereignty of Nations: 'Nothing contained in the present Charter shall authorize the United Nations to intervene in matters which are essentially within the domestic jurisdiction of any state or shall require the Members to submit such matters to settlement under the present Charter …'

73 Rhona K.M. Smith and Christien van den Anker, eds, *The Essentials of Human Rights* (London: Hodder Arnold, 2006) at 37; Freeman and Van ert, *International Human Rights Law*, op. cit. at 37; See also Amyrta Sen, *Poverty and Famines: An Essay on Entitlement and Deprivation* (Oxford: Clarendon Press, 1981) observing that no substantial famine in the last 50 years has occurred in a state with a democratic form of government and free press.

preambles of the ICCPR and ICESCR similarly decree that human rights 'derive from *inherent dignity*, a divinity of the human person'.[74] Protecting the *dignity* of the person is the essence of the living culture of our birth rights and is the common denominator for humanity across all cultures. Strong, internationally mandated IPRs, if extended to life, may be contrary to these provisions, to the HR reviewed in the last chapter, and to numerous other sovereign rights including the free and full use of natural wealth and resources (Article 47, ICCPR and Article 25, ICESCR). Also threatened might be the sustainable use and equitable benefit sharing of and access to genetic resources under the Convention on Biological Diversity (CBD). The CBD, discussed in greater detail in chapter seven, recognizes genetic resources as sovereign property and is integral to the state's role in mediating competing claims to the benefits of these resources.

The Declaration on the Human Genome and Human Rights (HGHRD, Article 1) confirms that the 'human genome underlies the fundamental unity of all members of the human family, as well as the recognition of their inherent dignity and diversity'. The HGHRD recognizes that the *human* genome at least is part of our common heritage.

In 1997, the Convention for the Protection of Human Rights and Dignity of the Human Being with regard to the Application of Biology and Medicine[75] provided (Article 1) that Parties to the Convention '*shall protect the dignity and identity of all human beings* and guarantee everyone, without discrimination, respect for their integrity *and other rights* and fundamental freedoms with regard to the application of biology and medicine'. Article 2 confirms the primacy of human beings: '*The interests and welfare of the human being shall prevail over the sole interest of society or science*' while Article 3 obliges Parties, taking account of health needs and available resources, to take measures with a view to providing *within their jurisdiction* equitable access to health care of appropriate quality. Article 15, articulates the general rule providing that '[s]cientific research in the field of biology and medicine shall be carried out *freely*' subject to any legal provisions *ensuring protection* of the human being. The 1998 Additional Protocol to the Convention for the Protection of Human Rights and Dignity of the Human Being with regard to the Application of Biology and Medicine on the Prohibition of Cloning Human Beings, further demonstrates the expansive and dynamic content of universal HR by extending these individual protections to prohibit the offensive threat of cloning people that new stem cell and cloning technology will soon enable. Laws imposing such a ban would be permissible to prevent commercialization even if clones or human chimeras *were* patentable.

---

74 Yet, the scholarly literature is ripe with contested foundational theories. See J. Donnelly, *Universal Human Rights in Theory and Practice*, 2nd edn (Ithaca, NY: Cornell University Press, 2003) 13–21.

75 See full text, <http://conventions.coe.int/treaty/en/treaties/html/164.htm>.

In October 2003, UNESCO adopted the International Declaration on Human Genetic Data as a standard-setting instrument designed to guide states in developing domestic legislation in recognition of our 'common genetic heritage' and in support of the use of human genetic data in a manner respectful of human dignity and HR including the freedom of research (Article 1) and benefit sharing (Article 19). These instruments reflect the emergence of a growing ethical standard that transcends the political, cultural and economic nature of patent law and obligates states to recognize the link between new genetic technologies and human *dignity*.[76] A 'development' approach to IPRs in their utilitarian and instrumental tradition would help governments foster greater appreciation of the potential costs of over-protection: strong IPRs may in fact slow the *development* that trade law tries to promote, by restricting the free pursuit of economic, social and cultural development that might otherwise occur through the comparative advantage in imitation enjoyed by poorer states, through liberal access to knowledge and through unencumbered technology transfer.[77] TRIPS monopolies consolidate new protectionism:

> The new highest expression of protectionism is, in the view of developed countries, a necessary condition to promote innovation and to stimulate technology and capital flows to developing countries. Their assumption is that people from developed and developing countries will benefit alike from this kind of intellectual property rights.[78]

Using foreign IPRs law and policy under political and legal pressure to spur domestic innovation makes for poor local governance; it belies the right of a people to determine their own fate without compulsion based on their own governance needs, socio-economic requirements, and cultural contexts rather than the monetary interests of the 'first world' and their multinational and transnational corporations. The UNDP reports:

> [t]he expected gains from TRIPS are unlikely without a range of complementary policies such as investments in tertiary education and research capabilities, reward mechanisms in research sectors and an appropriate investor climate – all of which depend on different government policies.[79]

76 See Brad Sherman and Lionel Bentley, 'The Question of Patenting Life' in Bentley and Maniatis, *supra* note 2, 111–25 discussing ethics and patent law with particular attention to the European experience.

77 See Chon, '"Postmodern Progress"', op. cit.

78 B.K. Keayla, 'Agreement on Patent Laws', op. cit. at 259.

79 UNDP, *Making Trade Work*, *supra* note 24 at 208.

The OECD also suggests a need to locate IPRs within a context of existing infrastructure:

> ... government policies, particularly those relating to science and technology, industry and education, will need a new emphasis. The central role of the firm, the importance of national innovation systems, and the requirements for infrastructures and incentives that encourage investments in research and training have to be acknowledged.[80]

Public education, free and open access to information and sharing in knowledge, and public health delivery are all public goods that create assets for the poor and are an essential obligation of the state to balance against its industrial policy in its right to self-determination. The World Bank underscores the importance of state agency to this end:

> Why does the state have a role in expanding poor people's assets? For two main reasons. First, markets do not work well for poor people, because of their physical isolation and because of market failures in financial, health, and insurance sectors, for example. Second, public policy can *reduce initial inequalities and increase the opportunities for poor people to benefit from growth.* Equity and efficiency considerations can be largely independent, but they generally overlap ... Among the most effective antipoverty policies are those that achieve more equity through redistribution and simultaneously enhance the efficiency of markets used by poor people.[81]

These *assets* help tackle asset inequalities only if state regulatory power is appropriately used to redistribute rather than to facilitate their private control. Any reservation or restraint on IPRs is often critically characterized as 'government interference' in the normal operation of the free market, which ignores the fact that the regulatory recognition of a property right in the first place is a 'government interference' that creates a subsidy for certain parties in the knowledge economy at the cost of others. Moreover, '[a]n active national

80 OECD 1996 Knowledge Based Economy Report, op. cit. at 237.

81 The Report adds, '[s]ome health services, such as mosquito and other pest control and health education on basic hygiene and nutrition are pure public goods ... Governments are responsible for infectious disease control on efficiency grounds. But such policies have important equity benefits as well. While the poor suffer more from almost all diseases than the non-poor do, the difference is greatest for infectious diseases. In India, the poorest tenth of the population is seven times as likely to suffer from tuberculosis as the richest tenth': World Bank Development Report, 'Expanding Poor People's Assets and Tackling Inequalities' (2000–2001), <http://www1.worldbank.org/prem/poverty/wdrpoverty/report/ch5.pdf>, c. 5 at 79 and Box 5.3 outlining 'win-win' policies in public health sector.

competence in basic research is … a necessary condition for benefiting from research undertaken elsewhere in the world; indeed it can be viewed as a national scientific intelligence system'.[82] That means domestic skilled labour is critical for determining the social value of basic research and its commercial potential. DCs and LDCs may not have the same kinds of public infrastructure necessary to become better innovators, offering strong IPR protection for foreigners without the potential for similar gains for themselves. Nuanced regulatory approaches are needed, contends Peter Drahos, because:

> [m]arkets are dependent upon a range of primary public goods that come in the form of rules and institutions (the rule of law, contract, property, banking, corporations, securities and stock exchanges). A flourishing capitalism equipped with such institutions allows entrepreneurs the freedom to act, to create the spontaneous ordering that is said to characterize markets. The regulation of public goods that serve as inputs into the exercise of skill by individuals also has important effects on these processes of spontaneous ordering. For example, over time societies have evolved different regulatory models for the intellectual commons (an information commons). An intellectual commons can be negative (open to individual appropriation) or positive (in the co-ownership or co-use of all and not open to appropriation) and inclusive (open to all) or exclusive (open to a select group).[83]

Countries where educational infrastructure *is* strong but policies supporting application development are not may become source countries for foreign-held patents, particularly given how important corruption-free governments and stable markets are to creativity and innovation. Then there is the issue of knowledge leakage. *Leakage* depends, 'as a first approximation, on the proportion of a country's corporate technological activity that is controlled by foreign firms, which reflects their capacity to monitor and absorb local basic research skills and knowledge'. By the same token, Keith Pavitt contends, 'the importance of the foreign technological activities of nationally owned firms will reflect a country's capacity to benefit from basic research undertaken in other countries'.[84] TRIPS would facilitate a western firm's absorption of leakage from countries with competitive and strong university systems, like

82 Keith Pavitt, 'What makes research economically useful?' (1991) 20:2 Research Policy 109 at 115; see also P. Patel and K Pavitt, 'Do Large Firms Control the World's Technology?' (mimeo) in *Science Policy Research Unit* (Brighton: University of Sussex, 1989), cited therein at 115–16.

83 See Peter Drahos, 'The Regulation of Public Goods' (2004) 7.2 J. of Int'l Econ. L. 321–39 [Drahos, 'Public Goods'] at 333 also published in Maskus and Reichman, *IPG*, op. cit. See also Drahos, *A Philosophy*, op. cit. at 57–60.

84 Pavitt, *supra* note 82.

China and India, and facilitate the means of foreign commercialization.[85] Considering their infrastructure and labour markets, DCs that adhere to stronger protection standards of Northern and developed countries may be providing another form of subsidy through leakage to an already highly subsidized (through trade related IPRs localized in knowledge goods) foreign export market. The effect is to domestically absorb the costs for external gains by large MNCs. IP scholar David Vaver explains how IPRs are domestically subsidized:

> True, the state may pay no money from general revenue, but it sheds this responsibility by dictating that one person or one class of people should pay another person or another class a fee – that is, subsidize them – ostensibly for the benefit of the community as a whole. It is a subsidy with a difference. Questions of who can benefit directly and who must pay directly or indirectly are constrained by the classification of intellectual property.[86]

Globally protected IPRs, through extrapolation, become globally subsidized by the people least likely to generate them (by western standards) or afford them (by any standards). While Professor Vaver's insights are well articulated, the characterization is too generous because in the developed world the (welfare) state *does* pay money from general revenue by supporting public education, subsidizing universities, public health, health and safety standards and other generally supportive and necessary infrastructure whereby a healthy and functioning labour force can 'create'.[87] Since this general revenue is our tax dollar and modern IPRs are so often conditioned on labour and investment, in some sense we are *all* 'contributing authors', reinforcing the argument that if IPRs are a HR, they are a right of participation, creation, use and enjoyment and not one of exclusion. From this perspective, we may reimagine the TRIPS

---

85 See e.g. Stone, 'An Islamic Science Revolution?', op. cit. Iran has burgeoning R&D in scientific sectors including biotech and stem cell research despite the lack of strong IPRS and their strict enforcement; it has not proven to be a significant impediment. University educated human resources is one of Iran's major exports according to a national Iranian newspaper, the *Iran Daily* which reports that some 180,000 who hold university degrees permanently leave Iran each year: 'The sad truth is that Iran ranks first in the world in terms of the phenomenon of brain drain' *Iran Daily*, 17 September 2005 at 2.

86 Vaver, *IPL*, op. cit.

87 'The returns from intellectual property systems, for instance, will be affected by other public goods, such as education. Strong [IP] regimes that deliver the public good of knowledge through the market have their efficiency improved if they are complemented by labour markets that allow skilled workers to move from firm to firm (restrictions on the use of codified information are counterbalanced by regulation that allows for the movement of skill-embodied information)'. Drahos, 'Public Goods', *supra* note 83 at 332.

dynamic as one in which the protections prescribed must be complied with but imports are nevertheless subject to a countervailing duty based on the foreign market subsidy in order to level the playing field for DCs and LDCs now that their comparative advantage in imitation has been eliminated. Such an approach would help correct the mismatch of institutions that domestically compound existing inefficiencies of TRIPS '[i]mposing higher standards of IP protection on developing countries that suffer from severe shortages of skilled labour', writes Drahos. For poorer member states, granting exclusive rights in the domestic market would have little impact on domestic innovation if the capacity to innovate is lacking; this 'means that intellectual property rights are less likely to achieve their incentive effects and may simply worsen a developing country's terms of trade'.[88]

A nation whose comparative advantage lies in imitation rather than innovation may stand to gain significant consumer and producer benefit from adopting *weak to no IP* protection. Trebilcock and Howse suggest that variation of standards may be efficient since

> [t]he level of intellectual property protection each country decides to afford will thus be rationally related to whether its comparative advantage resides more in innovation or imitation and adaptation of innovations made elsewhere, and the relative weight it gives to the interests of consumers (including its own producers who are consumers of inputs), imitators, and innovators.[89]

Best policies would move beyond accessing technology that could sustain development, under licence, to free *distribution* of knowledge to DCs and LDCs, recognizing that meeting individual HR and promoting development in target countries will have possibly greater positive externalities. There is actually little evidence to support the MNCs' claim that strong IPRs are good for DCs and the growth of their economy or that they *will* increase technology development, flows and transfer of capital by attracting innovation and foreign direct investment necessary to establish local economy-boosting enterprises. In fact, corporations are drawn to establish plants where the human resources are more educated, the political climate offers stability, the rule of law operates, and not just where labour is cheap and IPRs strong.[90] It is more likely that the most developed countries will just engage in *trade,* exporting goods to a country which in turn offers them strong IPRs without reciprocal economic benefit to the host. A UN based study confirms that innovatory companies would prefer to *sell* the products or services incorporating their innovation than establish a local presence that would facilitate transferring the technology

88 Ibid. at 333.

89 Trebilcock and Howse, *ITR*, op. cit. at 396.

90 Nayak, *Global Health Law*, op. cit.

or licencing agreements because uniform prices would 'avoid the downward ratchet effect of cross-country comparisons' and translates into 'more exports by developed countries, and less opportunit[y] for industrial and technological development for developing countries'.[91] The HR violations here are more tacit, as uniform IP standards erode all incentives for price differentiation. Once enforceable, firms seek maximum profits under patents without due consideration of poor consumers' buying capacity. Standard prices may be subsidized by foreign aid that allows for full profit but removes incentive for geographic price discrimination at less than 'full' market price. If TRIPS obligations are expansively interpreted, the impact on sovereignty is that not only will a state be unable to adopt policy preferences of its peoples towards HR and HD objectives, but actually may be compelled against it under threat of complaint within the framework for WTO dispute resolution.

The WTO Dispute Settlement Procedure

The binding dispute settlement process of the WTO is set out in the *Dispute Settlement Understanding* (DSU)[92] that is designed to provide 'security and predictability in the multilateral trading system' (DSU, Art. 3(2)). The 'rules based' system prescribes a judicial model for decision making and establishes an effective means of enforcing decisions. The Dispute Solution Body (DSB) administers the dispute resolution process (Article 2) if members are unable to resolve their disputes at first instance informally through consultations (Article 4), conciliation or mediation (Article 5). It is only after such procedures have failed that a formal complaint should be brought. A request may be made for a dispute resolution Panel to hear the complaint and make its ruling (Article 6). Because of the judicial nature of the system, appeals are allowed from Panel decisions to the Appellate Body (Article 17(6)) but the jurisdiction of the latter is limited exclusively to 'issues of law covered in the panel report and legal interpretations developed by the panel'. Much like domestic judicial systems in common law jurisdictions, the AB will not engage in the hearing or determination of new facts. A panel report shall be adopted by the DSB (unless there is notice of an appeal or consensus not to adopt the report as per Article 16). The final decision adopted by the DSB is binding on all parties to the dispute.

Prior to the Uruguay Round, GATT decisions had to be adopted by consensus, which meant that a losing party could effectively veto an adverse

91 Keayla, *supra* note 78.

92 See Understanding on Rules and Procedures Governing the Settlement of Disputes [DSU], 15 April 1994, WTO Agreement, Annex 2, art. II, Legal Instruments – Results of the Uruguay Round (1994) vol. 31, 33 I.L.M 1226 (1994), <http://www.wto.org/english/docs_e/legal-e/28-dsu.pdf>.

finding. However, as a result of this provision in the DSU, a final decision is *automatically* adopted unless the DSB decides unanimously against it. Once adopted, if the offending party does not comply with the ruling, trade sanctions may be imposed on other areas of export in respect of which the parties ordinarily trade.[93] Just as TRIPS provides transitional and special provisions regarding LDCs, so too does the DSU. For example, Article 24 articulates special procedures involving least-developed country members for both violation and non-violation complaints and provides that 'Members shall exercise due restraint in raising matters under these procedures involving a least-developed country'. The formality of the DSU makes it a very powerful institution. But, the TRIPS/DSU mechanism is not a court of last resort for corporations and private firms, as only member states have standing to bring a complaint; usually there would need to be a pattern of violation before a complaint will be made by a member's trade representative. However, just as with litigation, some complaints may be made for tactical advantage and strategically to leverage against other existing complaints between disputing parties. For example, the US Trade Representative might have agreed to bring a complaint based on Ontario's continuing alleged breach of MGL's BRCA patents as a political tool for resolving the US-Canada softwood lumber dispute that predated it and was continuing at the time. Moreover, trade regulation and its discourse are dominated by the comparative advantage of the industrialized world in law, politics, human resources, technological capacity, linguistic dominance, intellectual capital, and rhetorical eloquence, all of which allow for better representation of first world interests in the negotiation process and, therefore, the negotiated outcomes including dispute settlement mechanisms.[94]

The dispute settlement process is also captured, as Barfield reports: '[t]he United States has been the world's most active user of the new system – and the most successful. As of January 2001, the United States had brought fifty-seven complaints to the WTO … [and] lost only three cases.'[95] Under Article 64 of TRIPS, IP complaints are made subject to the DSU and since TRIPS covers both substantive standards and procedural implementation requirements, both types of complaints may be raised before the DSU.

Since these international norms are meant to legally bind nation states, they naturally derogate from the *full* sovereignty of these states. Jurisprudence professor Dennis Lloyd normalizes the relationship between internal sovereignty and international duty in order to derive a rational understanding of how this relationship fits with external rules of law. Lloyd contends:

---

93 *Canada – Certain Measures Concerning Periodicals*, 14 March 1997, WT/DS31/R (Panel); 30 June 1997, WT/DS31/AB/R (Appellate Body) [*Canadian Periodicals*], adopted 30 July 1997 [*Canada Periodicals*, AB].

94 See e.g. Drahos, 'Weak Bargain', op. cit. at 79–109.

95 Barfield, *Free Trade, Sovereignty, Democracy*, op. cit. at 32.

> [e]very rule of international law imposes a legal fetter on national states in the international sphere, for this is the very sense and meaning of an international legal community. But within its own internal sphere the national sovereign still retains domestic sovereignty and may legislate or act in disregard of international obligations. But by so doing it cannot alter, abrogate, or lessen the force of these obligations *vis-à-vis* other states, and *will have to take whatever consequences its default may entail, having regard to the existing state of international law* and the sort of pressures which can be brought upon it in the particular circumstances.[96]

Liability would be harmful to the state and its people but misgivings regarding the role of the WTO and its proper jurisdiction may prove more harmful as governments decide not to take HR promoting actions that would otherwise be a regulatory priority, in order to avoid risking the costs of defending a complaint. The graver risk is that the same kind of anti-competitive and predatory behaviour that is occurring in domestic patent practice will become global in scope and executed by the more powerful states that represent these firms in disputes. And, just as in patent law, the weaker party may simply rationally acquiesce to demands in order to avoid 'litigation' costs. Still, 'States *must* make rules for the effective functioning – in the social interest – of markets'[97] and should be able to defend these domestic measures at the WTO.[98]

Although TRIPS has been referred to as 'a "charter of rights" for the patent holders …',[99] respect for sovereignty will promote respect for cultural diversity; domestically nuanced preferences; differences in morality, ethics, economics, politics and the social objectives of any given 'public'; and the ability to respond with diverse regulatory measures to universal (HR) claims, essential for encouraging continued cooperation amongst nations and their peoples. Michael Trebilcock and Robert Howse question the more extreme and indiscriminate claims for policy convergence and international harmonization of 'within-the-border' domestic policy diversity. In their article, 'Trade Liberalization and Regulatory Diversity: Reconciling Competitive Markets with Competitive Politics',[100] they argue that the welfare implications of policy harmonization are ambiguous and highlight the importance of competitive governments and politics over competitive markets while acknowledging that mutual benefits

---

96 Lloyd, *The Idea of Law*, op. cit. at 190.

97 Ibid.

98 The Doha Declaration lists 'the relationship between existing WTO rules and specific trade obligations set out in multilateral agreements (MEAs)' as a topic for negotiation, see Para 31(i) Doha Declaration, *supra* note 726. See generally Pauwelyn, *Conflict of Norms, supra* note 31.

99 Keayla, *supra* note 78 at 263–4.

100 Trebilcock and Howse, 'Regulatory Diversity', op. cit.

may be gained in consensual forms of policy convergence through reducing the costs of divergence. The authors make the case for 'substantial national political autonomy in formulating domestic regulatory and related policies even though such politics will often have significant impacts on resulting trade flows[,]'[101] so long as the principle of non-discrimination is observed. That is, domestic regulatory diversity should be encouraged so long as the policies or practices do not violate the National Treatment principle and are not forms of disguised or unjustifiable discrimination offensive to trade. One of the benefits of regulatory competition is that it allows the possibility of making informed evaluation of different approaches to regulation (and their efficacy), which may in turn lead to more transparency and regulatory competition at both the domestic and international level. Whatever the approach taken to trade and HR issues, it necessarily must be a sovereign one.

Conclusion

Economic theories that directed the trajectory of domestic fiscal policy and accompanied the emergence of social welfare programmes *within* the State preceded the social and trade policy initiatives internationally marshalled in institutions like the World Bank, the IMF, the GATT/WTO and the UN in the postwar era. Trade liberalization proponents continue to recognize the neoliberal maxim that 'creating wealth is a precondition to redistributing it'[102] while offering capitalism's 'successful' marriage to the welfare state as a prime example. Despite growth in production and world wealth, however, '[p]overty amid plenty is the world's greatest challenge'.[103] Neoliberal trade policies increase the wealth of a nation and thereby *allow* for its internal redistribution but they do not *ensure* it without the attendant regard to welfare-promoting HR and its complement, HD. Liberalizing trade is a means of effecting *redistribution* of wealth *internationally* by encouraging free competition, transfer of goods and technology, services and even human resources but insofar as TRIPS consolidates monopolistic power, it fails to fulfil the mandate of the regime within which it is found and should be

101 Ibid. at 28

102 Michael J. Trebilcock and Robert Howse, *ITR*, op. cit., at 12. The authors observe that '[d]irect financial benefits paid to Canadians under various social welfare programmes amounted to $49, 136 million in 1985, compared with 1947 expenditures of $3,838 million ... Trade liberalization and trade expansion have not been inconsistent with these redistributional, social and cultural policies ... [O]nly relatively prosperous countries can afford generous social policies. Impoverished third world countries ... do not have the wealth to afford them'.

103 See World Bank Development Report 2000/2001, *Attacking Poverty*, op. cit. at v.

approached with cautious reserve. Neither the theory nor history of trade is incompatible with redistributive functions that underlie (social and economic) HR and ideally inform state agency for their protection,[104] but TRIPS may find itself realized or circumscribed by state agents and national and international decision makers. What is important is that the protection, promotion, and fulfilment of HR remain at the forefront of political and regulatory agendas both domestically and internationally when debating patenting life; it is HR instruments that require and enable redistribution to be effective *locally* and individually.[105]

As Louis Henkin aptly stated: '[E]very individual has claims upon … society – both claims to freedom from undue governmental intrusion and claims for governmental support for economic and social welfare'.[106] While the global capacity for food production has increased exponentially with the advent of new technologies, famine continues to exist despite the promises of biotech used to justify patent monopolies in food, plant, and seed sectors. While advances in new biomedical technology offer unimagined potential, most of the world's peoples continue to suffer from pandemics and disease that most readily can be cured by alleviating their dire conditions of poverty and improving standards of living in meeting basic needs for food, water, and shelter. These failures may be better characterized as *political* ones since the correction of market failure relies on the will of government(s) to intervene, and not just on behalf of property owners.[107] A state needs to be made accountable for its failures and able to defend its successes in terms of a just world order. Protecting human dignity from political failure is a dominant function of HR instruments and is further promoted by trade liberalization efforts which are premised on a reasonable assumption that greater economic interdependence and greater market competition would lead to friendlier and peaceful relations, global prosperity, and lower consumer costs. These were the goals of both institutions for global governance as the historical and

104 See generally Michael Trebilcock, 'Critiquing the Critics of Economic Globalization' (2005) 1 *J. of Int. Law and Relations* 213 ['Critiquing the Critics'].

105 Generalized indicators for growth and redistribution may be misleading. There is a notable difference in looking at growth of incomes between countries (and making the erroneous assumption that all people in one country earn the average income for that country) and using more accurate measures for income inequality that consider disparities of individual incomes within a country. See Branko Milanovic, 'True World Income Distribution, 1988 and 1993: First Calculations Based on Household Surveys Alone' (2002) 112 *Econ. J.* 51.

106 Louis Henkin, *The Rights of Man Today* (Boulder, CO: Westview Press, 1978) at 10.

107 See Rawls, *The Law of Peoples*, op. cit., referring to Amartya Sen, *Poverty and Famines* (Oxford: Clarendon Press, 1981) and Sen, with Jean Dreze, *Hunger and Public Action* (Oxford: Clarendon Press, 1989) – '*confirms these points and stresses the success of democratic regimes in coping with poverty and hunger*' (at 109).

institutional review shows but are also the goals of the future set out in the MDGs. Foreign-based subsidies of first-world intellectual objects through universal prescriptions under TRIPS increase the size of the pie but some parties continue to go hungry while others have secured a larger slice. The new WTO rules post-Uruguay have been criticized because their intentions regarding domestic and global welfare effects are more ambiguous than those of GATT:

> These rules could not be justified through the idea of Kaldor-Hicks efficiency – there is no particular reason to believe on the basis of economics that increasing intellectual property protection will increase aggregate domestic welfare. Some countries gain from increased patent protection and some countries lose; aggregate welfare may increase or decrease. And the issue of who gains and who loses *within* a given society rears its head and cannot be avoided or suppressed by any idea tractable to technocratic management of the trading system.[108]

While economic ideology has seemingly replaced the 'embedded liberalism' foundational to the origins of the trading system, it is a moral question we are left with because, as Thomas Cotter observes, 'economics may help us to predict the possible consequences of various legal rules, but the ultimate decision whether those consequences are good or bad, and therefore whether the rules should or should not be adopted, must rest upon moral principles to which economics does not speak'.[109] If the WTO remains insensitive to modernity's social demands, as an institution it would risk being undermined by a reasonable apprehension of bias in favour of private capital, corporate interests, and the dominant cultural and legal imperatives of developed members, shaping its culture and politics. Increasing public protest, gradual non-participation and disregard by the vast majority of less privileged members, and a Round that will neither conclude successfully nor die a quite death, demonstrate that social movements animating counter-narratives of anti-globalization activism (equating the politics of globalization and economic integration woes with the WTO as an institution) are gaining

108 See Howse, 'Technocracy', *supra* note 27 at 99 and 102: 'The belief that the removal of trade restrictions is Kaldor-Hicks efficient cannot be reduced to *just* blind ideological faith – in many situations the empirical evidence suggests that one could and should replace trade restrictions with other policy instruments, and make everyone better off.' See also Michael J. Trebilcock, Marsha A. Chandler and Robert Howse, *Trade and Transitions: A Comparative Analysis of Adjustment Policies* (London: Routledge, 1990).

109 Thomas F. Cotter, 'Introduction to IP Symposium' (2002) 14 Fla. J. of Int'l L. 147 at 151.

traction,[110] embracing in the debate as they often do vocabularies of HR and the discourse of cultural difference.[111] To make trade *work*, TRIPS's 'future will depend on the decisions taken by the dispute settlement body, which will determine to what extent the agreement is implemented *in line with the social and economic development objectives of member nations*'.[112] WTO jurists would do well to honour the prescient message inscribed on the plaque found on the stone outside the WTO: *Si vas pacem cole justitiam* – If you seek peace, cultivate justice.[113]

110 These criticisms follow the operations of the IMF and WB too. See 'Globalization and its Critics: A Survey of Globalization' *The Economist* (29 September 2001) 3 at 3–30.

111 Rosemary Coombe, 'Legal Claims to Culture in and Against the Market: Neoliberalism and the Global Proliferation of Meaningful Difference' (2005) 1 *Law, Culture and Humanities* 32, York University, <http://www.yorku.ca/rcoombe/publications/Coombe_Legal_Claims_to_Culture.pdf> offers: 'These new social movements express forms of aspirations that deploy the rhetoric of international human rights, indigenous rights, and environmental agreements in ways that nonetheless significantly exceed the modern categories that dominate rights claims. If these assertions are made in cultural terms, it may be because the area of cultural rights is the least interpreted of the human rights fields and thus most amenable to claims that combine but go well beyond our modern understandings of the political, the economic, the social, and the cultural. The specific paradigms ... that govern human rights laws are arguably too narrow to encompass the aspirations for social justice expressed by the world's poor in their resistance to the violence of modernization and the destruction wrought by state and capitalist development projects.'

112 UNDP, *Making Trade Work*, *supra* note 24 at 222, emphasis added. See also Article 28 of the Universal Declaration of Human Rights, op. cit..

113 Cited in Adelle Blackett, 'Whither Social Clause? Human Rights, Trade Theory and Treaty Interpretation' (1999) 31 *Colum. H.R.L. Rev.* 1 at 5 and n. 5.

Chapter 7
Reconciling Competing International Obligations: The Equitable Conduct Defence and the Stewardship of the WTO

> We can sit down and look at the realistic possibility of making the WTO work for the whole world. We should be realistic. We shouldn't be kidding ourselves that the WTO is all right at the moment. It's not all right.[1]

> The purpose of the law is to prevent the strong from always having their way.
>
> Ovid

Introduction

This chapter completes the prescriptions advanced in this text for appropriate state agency when regulating biopatenting and builds on the evidence put forward in prior chapters of the imperative to reserve sufficient national regulatory space for HR-related IPRs. The focus in this chapter is to consider how WTO jurists might 'cultivate justice' by re-uniting markets with universal principles and reconciling HR principles with trade objectives.[2] Accordingly, a human rights-approach to dispute settlement is advanced. To summarize the framework proposed for state agency in relation to patenting life, the first branch of the bifurcated approach seeks to address the domestic need for improving patent system efficiency and efficacy to better achieve cross-policy coordination and meet the public's interest in more optimal instrumental tools for encouraging inventions. At the domestic level, I urge states to strive to reconcile competing international obligations with domestic law by adopting what would be TRIPS-compliant regulatory measures. The second branch of my prescription is to encourage states to anticipate a WTO challenge to

1 Supachai Panitchpakdi (WTO Director-General 2002–2005), 'Keynote Address: The Evolving Multilateral Trade System in the New Millennium' (2001) 33 Geo. Wash. L. Rev. 419 at 432.

2 Kofi Anan's eloquent appeal was: 'Let us choose to unite the powers of markets with the authority of universal principles.' See UN Global Compact, <http://www.unglobalcompact.org/Portal/>.

their adopted measure whenever domestic patent reform is undertaken; this is not to curtail such reform but to explore and develop available defences in light of them, given a WTO member may become party to a complaint contesting its measure(s) as a violation of TRIPS regardless of the merits of such a claim. Developed and DCs alike will benefit their peoples by embracing such an approach. This chapter considers what a state can do to defend the regulatory choices made to better balance patents and public health, or other key trade/human-rights issues raised at the nexus of biopatenting, should the matter proceed to WTO dispute settlement. IPRs are 'in essence, government tools for regulating markets in information' write Drahos and Braithwaite, 'the problems of government capture … have led us to the view that there should be a presumption against the use of these rights to regulate markets rather than a presumption in their favour'.[3] Regrettably, that has not been the case; TRIPS tips the scale in favour of their use.

Issues at the intersection of trade and HR have typically been concerned with the propriety of a domestic *trade-impacting* measure.[4] Trade purists believe the system should remain insulated from extraneous considerations and that balancing policy concerns (the so-called 'social dimensions' of trade) is not within the WTO's mandate or jurisdiction. Critics conversely argue that fairness demands a purposive interpretation to be given to state obligations in order to integrate the international trade system and the human rights system into a *cohesive* and *coherent* system of public international law – the body of law governing state relations. Debate about a 'social clause', justice, and equity, and the burgeoning so-called 'trade-and ___' scholarly literature are testaments to the challenges of uncertainty the WTO faces. The institutional deficit for welfare maintaining mechanisms and constitutional protections coupled with the effective compliance-enforcing dispute settlement procedures has embroiled the WTO in controversy and myopically rendered it '"public enemy number one" on the anti-globalist agenda'.[5] Matsushita et al. identify some of the charges:

> 1. The WTO serves the interests of transnational corporations to the detriment of developing nations and the poor[;] 2. The WTO is anti-democratic and

---

3 Drahos and Braithwaite, *Information Feudalism*, op. cit. at 3.

4 See e.g. Blackett, 'Whither Social Clause', op. cit., focusing on trade and labour standards; McDonald, 'It's Not Easy being Green', op. cit., on trade and the environment and Stephan Bottomley and David Kinley, eds, *Commercial Law and Human Rights* (Brookfield, VT: Dartmouth Publishing Company, 2002) arguing that commerce and HR are not mutually exclusive.

5 See generally Matsushita et al., *World Trade Organization: Law, Practice, and Policy* (Oxford: Oxford University Press, 2003). The entire text is devoted to negating this judgment by explaining the law, institution, and function of the WTO and the ways in which it can be improved on its own terms. See also the rejoinder to such charges by Trebilcock, 'Critiquing the Critics', op. cit.

> violates national sovereignty[;] 3. The WTO elevates property interests over human rights and health. It particularly 'enshrines' intellectual property rights[;] 4. The WTO is hostile to protection of the environment, workers' rights, agriculture, food safety, and social welfare generally because it strikes down protection of such interests as trade barriers. In summary, to some, the WTO and globalization are forces for oppression, exploitation, and injustice.[6]

All of these complaints, in fact, also dominate the discourse surrounding patenting life forms given the many sectors of vital importance that are affected.

Economist and Nobel laureate Amartya Sen asserts, the 'compelling need in the contemporary world is to ask questions not only about the economics and politics of globalization, but also about the values and ethics that shape our conception of the global world'.[7] Trade liberalization and economic integration offer welfare gains for world citizens but 'the greatest challenge facing the WTO today is also the greatest challenge facing the world. That challenge is the ever-increasing inequality that characterizes the division of income and assets on our planet'.[8] Some scholars believe that the division between and within nations is growing and is exacerbated by trade liberalization;[9] a view increasingly shared by many NGOs and many academics. Mendes and Mehmet write,

> [t]he technical responsiveness to the changing picture of global trade shown in the establishment of the WTO is not matched by a sensitivity to existing and emerging trends in the social dimensions of trade, including labor, the environment, and human rights ... the impressive responsiveness to the growing importance of services and intellectual property by the trade

6 Matsushita et al., *supra* note 5 at vii.

7 See Amartya Sen, 'Globalization: Value and Ethics', talk given at Falcone Foundation, 23 May 2001, <http://www.fondazionefalcone.it/sen.html>.

8 Buckley, *Changing Face*, op. cit. at 3.

9 'The richest 1 per cent of people in the world earn as much as the bottom 57 per cent; in other words, less than 50 million of the richest people earn as much as the 2,700 million poorest people ... If we look at assets rather than incomes, the richest 20 per cent of humanity owned 60 times as much as the poorest 20 per cent in 1990 and 74 times as much in 2000, and three men, the two founders of Microsoft, Bill Gates and Paul Allen, and the investor Warren Buffet, own as much as the 600 million people in the world's 48 least developed countries'. Ibid. at 3–4 [footnotes omitted]. See also Milanovic, Development Research Group, World Bank (taking into account income disparities within each country as well as between), op. cit. and his new treatise, *Worlds Apart: Measuring International and Global Inequality* (Princeton, NJ: Princeton University Press, 2005); for global inequality trends, see 'World Inequality', *BBC News* 18 July 2001.

> experts has ignored the possible conflict between the rules of these new trade areas and certain categories of human rights.[10]

The treatment of economic liberalization as an *end* instead of a *means* has raised tremendous concerns for civil society. The Report of the Canada Standing Committee on Foreign Affairs and International Trade captures these concerns cogently:

> For example, health, consumer, student and other social advocacy groups raised issues with respect to: the right of entry, establishment and 'national treatment' for private health and educational services corporations; the extensions of patent rights for pharmaceutical drug companies; trademark rights in relation to cigarette packaging; the patenting by large transnational enterprises of genetically modified organisms and other life forms. Whose rights need to be protected in the public interest? What is consistent with international human rights norms?[11]

As a result, the WTO must find a means to

> manage globalization in such a way that its benefits are shared far more equally among individuals than is today the case[,] ... not merely desirable on grounds of fairness or ethics, though it is utterly defensible on those grounds, [but it] ... is essential if we are not to slide backwards into a less interdependent, and far poorer world.[12]

External liberalization has entrenched itself in the international community with agreed upon economic obligations becoming binding and enforceable through new dispute mechanisms developed to ensure compliance under the auspices of the WTO. Yet the effective means of acquiring and enforcing constitutional-like rights that are germane, by liberal democratic ideals, to the development and operation of any nation state and the dignity of its peoples are notably absent in our global institutions. Trade agreements are no longer confined to ensuring trade 'free from discrimination' but have expansively encroached through language, law, and ideology, into the sovereignty of members with a new monolithic mantra of trade essentially

10 Mendes and Mehmet, *Global Governance*, op. cit. at 72.

11 Report of the Standing Committee on Foreign Affairs and International Trade: Canada and the Future of the World Trade Organization, Advancing a Millennium Agenda in the Public Interest (Canadian Millennium Report), Chair Bill Graham, MP (Ottawa: House of Commons, Canada, 1999) at 11–17.

12 Buckley, *Changing Face*, *supra* note 8.

free from regulatory burden.[13] As our preceding discussion has established, since the state is the purveyor of IPRs, the state's role is central to the realization, or alternatively the violation, of HR obligations. Yet governments may legitimately believe that any derogation from patent rights would attract potential liability under TRIPS in the form of trade sanctions. Does the TRIPS text allow for concurrent compliance of members with HR instruments? Or will the global economy continue to be dictated by the coercive suasion of its political economy with the private power of corporations the puppeteer of public law? This is the question occupying NGOs,[14] public interest advocates, special interest groups,[15] academics,[16] government advisory committees and other institutional players analyzing issues surrounding the fundamental question of whether the rights to life can be owned. TRIPS obligations are not necessarily *antithetical* to human rights, though they provide a position of privilege through the power of an exclusive grant of monopoly. The provisions require further scrutiny of the obligations they impose on the rights holder and the *justifications* allowed for divergence in certain contexts, such as to prioritize HR. When competing interests are characterized as *human rights*, they are distinct in their persuasive claim on international arrangements from traditional moral and ethical arguments. As human rights, they have a separate line of authority as customary law and demand an almost sacred level of accountability for their subjects.[17]

Sovereignty and self-determination require discretion to regulate biopatenting domestically without threat of trade-related interference. These oblige the state as much as the substantive intersecting rights of health, food, and the right to benefit from scientific progress independently – even if only in terms of a negative right to be free from harmful government intrusion. State discretion to prioritize HR, consequently, should be re-imagined as a contribution to the *cooperation* paradigm in which the state promotes 'self-help' for its peoples, so long as the trade contested measure is adopted on a non-discriminatory basis. After a discussion of the potential forms of instrumental

13 See e.g. D. Driesen, '"What is Free Trade?" The Real Issue Lurking Behind the Trade and Environment Debate' (2001) 41 *Va. J. Int'l L.* 279, cited in McDonald, *supra* note 4 at 145, n. 1.

14 For strong advocacy in this area, see RAFI at <http://www.rafi.org/>, Third World Network (TWN) at <http://www.twnside.org.sg/> and GRAIN at <http://www.grain.org/publications/it-november-2001.cfm>.

15 See Debra Harry, Indigenous People's Council on Biocolonialism, <http://www.ipcb.org/pub/>.

16 See Amani and Coombe, *supra* note 78; Sell, *Private Power supra* note 407; and, Marjorie Cohn, 'The World Trade Organization: Elevating Property Interests Above Human Rights', 29:3 *Georgia Journal of International and Comparative Law* (2001) 427-441.

17 See e.g. Ryan Goodman and Derek Jinks, 'Measuring the Effects of Human Rights Treaties' (2003) 14.1 *European Journal of International Law*, 171–83.

conflict, the guiding principles of treaty interpretation, and the provision of both rights and obligations, this chapter concludes by advocating that any biopatent complaint made under TRIPS be interpreted in a conciliatory rather than a conflicting manner before matters proceed formally. Failing this, WTO jurists should recognize an 'equitable conduct defence' (ECD) based on the language of the TRIPS text. Accepting the ECD may elucidate an important ideological question about the role of the WTO that troubles many. Helleiner is not alone in asking:

> What sort of World *Trade* Organization is it, after all, that doesn't seriously concern itself with trends and fluctuations in its members' terms of trade particularly those of its weakest and most vulnerable members? Or with 'burdensome surpluses' (as the ITO Charter called them) in primary commodity markets? Or with restrictive business practices and abuse of dominant power in international goods and services markets? While it has so far paid scant attention to such obvious *trading* issues it has moved deeply into such *domestic* policy issues as intellectual property regimes, domestic investment and subsidy policies and some would even push it into labour standards and environmental practice, all of which may or may not, in fact, be 'trade-related'? On the basis of current practice it might better be called the World Market Harmonization Organization. [18]

A better, more effective and true-to-purpose WTO is possible. WTO jurists should strive to achieve a balanced contextual and purposive interpretation of texts through the DSU that seeks to reconcile trade with its social dimension and a state's HR obligations. '[M]arkets', economist Helleiner writes, have never been the 'primary basis for human interaction' and globalized markets 'operate within politically defined rules and governance institutions'.[19] The 2001 'Nightmare Report' of the Expert Group of the UN Subcommission on Human Rights qualified the WTO as the 'nightmare' of human rights.[20] To overcome strict WTO rule orthodoxy, respect for diversity is now more than ever necessary to ensure that TRIPS' interpretation does not dictate how governments can and should interact with the new global economy,[21] or undermine core human values. WIPO shares the role of norm setting and

18 Helleiner, 'Markets, Politics and Globalization', op. cit. at 12.

19 Ibid, at 3 and 16.

20 See Preliminary Report, by J. Oloka-Onyango and Deepika Udagama, submitted to the UN Sub-Commission on the Promotion and Protection of Human Rights, 15 June 2000 in accordance with Sub-Commission Resolution 1999/8 and UN Press Release, 'Globalization and Its Impact on the Full Enjoyment of Human Rights, E/CN.4/Sub.2/2000/13.

21 See Helleiner, *supra* note 18 at 3.

there have been attempts to strengthen the role of the UN,[22] but because the contentious issues of trade-related IPRs fall to the jurisdiction of the formalized dispute-settlement mechanism, the WTO is the preferred forum. The WTO membership of approximately 153 states is represented by more than two-thirds DCs and LDCs.[23] Given the disproportionately large population of these countries, notably that of India and China,[24] and the disproportionately large population of poor within these populations,[25] the future of multilateral trade governance is fragile. The apparent collapse of the Doha Round in 2003, and the failure to reach consensus on the July 2008 package to successfully conclude it, reflect the uncertainties the WTO must face. The July package had emerged from another renewed attempt to overcome intractability over agricultural modalities (to determine the future level of tariffs and reductions in farm subsidies) and non-agricultural market access. Although the Doha Round has been in progress more than eight years, Director General Pascal Lamy assures that no one is ready to throw in the towel yet. The Doha Work Programme states, '[p]olitical breakthrough requires consultation among a group of ministers representing *all* interests in the negotiations'. The interests are complex and require assuaging of main institutional criticisms over

22 See 'Revised draft outcome document of the high-level plenary meeting of the General Assembly of September 2005 submitted by the President of the General Assembly, 10 August 2005, Fifty-ninth session, Agenda items 45 and 55, Integrated and co-coordinated implementation of and follow-up to the outcomes of the major United Nations conferences and summits in the economic, social and related fields. Revised Draft Outcome Document (10 August 2005) [A/59/HLPM/CRP.1/Rev.2], in particular para. 14 (Values and Principles), <http://www.un.org/summit2005/documents.html>, which states: 'We pledge to make the United Nations more relevant, more effective, more efficient, more accountable and more credible ...' (para. 125) '... to strengthen the United nations with a view to enhancing its authority and efficiency, as well as its capacity to address effectively the full range of the challenges of our time ...' (para. 103): 'We resolve to strengthen the United Nations human rights machinery with the aim of ensuring effective enjoyment by all of the human rights, civil, political, economic, social and cultural rights, including ... to development.'

23 <http://www.wto.org/english/thewto_e/whatis_e/tif_e/dev1_e.htm>. The terms 'developed' and 'developing countries' (DCs) are not defined; states self-select this status subject to WTO acceptance.

24 'By 2001, India had 1,033m people against China's 1,272m. But China's national income per head, according to the World Bank, was $890, nearly double India's $890. Adjusted for purchasing power, the Chinese were still 70% wealthier than Indians were. In the ten years from 1992, India's GDP per head grew at 4.3% a year, China's twice as fast. Some 5% of Chinese now live below the national poverty line compared with 29% of Indians. Economic strength begets strategic clout.' 'Special Report India and China: A Tiger, Falling Behind a Dragon', *The Economist* (21 June 2003) at 21–3. See also 'India and China oppose attempts by some developed countries to 'divide' developing nations as WTO negotiations', *India Daily* 12 July 2005.

25 See Sachs, *The End of Poverty*, op. cit. at 32–9.

legitimacy, process, and architecture but also the content and relationship of substantive international norms. HR and trade obligations derogate from full unfettered sovereignty of a state but trade obligations are appropriately characterized as enabling the means to HR's universal ends;[26] they may have to surrender to means that have less of an impact on HR.

A Human Rights Framework for WTO Dispute Settlement: Human Rights as a Shield

While WTO literature analyzes globalization's impact,[27] the consequences of economic integration,[28] and the protection of HR in a trade regime,[29] often the focus is on the potential use of restrictive border measures on imports, such as quotas or tariffs,[30] to offensively elicit human rights compliance in another *exporting* nation and the legitimacy of doing so to affect foreign policy objectives.[31] But the conceptual framework for global cooperation needs to

26 Hastrup, 'Accommodating Diversity', op. cit. For a thoughtful discussion of two distinct approaches to the IP-HR interface, see Laurence Helfer, 'Human Rights and Intellectual Property: Conflict or Co-existance' (2003) 5 Minn. Intell. Prop. Rev. 47, <http://mipr.umn.edu/archive/v5n1/Helfer.pdf>.

27 See Robert McCorquodale and Richard Fairbrother, 'Globalization and Human Rights', 21 Hum. Rts Q. (1999) 735–66. Compare with 'Globalization and its Critics: A Survey of Globalization' *The Economist* (29 September 2001) 3. For a discussion on how globalization is resulting in increased income inequality and poverty and challenging the view that poverty is best reduced through growth oriented, rather than distributive policies, see Giovanni Andrea Cornia, 'Liberalization, Globalization and Income Distribution', Working Papers No. 157, March 1999, The United Nations University World Institute for Development Economics Research, at <http://www.wider.unu.edu/publications/wp157.pdf>.

28 See Michael Trebilcock, 'Post Seattle Reflections: A Qualified Defense of the WTO and an Unqualified Defense of the International Rule of Law' presented at Session 6, Corporate Power, National Sovereignty, and the Rule of Law in a Global Economy, NY University School of Law (2000), <http://www.loc.gov/bicentennial/abstracts_trebilcock.html>.

29 See e.g. Hoe Lim, 'Trade and Human Rights: What's at Issue?' (2001) 35(2) J. of World Trade 275.

30 Non-discrimination and reduction of quantitative restrictions are the two main pillars of the multilateral trade system. Tariffs (import duties) are a barrier to trade because they add costs to consumer purchase prices and thereby make foreign exported producers less competitive with domestic ones not subjected to this additional cost. If quotas are also not reduced or eliminated, quantitative restrictions would simply substitute for tariffs.

31 See e.g. Michael Trebilcock, 'Trade Policy and Labour Standards: Objectives, Instruments, and Institutions' (2001) University of Toronto Research Paper No. 02-01, <http://ssrn.com/abstract/_id=307219>; Sarah Cleveland, 'Human Rights Sanctions

be further developed. In particular, it must establish the means by which a dispute settlement body 'adjudicates' measures taken to protect domestic HR in alleged violation of TRIPS. According to Vazquez, the HR community is ostensibly threatened by the international trade community while the trade system does not appear to suffer from the same insecurities. In March 2005, the UN General Assembly issued a Report of the Secretary General[32] revitalizing HR and the role of the UN in the new era and while some commentators argue that trade sanctions can and should be used to induce HR compliance in trading partners, the position advanced here in support of the Article 8 ECD is a more modest proposal of HR as a *shield* against liability, not as a sword for intervention. At a minimum, trade obligations should *not interfere* with the domestic realization of internationally recognized HR obligations where a state gives these obligations priority in domestic policies in order to resolve issues arising from the patenting of life. The threat of trade-related costs should not form a disincentive to states and representative governments in forming policies for meeting, and remaining accountable for, HR duties owed to citizens.

The duty of national governments to their *own* citizens to ensure HR are met, or at least not interfered with at *home*, is of practical relevance as civil society attempts to *localize the global* in domestically contingent frameworks.[33] It is hypocrisy not to mind one's own house when committing to these values *globally* through the MDGs and Monterrey Consensus. If each state governed by the ethos that domestic HR are the primary obligation of the nation state, the aggregate status of individual HR protection worldwide would be vastly improved. Instead Canada, for example, while amending its national patent laws to address HR concerns in foreign countries related to access to essential medicines, continues to shirk its regulatory responsibility to legislatively address the growing contestability and incoherence of domestic biopatent law. Canada did not fare well in its 2006 country-specific HR Committee review of its ESCR's commitments and is also precariously listed on the United

and International Trade: A Theory of Compatibility' (2002) 5(1) *J. Int'l Econ. L.* 133; Robert Howse, 'India's WTO Challenge to Drug Enforcement Conditions in the European Community Generalized System of Preferences: A Little Known Case with Major Repercussions for "Political" Conditionality in US Trade Policy' (2003) 4 *Chicago J. Int'l L.* 385-405. See generally Vazquez, 'Trade Sanctions and Human Rights', commenting on the bluntness of economic sanctions as a mechanism for coercing HR compliance in trading partners given the potential ramifications of sanctions and withdrawal of economic engagement as potentially penalizing not the elite decision makers or governments but rather 'devastating an already much abused populace', See e.g. Craig Forcese, 'Globalizing Decency: Responsible Engagement in an Era of Economic Integration' (2002) *Yale Human Rts and Dev. L.J.* 1.

32 Report of the Secretary General, A/59/2005, <http://www.un.org/ga/59/hl60_plenarymeeting.html>.

33 See e.g. Study on Canada's Human Rights Obligations Report, <http://www.sen.parl.gc.ca/>.

States' Special 301 priority watch list for failing to adopt stricter IPRs. The Canadian government should respond, as called upon, legislatively but also with the non-legislative approaches available to it, defending its conduct with the ECD described below. The ECD essentially provides a HR approach to interpreting TRIPS obligations, using HR as a shield to allow governments the ability to defer to these obligations and defend them above those of trade while encouraging more efficient alternative regulatory responses. Recognizing such a defence would be an important step forward in allaying the legitimacy crisis of the WTO and is commited to Article 2.7 of the UNC.[34]

Robert Howse contends that (unilateral) trade sanctions can be used proactively to coerce HR compliance in trading partners.[35] Howse and Makau Mutua have argued that violators of HR tend to be WTO rule violators.[36] A more inclusive HR approach to trade advocated by some scholars, NGOs, and representatives of developed countries, as a *sword* to elicit compliance in a trading partner through trade sanctions, however, would graft the hegemony of trade values onto HR norms and thereby supplant the role and importance of the UN system.[37] Concessions conferred *ex gratia* (such as the GSP) can, nevertheless, currently be unilaterally withdrawn even in the absence of violation of international HR law.[38] But to impose trade sanctions or refuse trade with

34 '[n]othing contained in the present Charter shall authorize the United Nations to intervene in matters which are essentially within the domestic jurisdiction of any state or shall require the Members to submit such matters to settlement under the present Charter ….' UNC, <http://www.un.org/aboutun/charter/>.

35 See also Vazquez, *supra* note 31 for using trade to sanction HR in other countries and Gabrielle Marceau, 'WTO Dispute Settlement and Human Rights' 13.4 E.J.I.L 753.

36 Howse and Matua, 'Protecting Human Rights in a Global Economy', op. cit. Compare this with Jose Alvarez's reply, 'Trade and the Environment: Implications for Global Governance – How Not To Link Institutional Conundrums of an Expanded Trade Regime' (2001) 7 Widener L. Symp. J. 1 that 'many HR violators will routinely comply with international economic agreements and many prominent defenders of human rights, including the United States, have trouble adhering to and complying with some international agreements'.

37 Historically, trade sanctions or the suspensions of concessions replaced the use of force as a legitimate means of coercing compliance with international obligations owed to the nationals of another state. Such sanctions can now only legitimately be imposed by the UN's Security Council as a form of coercing HR compliance. In the multilateral trade regime, sanctions for enforcing a decision are to equal the amount of lost trade related to the impugning measure. The UN continues to play an important role in many respects, one of which is in interpreting the ICESCR and thereby serving in an influential institutional advisory capacity for a human rights approach to IPRs and trade. See e.g. discussion in Chapter 6, footnote 111.

38 But, if the concession was conferred by a treaty, it could only be withdrawn in accordance to that treaty on the basis of a violation of it or some other norm (as a countermeasure), Vazquez, *supra* note 31 at 799.

a partner, for violating the HR of those producing its goods for export, for example, rendering these products not 'like' products for the purposes of a GATT dispute, while laudable in its objective of reducing HR violations is paternalistic and invasive of a Member's sovereignty and would slow the global 'development agenda'. Moreover, HR as a sword in trade is more vulnerable to abuse as disguised protectionism and creates compounding HR complications because sanctions may lead to graver subverted forms of HR violations in the targeted state, with a shift to domestic supply industries.[39] Sir Anthony Mason's view is that

> [t]he correct balance between efficient economic outcomes and respect for human rights may depend upon the stage of development that a particular country may have reached. In the developed countries of the world, human rights are an established culture, centred upon judicially enforced human rights which are either entrenched or statute based.[40]

The legal justification and authority derived from the ECD offered in this chapter is a healthy compromise. The ECD shields sovereign discretion to give potentially trade-impacting HR priority within a limited domestic context without *requiring* this priority (a sword) within a multilateral trade setting that may be seen as *ultra vires* of the WTO. Most fundamental to the multilateral trade regime are its foundational pillars: the tenets of non-discrimination, non-unilateralism and the preservation of cooperation and conformity to the rule of law. These criteria should be the only qualification in assessing the ability to invoke an otherwise available defence and they are indeed retained by the proposed ECD. HR as a shield against the complaints of another member requires that the trade-contested measure be applied in a non-discriminatory manner; this is actually more important than deciding whether HR have any formal role to play in the trade regime. Robert Howse suggests:

39 For example, a ban (quantitative restriction) might be imposed on imported products made by child labour on the basis that these are not 'like products'. However, such a measure might shift domestic child labour away from export oriented industries (i.e. rug weaving) towards domestic service industries such as the child sex trade. Although this is a gross oversimplification, has the measure been effective in curtailing child labour or simply compounded it by adding another dimension that drives it further underground? Is a country taking such a measure obliged to cease trade with *all* other trading partners who continue to do trade with the human rights violating member on the basis of non-discrimination? But see Bottomley and Kinley, 'Human Rights as Legally Binding or Merely Relevant' in Bottomley and Kinley, *Commercial Law and Human Rights*, op. cit., where it is argued that sanctions help create pressure for a corporate culture sensitive to HR.

40 Sir Anthony Mason, 'Comment on Papers presented at the Commercial Law and Human Rights Conference' in Bottomley and Kinley, *supra* note 39 at 6.

> [r]ather than attempt once again to decide what is 'in' or 'out' of the WTO, we should try to mould the rules and their interpretation to structure the *interaction* of the trading regime with other powers and authorities, both domestic and international, in a legitimate manner.[41]

International law should be a process – a dialogue – of different but equal sovereign voices in chorus; recognizing it as such would reduce some of the current politics that detract from the legitimate role and function of the WTO. DCs have made significant strides in revisiting the issue of patents and public health, and revitalizing the discussion on trade and agriculture, but at the point where health and agriculture merge, with the patenting of life, bilateral and regional arrangements continue to erode TRIPS protections, suggesting that the industry lobby in the North will continue to push for the removal of sovereign discretion and narrow interpretation of member rights under TRIPS until they are completely eliminated. States should not assist such efforts by voluntarily surrendering these rights through their common law, by failing to exercise the scope of their TRIPS rights, or by committing to TRIPS-Plus agreements. But nor should WTO jurists effect the same in strictly approaching trade from a 'purist' perspective. The proliferation of international agreements adds to the sites of potential conflict. While a more comprehensive agenda is underway to revisit the equal participation by members in trade negotiations to ensure justice in mediated outcomes, conflict resolution requires a concurrent complementary framework for disputes.[42]

41 Howse, 'Technocracy', op. cit. at 112. Helleiner similarly proposes increased democratic participation in the global governance institutions, political gains of co-operation amongst the weaker and poorer WTO members, and attempts to address equity and redistributive justice issues, *supra* note 18.

42 Supachai Panitchpakdi in his address, *Prospects for the Millennium Round* states, 'The top priority facing the WTO today is first to identify and then redress many of its shortcomings … [W]e need to strengthen the organization and rebuild confidence among its Members after the Seattle Ministerial Conference. To this end, the WTO is currently working on the confidence-building measures as agreed to recently in Geneva. They include: … increasing technical assistance to developing countries and LDCs, addressing the implementation concerns, and improving transparency in decision-making process … *Most developing countries are facing the lack of human, financial, and institutional resources, and this inadequacy has hindered them from realizing the full benefits of the negotiating process*'. <http://www.moc.go.th/thai/dbe/global10.html>.

Regime Shift or the Desire for Conflict?

The state as 'subject' of international law[43] has given way to a plurality of subjects matched by a parallel pluralism in the *sources* of the law,[44] precipitating concerns over a 'disaggregated' if not 'dissolving' state given the growing fragmentation of international law.[45] Ostensibly, this is paired with the need to reclaim politics in the face of equal fragmentation occurring in that field. David Kennedy's insights raise some interesting issues related to the emasculation of participatory processes and how this is occurring:

> [T]he resignation about the demobilization of a vigorous public policy indicates that even as welfare states erode, the notion of public policy they exemplified is alive and well: pubic policy is territorial intervention by 'public authorities against a background of apolitical private initiative. This resignation refuses to treat as political, as public, as open to contestation, the institutions and norms which structure background market. If we think of the private domain as political, it is not at all obvious that the current situation is one of fragmentation rather than concentration … Where factors of production are relatively immobile, a locality or private actor may have more capacity to conduct global public policy than either the welfare state or the institutions of international economic law. The question, in other words, is not whether politics or where politics, but what politics. Internationalists

43 Art. 38(1) in *The Statute of the International Court of Justice* (ICJ) identifies the state as subject and sets out valid sources of law. Non-state entities have acquired some international personality and can maintain legal relations, enjoy rights, or assume prescribed obligation. See e.g. Kindred et al., *International Law Chiefly as Interpreted and Applied in Canada*, op. cit. at 11–91.

44 See Anne-Marie Slaughter, 'Breaking Out: The Proliferation of Actors in the International System' [Breaking Out] in Dezalay and Garth, *Global Prescriptions*, op. cit. at 12–36 argues that with the proliferation of actors and legal norms, international law can no longer be limited in application to the monolithic sovereign state – itself disaggregating into its components of courts, regulatory agencies, and legislatures legitimating the system through law and creation of legal authority for the legitimate set of actors who can *produce* it. But see Linda Weiss, *The Myth of the Powerless State* (New York: Cornell University Press, 1998) asserting that state capacities for domestic transformative strategies provide a competitive advantage by offering state-embedded institutions for economic governance.

45 The International Law Commission (ILC) established a Study Group on 'Fragmentation of international law' chaired by Professor Bruno Simma on 9 May 2002. See *Daily Bulletin*, 54th session of the ILC, at <http://www.un.org/law/ilc/sessopms/54/jourchr.htm>. See also Gerhard Hafner, 'Risk Ensuing from Fragmentation of International Law', ILC, Annex to Report on the Work of the 52nd session, General Assembly Official Records, 56th session, Supplement No. 10 (A/55/10), 321–39 at <http://www.un.org/law/ilc/reports/2000/english/annexe.pdf>.

> should care less about whether the State is empowered or eroded than about the distribution of political power and wealth in global society … Technocratic governance, a displacement of public by private, of political alignments by economic rivalries, the unbundling of sovereignty into myriad rights and obligations scattered across a global civil society – all this has transformed international affairs … The result is an intellectual class unable to develop viable political strategies for the world it has applauded into existence, ratifying the political choices that result from the arrangements of private power to which the State has handed its authority, while still celebrating the expansion of participation in an emasculated public policy process.[46]

The International Chamber of Commerce (ICC) reports that no less than '162 regional trade agreements notified under the GATT and the WTO are in force today'. It was anticipated that there would be between 100 and 200 new regional trade formations.[47] International law specialist Joost Pauwelyn, a former Legal Affairs Officer for the WTO and now professor of law explains the potential consequences that result:

> [m]odern international law, is indeed, composed increasingly of treaty-based sub-systems (such as that of the WTO … or the World Intellectual Property Organization) … said to have their own sector-specific 'international law', law-maker and law-enforcement mechanism. Like national laws within the discipline of 'conflict of laws', these sub-systems of public international law interact and may give rise to conflict … [T]o talk of these sub-regimes as being separate 'international laws' which may 'conflict' would give the wrong signal. First, it would lose sight of *general* international law in creating the impression that these sub-regimes are 'self-contained regimes' to be evaluated exclusively with reference to norms created *within* the particular sub-regime. Second, it could be understood by some as elevating what are basically treaty norms (say, WTO provisions) of a *contractual* nature to the status of 'law' in the strict domestic law sense of norms imposed by an independent

46 David Kennedy, 'The Forgotten Politics of International Governance' (2001) 2 E.H.R.L. Rev. 117, at 120.

47 ICC Policy Statements (2002). For a discussion of potential negative effects of regional trade agreements and the relationship with Articles XXIV and V of the GATT, see the ICC, 'Regional Trade Agreements and the Multilateral Trade System', Commission on Trade and Investment Policy, November 2002, <http://www.iccwbo.org/home/statements_rules/statements/2002/Regional%20trade%20agreements_multilateral%20trading%20system.asp>.

> 'legislator' on all subjects (i.e. states) of the sub-regime independently of their will.[48]

'For foreign policy', expounds Louis Henkin, 'perhaps the most important legal mechanism is the international agreement, and the most important principle of international law is *pacta sunt servanda*: agreements shall be observed. This principle makes international relations possible'.[49] Adhering to *pacta sunt servanda* grows more difficult the greater the number of international, transnational, multinational, and bilateral agreements imposing diverse obligations. A state may face conflict with disparities between *domestic laws* in a national legal system and the norms provided in treaty based sub-systems (vertical conflict) or between different international instruments (horizontal conflict). A counter hegemonic narrative has emerged from the neo-liberal rhetoric of TRIPS; Laurence Helfer describes it as a post-TRIPS regime shift that reflects efforts to (re)locate international IP governance to a more compatible space for DCs and LDCs:

> Intellectual property issues are now at or near the top of the agenda in intergovernmental organizations such as the World Health Organization and the Food and Agriculture Organization, in international negotiating fora such as the Convention on Biological Diversity's Conference of the Parties and the Commission on Genetic Resources for Food and Agriculture, and in expert and political bodies such as the United Nations Commission on Human Rights and its Sub-Commission on the Promotion and Protection of Human Rights. In some of these venues, 'intellectual property lawmaking' involves the negotiation of new treaties; in others, such lawmaking occurs through the re-interpretation of existing agreements and the creation of new nonbinding declarations, guidelines, recommendations, and other forms of 'soft law'.[50]

48 Pauwelyn, 'Conflict of Norms', op. cit. at 9. Pauwelyn notes how international law is *not* merely positive contractual law, however. For additional criticism of 'voluntarist' or 'auto-limitation theories' and their inability to adequately address *custom* derived norms and why as a 'genre' they are binding, see R. Higgins, *Problems and Process: International Law and How We Use It* (1994) at 15 asserting the view that international law is 'decision-making by authorized decision-makers, when authority and power coincide'.

49 Henkin, *The Rights of Man Today*, op. cit. at 19.

50 Laurence Helfer, 'Regime Shifting: The TRIPS Agreement and New Dynamics of International Intellectual Property Lawmaking' (2004) 29.1 Yale J. of Int'l L. at 6 and footnote 20: 'International law and international relations scholars have recently emphasized the importance of non-binding norms, or soft law, as a method to promote international co-operation and alter state behaviour'.

Helfer's delineation of a regime shift of IPR governance from 'WIPO to GATT to TRIPS'[51] and the subsequent post-TRIPS regime shifting *out* of the WTO into other international fora with non-binding soft law instruments, may be understood simply as a further *decentralization* and fragmentation of international law rather than a portent of a total regime *shift* per se. Helfer concedes this point despite his broader argument that orchestrated efforts by developing states and NGOs are deliberately taken to create IP norms that supplement, modify, or supplant TRIPS norms:

> But regime shifting may also spawn inefficient rivalries among actors or attenuate mechanisms for holding international institutions accountable to affected constituencies. And it increases the likelihood of conflicting or incoherent legal obligations for states and private parties – an especially grave concern for an international system with few hierarchical rules for resolving such inconsistencies.[52]

In the first decade of the WTO, civil society has begun to assume a critical role in influencing negotiated outcomes in IP protection.[53] It may be argued that the shifting is still occurring such that the transition of international IP out of the WTO and into subaltern 'international regimes concerning biodiversity, plant genetic resources, public health, and human rights'[54] is still incomplete. But considering the binding nature, political clout, and normative force of the TRIPS, as well as the developed world's attraction to the juridical quality of the WTO for its dispute settlement mechanism, developed countries would be reluctant to allow the shift to ever be completed unless moved into more protective TRIPS-Plus regimes; particularly since the more powerful WTO members enjoy a comparative advantage in information capital and familiarity with formal legal systems which allows them to dominate and succeed in the

51 Ibid. at 7. Helfer proceeds to explain how various actors use regime shifting 'to maximize desired policy outcomes, to relieve pressure for action in other international venues, to create treaties and soft law in tension with TRIPS and to lay the political groundwork needed to integrate new principles, norms, and rules of intellectual property protection into the WTO and WIPO'.

52 Ibid. at 81.

53 See e.g. Sell, *Private Power*, op. cit. at 121–62 and the role of civil society in negotiation of the Doha Declarations on TRIPS and Public Health, op. cit. See also Amani and Coombe, 'The Human Genome Diversity Project', op. cit. and the role of civil society in objecting to discriminatory genomics, the emergence of the *Indigenous Research Protection Act (IRPA)* of 2000 and UNESCO's 2003 adoption of the International Declaration on Human Genetic Data. See also Rosemary Coombe, 'The Recognition of Indigenous Peoples and Community Traditional Knowledge in International Law' 14 ST. Thomas Law Review 275–85.

54 Helfer, *supra* note 50 at 81.

WTO's judicial-like setting.[55] A countervailing regime shift that would match the magnitude of the one which occurred with TRIPS, or affecting the same scale of countries, may *never* transpire. But the norms, principles, and rules generated outside of TRIPS, Helfer explains, have their own utility:[56]

> Regime shifting allows state and nonstate actors, particularly those that have been ignored or marginalized in other international regimes, to experiment with alternative ways to achieve desired policy outcomes – such as different institutions, different decision-making procedures, and different compositions for actors with different types of subject matter or functional expertise. For NGOs, particularly those shut out of a forum by state-only access rules, regime shifting also offers the obvious advantage of greater access to lawmaking processes.[57]

Most UN reports released on this subject are critical of TRIPS' uniform standards and urge re-examination for DCs, LDCs, and development agendas.[58] Participation in norm generation in other legal fora has been successful but short of abandoning TRIPS or waiting for further amendments to take force,[59] it is doubtful whether individuals will become any greater beneficiaries of a *de facto* shift for the very reason that international IP governance under WIPO was unsuccessful: these other fora simply lack teeth. Furthermore, that is not to say that efforts undertaken to amend TRIPS are not worthwhile. Daniel Gervais argues that TRIPS may be the appropriate means for extending substantive IP protection to new fields in the interest of DCs and LDCs. In relation to prescribing protection for traditional knowledge he warns, however, of some potential problems:

---

55 Some states' legal systems may constrain government action and the ability to bring successful complaints. See Christina R. Sevilla, 'The Political Economy Model of GATT/WTO Trade Complaints' (1997) Jean Monnet Working Paper Series No. 5, <http://www.jeanmonnetprogram.org/papers/97/97-05.rtf>.

56 Helfer, supra note 50 at 9.

57 Ibid. at 55.

58 See e.g. UNDP, *Making Trade Work*, op. cit. at 221–2; Commission on Intellectual Property Rights, Integrating Intellectual Property Rights and Development Policy (2002), IPR Commission, <http://www.iprcommission.org/graphic/documents/final_report.htm>.

59 For an important discussion on this point and the impediment TRIPS has created for development, see the UNDP Report criticizing TRIPS for the developing world and encouraging these countries to 'begin dialogues to replace TRIPS … with alternate intellectual property paradigms' and in the interim to 'modif[y] … the way the agreement is interpreted and implemented'. There is no indication that TRIPS is being implemented appropriately and with balance. *Making Trade Work*, *supra* note 58 at 221, 222.

> First, while one is perfectly free to criticize TRIPS or suggest that it be amended, the prospect and extent of any such modifications are unknown and difficult to predict. In addition, if amendments do happen, they are likely to require several years to negotiate and enter into force. Third, one would have to be an extreme optimist to think that TRIPS will be amended in ways that respond to every need and concern of holders of traditional knowledge. For these reasons, finding solutions within the confines of the *current TRIPS* text seems a rational way to proceed – which does not mean that work on rewriting TRIPS is ill conceived. Both approaches are complementary.[60]

While abandoning or amending TRIPS may not seem a likely prospect to garner consensus, merely shifting the problems related to IP governance *outside* of the multilateral trading regime by creating *new* norms will not necessarily help solve them either unless, legally, we take a hard look at these other norms in dispute settlement at the WTO.

Professor Dunoff offers a different insight into the rationale behind Helfer's observations of a regime shift.[61] He contends that state and non-state actions in these other international venues are really about *creating* conflict in recognition and in protest at the binding yet flawed existence of TRIPS and the WTO's faulty 'regime architecture'. Effectively, the impetus for shifting is remedial but not exactly in the way Helfer envisions it. Rather, conflict creation (*within* the WTO and domestic contexts) is a means to bring awareness of, while fostering a demand for, change:

> [t]hat is, one significant component of civil society's response to the frustration resulting from WTO litigation is agitation in other international fora. How should we understand this response? Maybe it is simply forum shopping. But maybe something more significant is at stake. Perhaps civil society, unable to obtain a voice at the WTO, has embarked on a form of exit, in a way that constitutes a deliberate attempt to create conflicts with TRIPS norms. The resulting legal inconsistencies create pressures on states to try to address these conflicts, hopefully in a way more amenable to

60 Daniel Gervais, 'Traditional Knowledge and Intellectual Property: A TRIPS-Compatible Approach' (2005) Spring Issue 1 Mich. St. L. Rev. 137 ['TK'] at 139, footnotes omitted, emphasis added.

61 In addition to forum shifting as a means of forum shopping, and providing the opportunity for broader participation by state and nonstate actors (an opportunity to have 'a voice'), Helfer outlines how IP norm-creation outside of TRIPS serves as a kind of pressure valve that allows industrialized states that hope to defeat developing countries' proposals to amend TRIPS to resort to the WIPO or other soft law 'to shunt issues away from the WTO and thereby reduce pressure from developing countries to address those issues in the WTO'. Helfer, *supra* note 50 at 79.

> developing state interests. From this perspective, forum shifting represents a strategic exit and an effort to create strategic legal inconsistency, as part of a larger effort to force change by generating rules in one regime that are at least arguably inconsistent with those in another.[62]

Instead of 'forum shopping', Dunoff relates this 'forum shifting' to its utility in positively contributing to the legitimacy crisis of the WTO in order to rouse support for increasing equity, fairness and civil participation, that may improve regime architecture and help induce future TRIPS amendments; it seems to be working.

Two important developments have occurred under WIPO's auspices. In October 2000, the Intergovernmental Committee on Intellectual Property, Genetic Resources, Traditional Knowledge and Folklore (IGC)[63] was formed to address the challenges and claims of indigenous communities over the protection of their genetic resources, traditional knowledge (TK) and traditional cultural expressions – often as against their own governments. The IGC created a forum to further the dialogue on the relationship of these areas and IPRs and is supported by a voluntary fund, of contributions from member countries, used to finance the participation of indigenous communities. At the eleventh meeting of the IGC in Geneva in 2007, 24 new NGOs were admitted to the IGC bringing the total up to 180 and '[m]ost countries, especially developing countries, repeated that the existing intellectual property system is not sufficient to properly protect indigenous heritage, but that *sui generis* rights for collective owners would be needed'.[64] The discussions followed from the desire to customize IPRs to work for DCs and their agendas under WIPO and proved instructive, empowering, and enabling. One positive externality has been to encourage similar efforts under

---

62 Dunoff, 'Public Participation in the Trade Regime', op. cit. at 967. Dunoff adds, '[i]n short, it seems that a diverse group of participants, including grassroots activists, farmers' groups, environmentalists, development groups, human rights groups, consumer organizations – in a phrase, civil society – is the engine here. So after finding little or no voice at the WTO, it seems that civil society groups have started to agitate in other international regimes where they are finding that they can have a meaningful voice'. At 968.

63 IGC was created by WIPO's General Assembly (WO/GA/26/6) calling on a wide range of participants to examine issues and debate policy responses relating to IP, TK, and TRIPS in an international forum. For information on the IGC's mandate and related documents, see <http://www.wipo.org/globalissues/index.html>; in particular, Section II.D discusses why integrating counter-regimes norms into WIPO and the WTO is important for DCs and WIPOs role to this end.

64 Silke von Lewinski, 11th Intergovernmental Committee on Intellectual Property and Genetic Resources, Traditional Knowledge and Folklore, Geneva, 3–12 July 2007 (WIPO) Report on the Discussions on General Questions and the Protection of Folklore,<http://www.alai.org/documents/Report%20IGC%2007%20July.doc>.

the WTO forum where there is a now a strong push from NGOs, DCs and LDCs for another TRIPS amendment. Meanwhile, the February 2008 meeting of WIPO's IGC committee resulted in the commitment to a secretariat analysis and member state review, possibly in consultation with the Committee Chair, before the next IGC meeting anticipated for October 2008. Many developed countries prefer to deal with PGRs under WIPO (perhaps because of the ambiguities and flexibilities of TRIPS in this area) and may welcome the development of standards under its framework rather than that of the WTO. Still, there are concurrent efforts underway for a TRIPS amendment as part of the Doha Round that would help insulate DCs and LDCs resources against legal appropriation commonly referred to as 'biopiracy'. Its fate is not yet settled but demonstrates that WIPO's norm setting efforts continue to create a feedback loop for TRIPS and that conflict *is* useful in generating resolution. In the interim we are left with the task of resolving disputes in an era when the potential for disputes has substantially increased and the power for engagement and enforcement is inequitably disparate. To summarize, we can conclude that creating conflict is functional in generating the imperative for resolution in three noteworthy ways by: 1) articulating the conflict in concrete terms and thereby making the conflict more pronounced; 2) creating a more immediate need for resolution; 3) inviting the use of procedural principles of custom and law to interpret treaties, giving preference to the substantive content articulated in subsequent instruments and additional legal norms as required by the Vienna Convention on the Law of Treaties (VCLT).[65] Developing a framework within the WTO for working through contested operability and the juridical interaction of hard and soft law,[66] as well as different sub-regime obligations, becomes a critical pragmatic need as well as one of social and redistributive justice,and calls on established rules of treaty interpretation.

Conflict of Norms: The Matrix of Conflict

A conflict of norms arises when

> *it is impossible to comply with all requirements of two norms.* The impossibility of complying with two norms implies that the norms are mutually exclusive; they cannot coexist in a legal order. Compliance with one norm entails non-compliance with the other.[67]

---

65 VCLT, <http://untreaty.un.org/ilc/texts/instruments/english/conventions/1_1_1969.pdf>. See also discussion in text below.

66 See Helfer, *supra* note 50 at 77–9; these conflicts are not the subject of this analysis.

67 Seyed Ali Sadat-Akhavi, 'Methods of Resolving Conflicts Between Treaties' (Boston: Brill Academic Publishers, 2003) at 5. Sadet-Akhavi outlines the sources of

Some norms are mandatory and impose firm obligations on a party to do something (obligatory norms) or refrain from doing something (prohibitive norms) – identified by 'shall' in the norm. Some norms are permissive with discretion/freedom provided to do or refrain from doing something, using the language 'may'.[68] Technically, 'compliance' can only be required by mandatory norms but may extend to include the permissive exercise of discretion, recognizing that 'two norms are reconcilable if there is at least one way of complying with both of them'.[69] What reconcilable 'conflict' means is that the norms compete until the preferred reading (compliance) is achieved; if there is only one means of reconciliation, it effectively transforms the permissive norm into a mandatory one.

Conflict is '*inherent* in a system of law', according to Pauwelyn, because it is impossible to comprehensively foresee and anticipate the interaction of a norm with existing norms and future norms. In addition, to remain adaptable (and in international law, to be acceptable to the range of parties), norms are phrased in general and vague language.[70] Pauwelyn provides a detailed and comprehensive study of conflict of norms, offering a taxonomy of rights and obligations, a framework of how the different norms of international law interact, a hierarchy with WTO norms, and the supporting WTO jurisprudence. His classification of norms has two notable distinctions. First, he considers the mandatory norms to create *obligations*: the obligation to do something (a prescriptive norm) imposes a positive obligation and is a 'Command'; the obligation to refrain from doing something (a prohibitive norm) imposes a negative obligation and is a 'Prohibition'. The intricate system of regimes and sub-regimes are also *rights* creating. Discretionary norms create rights: the right not to do something is an 'Exemption'; the right to do something is a 'Permission'.[71] Two additional norms he identifies are those which empower an organ, institution or individual (other than states) with legal capacity under international law, such as Article 2.1 of the DSU which establishes the DSB or WTO committee decisions appointing a committee chairman, and norms that regulate the creation, application, interplay, suspension, termination, breach or enforcement of other norms of international law regardless of whether they additionally impose obligations on states or grant them rights

treaty law, the basis of real and false conflicts, and the resolution of conflict under international laws.

68 This is permission *stricto sensu* and is the meaning intended by the use of the term in this chapter. However, another possible use of the term 'permission' may be *lato sensu* which means that the act is not forbidden; that is, it is permitted. See discussion and ensuing footnotes in Ali Sadat-Akhavi, ibid. at 6.

69 Ibid. at 42

70 Pauwelyn, *Conflict of Norms*, *supra* note 48 at 12.

71 Ibid. at 158.

(secondary norms). The norms under the Vienna Convention would fall into this category.

The basic distinction he highlights between rights and obligations conferred in a single text is insightful and sufficiently nuanced to bring balance to various interests under TRIPS and the WTO Agreements. That TRIPS contains both *rights* and *obligations* may create intra-textual contestation 'not only between two contradictory obligations but also as between an obligation and an explicit right … [I]n practice a conflict of norms will always boil down to a conflict of *rights* and/or *obligations* resting on one or several states'.[72]

Using Pauwelyn's structural conflict taxonomy and assisted by insights on treaty interpretation offered by Seyed Ali Sadat-Akhavi, we may discern three primary (meta) forms of relational conflict: let us call these *Inter-conflict, intra-conflict, and apparent conflict* (reconcilable norms). Inter-conflict refers to competing norms found under different regimes or sub-regimes such as, for example, a potential conflict between the CBD and TRIPS Agreement or between TRIPS and human rights instruments, customary international law, or the concept of *jus cogens* which are peremptory norms of general international law governing state behaviour. An intra-conflict of norms is one internal to the trade system and might exist between *rights* and *obligations* within a single WTO text such as TRIPS or within different WTO agreements such as GATT Article XX and TRIPS. Both of these two categories raise issues of priority or hierarchy within conflicting norms.

Most conflicts under these two categories are merely apparent, however, in that the norms may compete for priority but do not actually conflict since they can be reconciled. Upon further scrutiny the relationship of these norms would better be classified under a third category: apparent but reconcilable conflict for conflicts that can in fact be avoided. Avoidance can take many forms: recourse to treaty interpretation principles and reference to the Vienna Convention on the Law of Treaties (VCLT), further examination and liberal interpretation of the language giving rise to the perceived conflict including any explicit exclusions ('this provision does not apply to'), permissions ('members may'), special provisions (such as transition periods or the allowance of regional arrangements which would appear in conflict with the non-discrimination/MFN principles of the WTO), 'statutory' treaty-based defences (such as Article XX of GATT or, as suggested here, Article 8 of TRIPS), an equitable defence similar in principle to that historically raised in common law jurisdictions to provide relief from law's formalism in pursuit of justice, or any combination of these. A text provision, for example, may have gained international recognition and currency as a customary norm *erga omnes*, that is, in relation to the international community as a whole. *Erga omnes* is closely linked with and often derivative of *jus cogens* (the body of peremptory norms and principles from which no derogation may be made). The right to

72 Ibid. at 10.

self-determination, for example, is another in which all states have a legal interest. These examples, some of which will be examined shortly, are merely illustrative and not intended to be exhaustive. So while HR *compete*, they do not necessarily *conflict* with trade obligations because conflict, by definition, requires that the norms be mutually exclusive. The obligations arising under HR instruments will *not conflict* with trade obligations for several reasons. First, though binding as treaty obligations, the substance and justiciability of HR obligations are still a contentious issue and in relation to ESCR, may be without the precisely defined content necessary as a substantive norm for the creation of conflict between international instruments. They can clearly lead to conflict between a state's implementation and operationalization of HR norms domestically and its international trade obligations (vertical conflict), something, if contested, the ECD – devised and advanced hereunder – would remedy in relation to patenting life. Furthermore, there should be no conflict if proper interpretation is given to TRIPS provisions in a conciliatory manner. At most, these obligations *compete* in the sense that there may be less *market* force or political pressure for states to give priority to the ethical issues and HR consequences of trading than there will be to respond compliantly to perceived trade obligations (because of costs associated with WTO complaints and their enforcement), even if a state may be liable for HR obligations in domestic law (the extent of which would be significantly less costly). Finally, these rights will not conflict because HR should always *trump* trade values, either as having achieved the status of customary law or due to their normative moral force.[73]

TRIPS Conflicts with Other Instruments: Inter-Conflict with the CBD

The BRCA1 example demonstrates that a state, in attempting to comply with one set of international obligations (human rights) through domestic implementation (i.e. delivering on the right to health) may find it has created a vertical conflict with a different set of internationally prescribed norms (i.e. TRIPS patent rights) whether or not a concurrent horizontal conflict between the competing international norms exists (as between ICESCR or CBD and TRIPS for example). Some key international HR obligations, such as the right to health and food, creating potential vertical conflict were discussed in Chapter 5 and are important in the relationship between the state and its citizens. In this section, I consider the relationship between two international texts and the question of whether there is horizontal conflict between TRIPS

73 The UN Sub-Commission on the Promotion and Protection of Human Rights argues for the primacy of HR over economic policies and agreements in order to avoid conflict at the intersection of IPRs, trade and HR, <http://www.unhchr.ch/html/menu2/2/sc.htm>.

and the primary non-WTO instrument that specifically addresses ownership over genetic resources and engages sovereign claims to self-determination.

The patenting of genetic material was debated in the biodiversity field well before TRIPS, resulting in the 1992 United Nations Convention on Biological Diversity (CBD) concluded at the Earth Summit in Rio de Janeiro.[74] The CBD was ratified by over 180 nation states, displaying a significant level of commitment by its signatories. It became an important locus for negotiations by developing countries and international NGOs seeking to offset the corporate lobby to extend patent provisions to life within the multilateral trade regime during the Uruguay Round. The tension between the CBD and TRIPS has continued since that time.[75] The CBD has three main objectives set out in Article 1: '… [t]he conservation of biological diversity,[76] the sustainable use[77] of its components and the fair and equitable sharing of the benefits arising out of the utilization of genetic resources including by appropriate access to genetic resources and by appropriate transfer of relevant technologies …' [footnotes added]. The CBD acknowledges and confirms that the UN Charter recognizes a sovereign right of states to exploit their own resources under domestic environmental policies (echoed also under Article XX(g) of GATT 1947).[78] States also have a duty to ensure the activities within their jurisdiction

74 United Nations Convention on Biological Diversity, 5 June 1992, S. Treaty DOC. No. 103–20 (1993) [hereinafter *CBD, Convention, or Treaty*] at <http://www.biodiv.org/convention/articles.asp?lg=0&a=cbd-08>. The Convention came into force on 29 December 1993 – 90 days after the 50th ratification pursuant to Article 37 [CBD]. Canada was the first industrialized country to fulfill the requirement to also ratify the Convention through its domestic national system. Chairman of the Senate Foreign Relations Committee, Senator Jesse Helms, refused to present the Treaty for a vote and therefore the United States has signed but not yet ratified the Treaty (it requires a majority vote by the US Senate).

75 See generally N.S. Gopalakrishnan, 'TRIPS and Protection of Traditional Knowledge of Genetic Resources: New Challenges to the Patent System' [2005] EIPR 11–18.

76 '"Biological diversity" means the variability among living organism from all sources including, *inter alia*, terrestrial, marine and other aquatic ecosystems and the ecological complexes of which they are part; this includes diversity within species, between species and of ecosystems'. Article 2, CBD, *supra* note 74.

77 Article 2 on the Use of Terms in the CBD defines this exhaustively to mean 'the use of components of biological diversity in a way and at a rate that does not lead to the long-term decline of biological diversity, thereby maintaining its potential to meet the needs and aspirations of present and future generations'. Ibid.

78 '… nothing in this Agreement shall be construed to prevent the adoption or enforcement by any contracting party of measures … (a) relating to the conservation of exhaustible natural resources if such measures are made effective in conjunction with restrictions on domestic production or consumption.' It is worth noting that the provision applies to 'exhaustible natural resources' and biological resources may be finite insofar as continuous use is exhausted where it has led to extinction. Ibid.

or control do not cause environmental damage 'beyond the limits of national jurisdiction'.[79]

The Parties to the CBD undertake to cooperate to achieve the objectives of the CBD and are accordingly obligated to take general *national* measures for the conservation and sustainable use of biological diversity.[80] Whether TRIPS can protect biological and cultural diversity through its various existing IP provisions in the absence of further amendment has been explored comprehensively by Graham Dutfield.[81] Patenting life raises issues related to the degree of access and benefit sharing under the CBD that may be attainable with regard to non-human genetic resources (plant, animal, soil etc)[82] and the availability of the *sui generis* option for plant varieties offered in Article

79 Article 3 CBD. See also Article 4 for the scope of the CBD to protect biodiversity within national jurisdiction. Article 4(b) deals with *processes and activities*, within or beyond the limits of national jurisdiction *regardless of where their effects occur.* Ibid., emphasis added.

80 Article 5 and Article 10 CBD, *supra* note 74. See also Article 8 for In-situ Conservation and Article 9 providing for Ex-situ Conservation obligations of the Parties. For a discussion of the CBD and TRIPS, see Charles R. McManis, 'The Interface between International Intellectual Property and Environmental Protection: Biodiversity and Biotechnology'. (1998) 76 *Washington University Law Quarterly* 255–79.

81 Dutfield explores the potential use of trademarks for this purpose, including the possibility of certification marks. The potential extension of geographic indications to protect such cultural community based commodities such as Basmati rice cultivated in Northern India and Pakistan (a variety of which has controversially been patented by a foreign firm) will remain a contested issue according to Dutfield until the Parties to the CBD resolve the issue of genetic resources in *ex situ* collections which cannot be protected under plant variety protection where available because they are not a new variety. Trade secrets/undisclosed information protection provided for under TRIPS Article 2 on Unfair Competition can be used to protect the knowledge or know-how of an individual or a whole community as long as it has commercial value and provides a competitive advantage. This would allow a share of profits to be obtained from a company that misappropriates the information. See generally Graham Dutfield, *Can the TRIPS Agreement Protect Biological and Cultural Diversity?* Biopolicy International Series No. 19 (Nairobi: ACTS Press African Centre for Technology Studies, 1997).

82 See Graham Dutfield (2002), 'Sharing the Benefits of Biodiversity: Is there a Role for the Patent System?' 5(6) Journal of World Intellectual Property 899–931 ['Sharing the Benefits'], reprinted in Keith Maskus, ed., *The WTO, Intellectual Property and the Knowledge Economy: Critical Perspectives on the Global Trading System and the WTO* (Brookfield, VT: Edward Elgar, 2004) at 292. For a discussion of the informing politics, see Rosemary Coombe, 'Works in Progress: Traditional Knowledge, Biological Diversity, and Intellectual Property in a Neoliberal Era'. In R. Perry and B. Maurer, eds, *Globalization Under Construction: Governmentality, Law and Identity* (Minneapolis: University of Minnesota Press, 2003) and J.H. Vogel, 'Know-how Licenses: Recognizing Indigenous Rights over Collective Knowledge' (1997) 4 Bulletin of the Working Group on Traditional Resource Rights, 17–18.

27.3(b) of TRIPS. In recognition that indigenous communities play a critical role in the preservation of biological diversity as guardians of TK, Article 8(j) of the CBD provides:

> Each country *shall*, as far as possible and as appropriate … subject to its national legislation, respect, preserve and maintain knowledge, innovations and practices of indigenous and local communities embodying traditional lifestyles relevant for the conservation and sustainable use of biological diversity and promote their wider application with the approval and involvement of the holders of such knowledge, innovation and practices and encourage the *equitable sharing of the benefits arising from the utilization for such knowledge, innovations and practices;*[83]

This *command* norm creates a potential inter-norm conflict with the monopolistic rights of patent holders prescribed under TRIPS Article 28; not only might the benefits arising from such knowledge not be shared, but the the community from which said knowledge, innovation, and practice has been derived might also be excluded entirely. The 'putative conflict between biodiversity and biotechnology', posits Jim Chen, 'arises from a fundamental difference in factor endowments. The global north is rich in financial capital and industrial technology but poor in genetic resources. The global south is the precise opposite: biologically rich but economically poor'.[84] The issue is not, however, just about inputs and outputs in innovation but rather equally about the inclusiveness or exclusiveness of global IPRs regimes and the disparate, culturally imperialist treatment given to TK, oral culture, and ethnobiological know-how.

Drahos and Braithwaite contend that the incongruence of the legal system and the scientific process may also share the blame:

> Because intellectual property relates to information and knowledge, and because information and knowledge is built up over time by many people, it is hard to work out just what any given individual is truly responsible for. Ideas are triggered by related ones. All ideas have fuzzy boundaries. Working out where the fence of intellectual property ownership should go is

83 CBD, *supra* note 74, emphasis added.

84 See Jim Chen, 'Intellectual Property and Biological Diversity: A Misunderstood Relation', <http://www.law.ufl.edu/faculty/pdf/chen.pdf> concluding that all stakeholders in the biodiversity and biotechnology debate have exaggerated the nature of their claim and that commercial exploitation of genetic resources holds the key to biodiversity conservation, helping overcome perverse incentives to consume scarce natural resources by fencing them off.

> very difficult. In the world of commerce it is legal muscle more than moral entitlement that determines the fenceline.[85]

In short, 'indigenous communities tend not to enjoy legal property rights over their valuable (or potentially valuable) knowledge' writes Dutfield. 'It would certainly help if they did', he adds, 'though the fact that they do not is really a matter of social injustice rather than one of economic inefficiency'.[86]

The Turmeric Example

The conflict over the turmeric patent is an illustrative example. In March 1995, the US patent on 'use of Turmeric in Wound Healing' was awarded to the University of Mississippi Medical Center[87] despite the treatment forming part of the traditional knowledge of Indian communities. By 1996, the Council of Scientific and Industrial Research of India (CSIR) requested that the patent be revoked on the basis that it failed to meet the novelty and non-obviousness requirements for patentability. The patent description listed other traditional uses for turmeric, some medicinal, derived from the plant *Curcurma longa* but failed to disclose the prior knowledge of its wound healing properties for which the patent was sought. Nor did the patent disclose any real 'invention' by strict standards, failing to provide a method of extracting the active therapeutic quality described. Instead, Dutfield writes, the patent simply declared that 'turmeric is a natural product that is readily available in the food store'. The patent was ultimately revoked on domestic patentability criteria but not on India's claim that it formed part of their TK. That is, the US patent examiners are not required by law to accept the evidence of such knowledge held outside of the US as prior art unless it has been reported and *published* in scientific journals or databases or made available to the public in some tangible medium. Coordinated efforts are now taken to help ensure that such misappropriation of TK no longer occurs. The People's Biodiversity Registers Programme is an initiative originally sponsored by the WWF India, and coordinated with the Centre for Ecological Sciences of the Indian Institute of Science and the Foundation for Revitalization of Local Health Traditions (FRLHT) and many NGOs since then. Its mandate is to provide a record of local knowledge for present and future use within those communities, to preserve the knowledge, and to protect it from foreign misappropriation.[88]

---

85 Drahos and Braithwaite, *Information Feudalism*, at 26.

86 Dutfield, 'Sharing the Benefits' *supra* note 82 at 905.

87 Specifically, 'a method of promoting healing of a wound by administering turmeric to a patient afflicted with the wound' was claimed including surgical wounds and body ulcers. See Dutfield, *IPRS*, *Trade and Biodiversity*, op. cit. at 40–74.

88 See e.g. Madhav Gadgil, 'People's Biodiversity Registers: Lessons Learnt', <http://www.etfrn.org/etfrn/workshop/biodiversity/documents/gadgil/bioreg.

Because of the public controversy the turmeric patent generated, the potential conflict was resolved at the state level through patent revocation[89] without further recourse to international law. Like the MGL example, the turmeric case highlights the importance of maintaining strict patentability standards, broad prior art considerations, a common sense approach to avoiding bad patents, and the use of attested domestic solutions for producing TRIPS-compliant regulations and as a means of minimizing or reconciling potential conflict before it escalates. But revocation does not solve broad policy issues. TRIPS does not yet protect TK; India would have had a difficult time bringing a trade complaint against the US for this act of biopiracy if the patent was not revoked. Moreover, Dutfield underscores, 'Indians in the US using turmeric to treat their children's wounds were therefore infringing the patent [and if] the University of Mississippi had been awarded a similar patent in India, tens of millions of people would then have become patent infringers!'[90]

To further play out this hypothetical scenario, India could have been liable under TRIPS for lack of enforcement. But one would expect as a matter of justice, if India did not grant the patent for a public policy purpose (rather than on the basis that it failed to meet strict patentability requirements, which is more difficult to challenge), or if India granted the patent but failed to enforce it, that it should be able to defend these decisions if taken to task before a DSB of the WTO. Whether it could successfully defend its conduct requires further scrutiny of the rights and obligations of states pursuant to TRIPS and its text. Dutfield suggests that Article 27.1 TRIPS contains positive commands in providing that 'patents shall be available for' but also *negative* commands (prohibitions) – what he calls 'compulsory exclusions' – which provide that inventions that are not new, do not involve an inventive step, or are incapable

pdf>. For greater discussion of these issues from various national perspectives, see The WIPO Intergovernmental Committee on Intellectual Property and Genetic Resources, Traditional Knowledge and Folklore (IGC) Draft Report, Tenth Session 30 November–8 December 2006, WIPO/GRTKF/IC/10/7 Prov., 26 January 2007, <http://www.wipo.int/edocs/mdocs>. The IGC's mandate relates more broadly to policy debates and governance issues over genetic resources, and traditional cultural expressions (folklore) with consideration to the demands of developing countries and indigenous peoples. Issues within its scope include creating searchable databases of contractual clauses in material transfer agreements, specifying conditions of access to genetic resources and benefit sharing, disclosure of biodiversity information/TK on patent applications, creating a TK database and finding means to document TK in the public domain, and providing rules for appropriate protection (i.e. through *sui generis* protection). See WIPO Press Release PR/2002/317: 'IGC Moves Ahead on Traditional Knowledge Protection', 25 June 2002.

89 TRIPS allows for revocation or forfeiture (a regulatory right of the state). However, TRIPS Article 32 commands that opportunity for judicial review of such a decision be made available.

90 Dutfield, *IPRS*, *Trade* and *Biodiversity*, *supra* note 87 at 65.

of industrial application or fail to disclose the invention in a manner clear and complete enough to be carried out by a person skilled in the art *must* be excluded from patentability. He contrasts these with the permitted exclusions in Article 27.2 and Article 27.3[91] – these, it will be recalled, were permissive discretionary norms that allow states to provide exclusions specifically related to life, health, morality, and *public ordre* (public policy) in domestic patent law.

Domestic law *can* impose *mandatory* exclusions from patentability by requiring *additional* pre-conditions to those required by Article 27.1 for granting a patent. But, there is an error in inferring that these compulsory exclusions derive from TRIPS Article 27.1 which provides that if certain conditions precedent are met (the patentability requirements), members are obliged by a command norm to issue the inventor a patent. If we were to reduce this to logical equation, it would be: 'if A then B', where A represents the criteria for patentability and B represents a patent grant. The claim that Article 27.1 creates *compulsory exclusions* is to assert that the extended logical truth from this syllogism is 'if not A, then not B'; this is not the logically deduced argument. Put another way, nothing in the command norm of Article 27.1 legally *prevents* countries or prohibits them from providing a patent when these conditions precedent are *not* met – so long as they are additionally available when they *are* met. In fact, what should be obvious by now is that lax patentability standards and the proliferation of bad patents sadly suggest that patents *are* granted without the TRIPS requirements *actually* being met even though this is, as Dutfield suggests, poor practice and a betrayal of the bargain with the Crown.

Since TRIPS prescribes universal *minimum* standards, the language of Article 27.1 appears not to prohibit the adoption of other – additional – criteria for granting patents so long as patents are available, at a minimum, where its conditions precedent are met; this is supported by the language 'patents shall be *available*' rather than 'patents shall be *granted*'. Thus, in terms of inter-conflict norm, the conflict may be resolved domestically in a conciliatory manner with both CBD obligations and TRIPS obligations by giving interpretation to the content of the patentability criteria which are undefined in TRIPS and by allowing states, India in this example, to create additional standards if necessary to reconcile CBD obligations, such as evidence of informed consent, benefit sharing contracts, and other pre-conditions to patentability, ownership, or commercialization.

In terms of intra-norm conflict, the requirements of Article 27.1 could possibly be resolved with reference to the exemptions in Article 27.2, 'Members may exclude from patentability inventions ... necessary to protect *ordre public or morality, including to protect* human ... health'. The grounds for invoking this right are not exhaustive but inclusive based on the language. Members

91 Ibid. at 19–20, 24.

also have rights of exclusion under Article 27.3(a) for therapeutic methods for the treatment of humans or could rely on clause (b) for plants (though *sui generis* protection is required as an alternative to plant patents).

Sui generis regimes may provide more or less protection than patent rights and can use different standards as criteria. Developed countries usually advocate UPOV 1991 as the template for such regimes while DCs prefer UPOV 1978 due to the greater flexibility, recognition of farmers' rights, and weaker scope of protection conferred. The latter upholds the farmer's privilege to save seeds from year to year and thereby would lessen intrusion on customary practices of plant variety breeding, but countries now joining UPOV are required to accede to the 1991 version.[92] Neither the nature nor scope of a *sui generis* regime, nor the period for protection, is prescribed by TRIPS, which is one reason that the United States has actively and consistently advocated for the removal of Article 27.3. IPRs have influenced the evolution of other international regimes governing plant genetic resources (PGRs) initially resisted and contested by developing country governments who considered them public goods defying private proprietary interests regardless of whether these resources were in their natural state (in situ), in global seed banks (ex situ), or modified through human intervention towards the creation of new seed or plant varieties.

Negotiations have usually occurred in the context of the Commission on Genetic Resources for Food and Agriculture (CGRFA) under the Secretariat appointed by the UN's FAO. Since 1983, the UN Food and Agriculture Organization (FAO) has been developing its Global System for the Conservation and Sustainable Use of Plant Genetic Resources for Food and Agriculture with the objective of promoting safe conservation and unrestricted availability and intergenerational sustainable use of PGRs. The CBD, while addressing genetic resources generally, did not deal with PGR for food and agriculture specifically or the access rights to PGR ex situ vis-à-vis international seed banks.[93]

TRIPS Articles 27.1 and 27.2 exemptions are discretionary and create *rights* for members. Given the connection between health and food, these can

---

92 For a discussion of UPOV and recommendations reconciling potential UPOV and TRIPS obligations or CBD and TRIPS obligations through *sui generis* protection within patent legislation, within plant variety protection or possibly under biological conservation law, see Dutfield, *IPRs*, *supra* note 87 at 75–85, 128–9.

93 See generally <http://www.fao.org>. Much of the soft law declarations developed to protect farmer's rights, national sovereignty and public domain treatment of CGIAR seed holdings in the non-binding *International Undertaking on Plant Genetic Resources* (IUPGR) were finally codified into the International Treaty on Plant Genetic Resources for Food and Agriculture (ITPGR), <http://www.fao.org/ag/cgrfa/IU.htm>. For a broader discussion on potential conflicts between this instrument and TRIPS, see Helfer, *supra* note 50 at 37–42. See also Dutfield, *IPRs*, *supra* note 87 at 100–105.

be said to extend to the turmeric example. The rights under these provisions pre-empt both intra-norm and inter-norm conflicts with TRIPS. That is not to say that inter-norm conflict between TRIPS and its predecessor, the CBD, is not possible. For example, it is not clear whether plant, animal, and human genetic resources fall within the language of Article 27.3(b) as 'plants' or 'animals' since they are merely a subcomponent of these. But, given the SCC interpretation, which would confer rights to the full higher life, perhaps there is now scope for Canada to argue plant genetic patenting is captured by this right in amending domestic laws for their exclusion. Unless a member invokes this exemption right under Article 27.2 or a liberal interpretation is given under 27.3(b) when a member contests a measure pursuant to it, then genes would fall within the command of Article 27.1 as patentable. Yet, Article 15 of the CBD affirms the sovereign rights of states over their natural resources and recognizes that the authority to determine access to genetic resources (coming from that country as a country of origin) rests with the *national* governments subject to *national* legislation. This is consistent with the permissive rights-creating norms of Article 27.2 and Article 27.3 but may be inconsistent with the interpretation that developed countries would seek.

Article 15.4 and Article 15.5 of the CBD provide for access and informed consent respectively; Article 15.6 supports the interpretation given earlier that any HR to IPRs is a right of *participation* and *access*: 'Each Contracting Party shall endeavour to develop and carry out scientific research based on genetic resources provided ... with the full participation of, and where possible in, such Contracting Parties'. Article 15.7 provides another command norm potentially in conflict with TRIPS:

> Each Contracting Party shall take legislative, administrative or policy measures, as appropriate and ... with the aim of *sharing in a fair and equitable way the results of research and development and the benefits arising from the commercial and other utilization of genetic resources with the Contracting Party providing such resources*. Such sharing shall be on mutually agreed terms.[94]

Provision for access to (and transfer of) technology is found in the command norms of Article 16.3 of the CBD:

> Contracting Parties ... which provide genetic resources are provided access to and transfer of technology which makes use of those resources, on mutually agreed to terms, *including technology protected by patents* and *other intellectual property rights*, where necessary, through provisions of Article 20 [financial resources] and 21 [financial mechanisms] and in accordance with international law ...

94 CBD, *supra* note 74.

Article 16.5 'Recognizing that patents and other intellectual property rights may have an influence on the implementation of this Convention, [Contracting Parties] *shall cooperate in this regard subject to national legislation and international law in order to ensure that such rights are supportive of and do not run counter to its objectives*'. These provisions create tensions but they seem reconcilable with TRIPS – for example, with the provision for exhaustion (Article 6), exception to the rights conferred (Article 30), compulsory licensing (Article 31), or even the revocation of such patents on the basis, for example, of strict domestic patentability standards not being met. The implication for (plant or animal) genetic resources unique to a particular state is controversial but less so than potential conflict over access to patented human genetic resources given the continued recognition that these are to be conceived as the *common* heritage of humankind.

Another TRIPS Amendment?

Some scholars believe there are *inherent conflicts* between TRIPS and the CBD that cannot be resolved without amendment to TRIPS because the latter allows patentability of genetic material which may encourage biopiracy of this material by private parties contrary to the CBD's confirmation of sovereign rights over genetic resources.[95] The envisaged amendment would expressly exclude the patentability of life and parts thereof, the patentability of inventions contrary to Article 15 of the CBD, and those based on traditional or indigenous knowledge. While express language would eliminate any doubt, there is nothing *inherently* conflicting about TRIPS if given proper interpretation. If we accept the Dunoff thesis that (apparent) conflict is useful as a system of generating improved measures for their resolution, then we need not attempt to solve the problem by resorting exclusively to international governance instruments. Rather, as suggested in the first branch of appropriate state agency prescribed, states should strive to resolve any potential conflict at first instance by adopting any of the permissible measures suggested such as stricter domestic requirements to patentability[96] or otherwise rely on member rights provided by the discretionary (exemption) norms of Article 27.

---

95 See e.g. V. Tejera, 'Tripping Over Property Rights: Is it Possible to Reconcile the Convention on Biological Diversity with Art. 27 of TRIPS Agreement?' (1999) 33 New Eng. L. Rev. 967 at 983–5. See also 'NGOs, Intellectual Property Rights, and Multilateral Institutions' Queen Mary Intellectual Property Research Institute, <http://www.ipngos.org/casestudies/agriculture/index.html>.

96 Although patent law's traditional distinction – between inventions, which are patentable because of the value added by the innovator and discoveries, which are to remain in the public domain because that which is discovered has always been there – is gradually being eroded in the biotech sector. See Peter Drahos 'Biotechnology Patents, Markets and Moralilty' (1999) 21 Eur. I.P. Rev. 441.

TRIPS Article 27.3(b) has an internal provision for review with respect to the patentability of plant varieties, animal inventions and the production processes for these, 'four years after the date of entry into force of the WTO Agreement'. Unlike other TRIPS provisions, this Article does not expressly state by *whom* that review is to occur and therefore the various UN reports outlined herein on the interplay of TRIPS, HR, and HD are even more relevant in tacitly 'regime shifting' in order to inform the dual process of negotiation and dispute settlement. Nevertheless, the TRIPS Council commenced its own review in 1999 which was broadened by paragraph 19 of the 2001 Doha Declaration to include consideration of the relationship between the TRIPS and the CBD, traditional knowledge and folklore. Members have contributed significantly by circulating opinion documents on their interpretation.[97] The Declaration also provides that the Council's work programme, including overseeing implementation issues pertaining to the whole of TRIPS under Article 71.1 and the review of Article 27.3(b), is to be guided and informed by the objectives and general principles of the Agreement as provided for under Articles 7 and 8 respectively, taking development issues fully into account.[98] The review of Article 27.3(b) has led to extensive discussions on substantive and procedural issues. As part of the Council's process of information gathering, many members have submitted formal communications to the Council and some informal ones. The Secretariats of the FAO, CBD, UPOV, and WIPO also contributed. In October 2000, the Special Session of the General Council (SSGC) on Implementation requested the Council continue its ongoing work to clarify the relationship between TRIPS and the CBD.[99] The SSGC urged the TRIPS Council to grant observer status to the CBD Secretariat but no consensus was reached on this and the matter was left for the Chairperson to report to the SSGC on his responsibility.[100]

What is emphasized in the CBD is not that such information *cannot* be patented but that cooperation amongst states for equal access and benefit sharing be ensured; strict enforcement of TRIPS-mandated exclusive proprietary 'rights' may unduly impose on these substantive distributive allocations. Also, the CBD requires that relevant communities obtain informed consent before they go prospecting for genetic resources. Accordingly, as Dr Bodeker

97 Papers at <http://www.wto.org/english/tratop_e/TRIPS_e/art27_3b_e.htm>. Other provisions requiring notification of Council by Members are Articles 63(1), 1(3), 3(1), 4(d) and 69. For a detailed history and discussion of the Council's formal meetings since 1999 and his view on the review provisions, see Gervais, *TRIPS*, op. cit. at 27–67, 227–32.

98 See para. 19, Doha Ministerial Declaration, op. cit. See also WTO's reference to this relationship at <http://www.wto.org/english/tratop_e/TRIPS_e/art27_3b_e.htm>.

99 See Document WT/CG/M59, para.43 and discussion of these proceedings in Gervais, *supra* note 97 at 38–40. (See Document IP/C/21.)

100 Document IP/C/21.

suggests, if greater use of TK (especially the health benefits of such traditional medicinal knowledge) is to be made for the benefit of all interested stakeholders, it would be prudent for 'WIPO, the CBD and the WTO to coordinate their policies and instruments in partnership with traditional knowledge holders as well as with conventional stakeholders such as governments and industry'.[101] Patenting life, preserving biodiversity, equitable access, and benefit sharing formed formidable challenges to TRIPS' inclusion in the Uruguay Round.[102] The relationship between the CBD and TRIPS was unresolved to the end and left for another day – perhaps another document. There is some discussion of the possibility of a TRIPS II emerging from the general nature of the review under Article 71.1 to address issues from the built-in agenda – such as whether the only acceptable *sui generis* regime under Article 27.3(b) would be a UPOV system. New issues such as stronger substantive requirements (TRIPS Plus) for protection and a strengthened enforcement section allowing, for example, trade associations to act before judicial authorities, freezing of bank accounts and assets and winding up companies in serious commercial infringement cases, are all emerging concerns for exploration. Daniel Gervais believes that a separate agreement is more likely, however, than TRIPS modification.[103] By other accounts, the proposed TRIPS Biodiversity amendment has secured support from more than 77 of the 153 WTO member states and would require domestic provision for securing prior informed consent, mandatory disclosure of genetic resources, and benefit sharing if adopted, but its fate has not yet been secured.

The Doha Round has succeeded in achieving limited improvement for HR through the public health amendment, which can serve as an instructive precedent for similarly coordinated efforts. The specific reference to biodiversity, TK and folklore protection in the 2001 Ministerial Declaration promises to level the gains of IPRs for DCs and LDCs, correcting the hitherto parasitic relationship of 'First World' inventions and 'Third World' resources.[104] For now, India may have resolved the potential conflict with s.3(d) of the 2005 *Patent (Amendment) Act* which excludes from patentable subject matter any

> new form of a known substance which does not result in the enhancement of the known efficacy of that substance or the mere discovery of any property or new use for a known substance or the mere use of a known process or

101 Gerard Bodeker, 'Traditional Medical Knowledge, Intellectual Property Rights and Benefit Sharing' 2 Cardozo J. of Int'l and Comp. Law 785 at 812.

102 Gervais, *supra* note 97.

103 Ibid. at 48.

104 Doha Declaration, *supra* note 98, para. 19. See generally Gervais, 'TK' *supra* note 60 at 146.

> apparatus unless such known process results in a new product or employs at least one new reactant.[105]

And China, another country with a strong history of TK and cultivated PGRs, has a broad set of exclusions from patentable subject matter that extends human and genetic molecular manipulation techniques and their products. Article 4 of the *Modern Patent Law* provides that 'inventions-creations' that are contrary to the laws of the state, contrary to social morality, or detrimental to public interest' are excluded from patentable subject matter. The Patent Examination Guidelines further instruct that this is to be interpreted to exclude from patentability:

> The process for cloning humans or cloned humans; the process for changing the inherent identity of the human reproductive system; the exploitation of human embryos for industrial or commercial purposes; the process for changing animals' inherent identity that may result in animal suffering but has no substantial benefit to human or animal medical treatment, and the animal obtained thereby; invention that may result in personal injury, property damage or environmental pollution, or the characteristics or device of which involve nationally important political events, people's feelings or religious beliefs.[106]

China's and India's exclusions reflect an intelligent and progressively coherent public policy perspective and demonstrate refreshing foresight in the regulatory space created for their governments to achieve broader social objectives yet minimizing potential legal costs.

Conflict Avoidance: Vienna Convention and Interpretation Issues

Most apparent intra or inter conflicts can be avoided through interpretation principles applied to treaties if the obligations set out in the treaties are *performed* in good faith as required by the principle of *pacta sunt servanda* (Article 26) codified in the Vienna Convention on the Law of Treaties (VCLT).[107] The VCLT defines 'treaty' exhaustively by stating in Article 2.1(a) that 'treaty' 'means an international agreement concluded between states in written form and governed by international law, whether embodied in a single

105 *The Patents (Amendment) Act*, 2005 No. 15 of 2005.

106 See the *Modified Patent Law* discussion in Herder and Gold, 'Intellectual Property Issues', op. cit. at 28. 'If China is to benefit from the bioeconomy, it may, ironically, be better off exploiting its cultural differences by building a model on a less rather than a more proprietary basis.'

107 VCLT, *supra* note 65.

instrument or in two or more related instruments and whatever its particular designation'. Treaties must also be '*interpreted* in good faith in accordance with the *ordinary meaning* to be given to the terms of the treaty *in their context* and in the light of *its object and purpose*' (Article 31.1). The WTO Agreements are treaties for the purpose of international law and subject to the VCLT's rules on treaty interpretation.[108] The WTO Appellant Body (AB) has held in several cases that the rules of the VCLT enjoy the status of customary international law and that terms of an agreement must be interpreted in light of their object, purpose and context.[109] Under Article 31.2, the 'context' includes, the text, preamble, annexes, subsequent agreements or instruments between the parties in connection with the conclusion of the treaty, as well as subsequent agreements on interpretation or practice or special meaning if intended by the parties. Supplementary means of interpretation may be made to confirm the interpretation under VCLT Article 31 or resolve any ambiguous or obscure meaning from the interpretation given (Article 32). However, there are general rules for priority of treaty obligations set out in VCLT Article 30, which provides that earlier obligations between the same parties of a later treaty will only apply to the extent that they 'are compatible with those of the latter treaty'.

TRIPS Conflicts Internal to the WTO

The global governance institutions have been instrumental for *a priori* conflict avoidance amongst nations despite the increased conflict potential that legal fragmentation has wrought. The mode of analysis adopted in this section is to consider the kinds of conflicts that may arise and prescribe a sensible legal approach (a human rights framework) using familiar legal tools, such as legal defences so that members may predict the outcome of a WTO dispute. A rational acceptance of a 'statutory' defence may lead some states to choose *not*

108 A technical distinction within domestic American law exists between treaties and trade agreements in the different approval procedures each requires for implementation. As a trade agreement, the WTO would require approval by a simple majority of both Houses of Congress, while treaties require approval by two-thirds of the Senate. See Jon R. Johnson, *International Trade Law* (Toronto: Irwin Law, 1998) at 40.

109 See *United States – Standards for Reformulated and Conventional Gasoline,* WT/DS32/R (29 January 1996) [*Reformulated Gasoline*, Panel Report] and *Japan – Taxes on Alcoholic Beverages,* WT/DS8/AB/R, WT/DS10/AB/R, WT/DS11/AB/R, Appellate Body Report (10 April 1996) at 11. The *Japan – Taxes* AB noted that Article 3.2 of the DSU requires that the provisions of 'covered agreements' and the GATT 1994 be clarified 'in accordance with customary rules of interpretation of international law'. For a more detailed discussion of a number of decisions, see generally, Terence P. Stewart and Amy S. Dwyer, *Handbook on WTO Trade Remedy Disputes: The First Six Years (1995–2000)* (Indianapolis, IN: Transnational Publishers Inc, 2001), 78–85.

to complain – even if it is suspected that there is a TRIPS violation – on the same principle of *pacta sunt servanda*: agreements shall be observed. It would be tantamount to evaluating the merits of bringing a civil action in domestic law in light of the particular facts, issues, *and* the availability and applicability of *defences*. It would also reduce the number of cases that do get 'litigated' and the uncertainty of having the matter resolved by a DSB seized by an issue within the broader debate of WTO jurisdiction.[110]

The best approach for a state to defend against a WTO complaint is to reduce the matter to intra-conflict (between one member's rights and another's obligations) or, I suggest, between a hierarchy of a member's *own* rights and obligations under TRIPS, narrowing the issues to an interpretive exercise of a single text. In *Japan – Taxes on Alcoholic Beverages,* the AB emphasized that the principal guide to treaty interpretation is the text of the treaty itself while refusing to interpret these agreements in 'clinical isolation' from other areas of international law.[111] The question is,

> how WTO jurists will respond to claims that TRIPS is inconsistent with the treaty commitments and soft law norms of other international regimes. Although such claims of inconsistency may have substantial persuasive force when used to support treaty amendments or promote soft lawmaking, they are likely to be viewed quite differently when raised during a WTO dispute settlement proceeding.[112]

Helfer argues that soft law can be effectively used to interpret TRIPS obligations even if it cannot be used to justify a violation of TRIPS obligations. He offers,

> because implementing TRIPS does not compel a violation of these agreements or standards (even though it may narrow the discretion or options available to states), WTO jurists are likely to reject claims that violating TRIPS is necessary to avoid conflict with other treaty commitments or regime objectives. The fact that noncompliance with TRIPS cannot be excused on

110 For a long list of recent articles on the trade/human-rights relationship, see Marceau, *supra* note 35, footnote 1. Some scholars believe that conflicts between trade and HR are really about conflict in systems of values unlikely to be resolved by dispute settlement alone as they need to be addressed as policy issues in political arenas, see e.g. Alvarez, *supra* note 36.

111 *Japan – Taxes on Alcoholic Beverages*, WTO, 11 July 1996, <http://www.worldtradelaw.net/reports/wtoab/japan-alcohol(ab).pdf> at 13 and 11 respectively. See also Gabrielle Marceau, 'A Call for Coherence in International Law: Praises for the Prohibition Against "Clinical Isolation" in WTO Dispute Settlement' (1999) 33 *J. World Trade* 87, cited in Helfer, *supra* note 50 at 68.

112 Ibid.

> these grounds does not, however, preclude states from arguing that TRIPS should be interpreted in a manner that avoids such conflicts and harmonizes international objectives. Soft law will be an important tool for WTO panels to use in resolving such arguments.[113]

In addition, Helfer argues that soft law may also 'serve as foundation for the evolution of binding international custom, which *is* accepted as a rule of international law'.[114] Helfer's insights are useful in the interpretive phase of treaty provisions for conflict avoidance. The scholarly focus on appropriate interpretation of treaty obligations in the literature often neglects the utility, as Pauwelyn suggests, of referring to TRIPS *rights* – to use his matrix – to inform legal understanding of the issues in a conflict-minimizing manner.

TRIPS confers onto WTO members specific *rights* by providing for a number of Exemption and Permission norms compliant as domestic solutions that give a state the right to exercise discretion in meeting other command (obligatory) norms, such as the legal taking provision of Article 31 which is not limited to compulsory licensing for government use but can be extended to any third party use so long as it is provided for in domestic law.[115] The permission norm of Article 31 helps states pre-empt conflict. Paragraph 5(b) of the Doha Declaration on TRIPS and Public Health recognizes that the TRIPS does not limit the grounds for issuing compulsory licences and paragraph 5(c) confirms the member's right to determine what constitutes national emergency or extreme urgency.[116] Thus a member cannot complain that another state invoked a compulsory licence or that the case was not a national emergency or extreme urgency, although there may be possible complaint on other vague language such as 'reasonable commercial terms' or even what constitutes permissible 'public non-commercial use' under TRIPS Article 31.[117] Article 31's remedial domestic response is enhanced by the precedent for a TRIPS amendment to deal with the trade/public-health conundrum of production of generic drugs for export. Export facilitating measures introduced by amendment to domestic

---

113 Ibid. at 71, footnotes omitted.

114 Ibid. Emphasis added.

115 The opening paragraph to Article 31 provides 'where the law of a Member allows for other use of the subject matter of a patent without authorization of the right holder, including use by the government or third parties authorized by the government, the following shall be respected …' and proceeds to list conditions.

116 *Supra* note 98.

117 Trebilcock and Howse argue that what should be taken account in making this determination is more than the economic interests of the private IPRs holder to just compensation, but also the economic and social interests that TRIPS provides for, including the transfer and dissemination of technology to the mutual advantage of producers and users under Article 7 and the need to prevent abuse of IPRS (under Article 8) all of which should be considered in interpreting the rights and obligations under Article 31. See Trebilcock and Howse, *ITR*, op. cit. at 412.

patent law are arguably significantly more trade-related than domestic patent reform directed at addressing domestic health priorities, yet these have proven acceptable to the international community. One would expect little reason to contest the more innocuous forms of patent reform that prioritize the human rights of nationals.

The nature of the member's right in Article 31 is such that a measure that falls within the language of the provision is *prima facie* legally justified. It would not be enough for the complainant to argue, for example, that the exclusive rights conferred under patent in Article 28 are being violated by a government grant of a compulsory licence under Article 31 because that licence is well within *the right-creating* norm of TRIPS and takes priority over any obligation. Article 3 of the DSU provides that a proper balancing between rights and obligations of members is to be achieved through dispute settlement. Consequently, a complainant would have the burden of proving that the alleged violating member's action did not *comply* with the text of the TRIPS compulsory licence provision and the text may subsequently be opened for interpretation. The AB has recognized that the burden of proof requires that the party who asserts a fact,

> whether the claimant or the respondent, is responsible for providing proof thereof. Also, it is a generally-accepted canon of evidence ... that the burden of proof rests with the party, whether complaining or defending, who asserts the affirmative of a particular claim or defence. If that party adduces evidence sufficient to raise a presumption that what is claimed is true, the burden then shifts to the other party, who will fail unless it adduces sufficient evidence to rebut the presumption.[118]

The complainant must establish that the *prima facie*-compliant measure does not in fact comply (perhaps because its application is discriminatory) or it fails to meet an internal requirement (a pre-condition) for the exercise of the right. Interpreting Article 31 as a TRIPS norm-creating right, instead of as a defence, gives the member who operates under this right the benefit of the burden of proof by imposing on the complaining party the burden of proving that the contested measure exceeds the scope of the member's right. Article 31 as it currently reads is not a complete solution, however, for a member as the restrictions on production for domestic consumption apply to other areas of technology and vital sectors, such as food, outside of the 'essential medicines' allowance made during the Doha Round. Moreover, certain textual ambiguities exist (what does 'predominantly' in Article 31(f) mean in quantifiable terms?).

---

118 *United States – Measures Affecting Imports of Woven Wool Shirts and Blouses from India*, WT/DS33/AB/R (23 May 1997) [*Shirts and Blouses*, AB Report], at 14, footnotes omitted.

Many conflicts of norms may be avoided by the recognition of member rights but some may need to proceed to dispute settlement.

Other permission norms provided in the TRIPS text include the right a member has to provide limited exceptions to the exclusive patent rights conferred (Article 30) but this is limited by the proviso that the exception does not 'unreasonably conflict with a normal exploitation of the patent' and does not 'unreasonably prejudice the legitimate interest of the patent owner, taking account of the legitimate interest of third parties'. Article 27.2, Article 27.3, and Article 8 also give the state regulatory discretion over certain domestic matters. Again, these exemption and permission norms preserve state domestic regulatory sovereignty by creating *rights* for the member which *prima facie* justify an action within their scope. They place the burden on the complainant to prove a violation, sufficiently, at least, to give rise to a rebuttable presumption. The language of these provisions defines the scope of a *positive right* of a member rather than stating an *affirmative defence* to the allegation of a breach of some other *positive obligation* of the member.

The exemption and permission norms should be effective in avoiding conflict by signalling that the member is well within its rights to adopt a given measure. However, it is foreseeable that an intra-textual, even intra-provisional, conflict dispute may arise on whether a measure does actually fall within the *scope* of the right and the task for the panel or AB will be to balance the interests of the parties taking into account TRIPS rights and obligations. The AB or panel when deciding a case are effectively acting as law makers vis-à-vis interpretation of textual and customary international obligations much like domestic courts do, with two notable differences: decisions are only binding on the parties and the DSB cannot add or diminish TRIPS rights or obligations (Article 3, DSU).

To date, the balance in patent complaints seems to be skewed in favour of property rights holders (or states acting on their behalf) with a rather conservative interpretation being given to TRIPS rights.[119] Despite the

119 In *India – Patent Protection for Pharmaceuticals and Agricultural Chemical Products*, WT/DS50/AB/R, adopted January 1998, the AB upheld a decision of the panel that found that the administrative means of maintaining patent applicants' priority cue during a transition period for implementing TRIPS obligations did not comply with its 'mail-box' condition which would require legislative oversight to secure the applicants' interest despite members' implementation discretion under Article 1.1 which places the burden on complainants to establish that the non-legislative method chosen was non-compliant. In *Canada – Generic Medicines*, op. cit., the AB rejected Canada's argument that the 20 year period would not have to be extended to pre-TRIPS patents granted. The panel was split on two issues. Regulatory testing before patent term expiry was permissible but manufacture and stockpiling of generics was not, discussed in Trebilcock and Howse, *supra* note 117 at 417–23.

permissive exemption norm of Article 30, its interpretation in *Canada – Generic Medicines*[120] has all but vacated its utility. Trebilcock and Howse write:

> [The panel] considered the meaning of the expression 'limited' solely from the perspective of the rights holder, and without regard to the policy goals or purposes of the exception (para. 7.13). Even though dealing with an explicit 'exceptions' provision, comprehensible only if there are legitimate competing policy interests, *the panel was only interested in how much the rights-holder might lose, not in how much society might gain, from a given exception. It never asked what scope the exception might require to achieve the social purpose at issue.* It also failed to consider the scope indicated by the expression 'limited' in light of the protection of public health, an objective explicitly affirmed in Article 8.1 of the TRIPS Agreement.[121]

Regrettably, the same normative force of property fundamentalism that ushered in the judicial extension of patenting life forms through domestic law in developed and common law jurisdictions continue to infuse the reasoning of WTO decisions relating to the balance of interests under TRIPS disputes. But the TRIPS text does allow a state to tighten patentability standards (Article 27) or to apply more stringent disclosure criteria (Article 29) and even specify licensing practices to address anti-competitive and predatory practices (Article 40) and thereby achieve desired policy outcomes in health. Once grounds for granting a patent have been met, the exclusive rights attached to the patent, set out in TRIPS Article 28.1, extend to 'owners' of a patent but no definition is offered. Nor are there any restrictions internal to TRIPS on states adopting specific criteria for the 'owner' status (to neutralize the corporate 'person' for example), other than the cornerstone criteria of NT and MFN.[122] Trebilcock and Howse contend this has

> major implications for the ability of developing countries, for instance to ensure that patent protection does not conflict with their development needs.

---

120 Ibid.

121 Ibid. at 418–19, emphasis added. See also Robert Howse, 'The Canadian Generic Medicines Panel: A Dangerous Precedent in Dangerous Times' (2000) 3 J. *World Intell. Prop*. 493 critiquing the Panel Report.

122 See discussion by Trebilcock and Howse on the implications to other areas of IP of the AB's decision and analysis of 'ownership' in the *Havana Club* case, *United States – Section 211* Omnibus Appropriations Act of *1998*, WT/DS176/AB/R, adopted 1 February 2002, where the EC challenged US legislation denying ownership and legal protection to trademarks associated with businesses and property confiscated by Castro's regime. The AB found that it was a legitimate basis for doing so. See also Trebilcock and Howse, *supra* note 117 at 425.

> Technology transfer, acceptance of price controls, etc. could all be imposed as conditions for an entity to be recognized as an 'owner' of a patent ... [123]

Remapping WTO Cartography with the GATT Article XX Defence

Unlike the exemption and permission norms giving members certain discretionary rights to be weighed against a member's TRIPS obligation, there are some WTO discretionary norms that may be better characterized as 'affirmative defences' to be invoked by the party complained against and considered by the panel only *after* the complainant has discharged its burden and proven a violation of some *other* provision of the covered Agreement.[124] Then, as with regular civil litigation in most common law jurisdictions, once a violation of the other provision has been established, the legal burden shifts to the party complained against to prove the affirmative defence in order to escape liability. The difference is not only technical but legal as it affects the burden of proof on the respective parties.

On the one hand, the legal burden of proof on a defending party for an affirmative defence is more onerous than the evidentiary burden of rebutting a presumption of non-compliance that a complaining party may successfully raise. The complainant still carries the burden of proving that what is claimed is true in a dispute and whether, if true, the allegedly violating act exceeds the scope of a *right* and thereby gives rise to the complaint that the measure breaches an obligation (the obligation to conform domestic measures to treaty rights). Conversely, with affirmative defences, the burden is on the defending party with respect to *all legal and factual tests in that defence*[125] but only *after* the complainant has successfully established a breach of some *other* obligation.[126] Having a formally recognized affirmative defence offers the defending party additional procedural recourse to negate liability once the complaining party has succeeded in making its case on some other provision and the defending party has already been determined unable to discharge the shifting evidential burden to rebut it.

Trebilcock and Howse note that while in '*Shirts and Blouses*, the AB gave the obvious example of GATT Article XX as an instance of an affirmative

123 Ibid. at 426.

124 *United States – Standards for Reformulated and Conventional Gasoline*, Report of the Appellate Body, WT/DS2/AB/R (adopted May 20, 1996) at 16 [*Reformulated Gasoline*, AB].

125 Subject to specific language in the text to the contrary. See *Shirts and Blouses*, AB Report, *supra* note 118 at 16. See also Trebilcock and Howse, *supra* note 117 at 128-9.

126 *Reformulated Gasoline, Panel*, *supra* note 124 at 16.

defence, what constitutes an affirmative defence may not be so obvious'.[127] The AB in *Reformulated Gasoline*[128] found that the burden of proof for showing that a measure was not an abuse of an exception (that it does not exceed it) falls on the defending party even after proving that the disputed measure falls within the enumerated exceptions under the paragraphs of Article XX[129] and this has a regrettable implication, Trebilcock and Howse argue:

> The AB, however, gave no explicit justification for this finding – one could as easily have argued that once a measure is shown to fall within an exculpating category, the burden of demonstrating that it is being used abusively shifts back to the complaining party, based upon the notion that Members of the WTO should not generally be subject to a rebuttable presumption that they are abusing rights acquired under the WTO law.[130]

The implication of the AB's decision is to create a presumption of bad faith contrary to the principle of *pacta sunt servanda* codified in the VCLT. The analysis that ensues strives to operationalize the 'correct' approach suggested by Trebilcock and Howse.

Article XX Jurisprudence for Interpreting the Affirmative Defence

Article XX(d) of GATT 1947 allowing Contracting Parties to adopt or enforce measures to ensure compliance with domestic IP laws[131] recognized the historically territorial quality of IPRs and meant that, as an enumerated exception to the basic GATT trade obligations, domestic regulation of IPRs were largely excluded from GATT disputes.[132] However, jurisprudence under

---

127 Trebilcock and Howse, *supra* note 117 at 128.

128 *Reformulated Gasoline, Panel*, *supra* note 124.

129 *Reformulated Gasoline* AB, *supra* note 124 at 22. But see *Japan – Measures Affecting the Importation of Apples*, WT/DS245/AB/R (adopted 10 December, 2003). The AB, para. 157, noted '[i]t is important to distinguish, on the one hand, the principle that the complainant must establish a *prima facie* case of inconsistency with a provision of a covered agreement from, on the other hand, the principle that the party that asserts a fact is responsible for providing proof thereof. In fact, the two principles are distinct'.

130 Trebilcock and Howse, *supra* note 117 at 129.

131 Specifically, measures 'necessary to secure compliance with laws or regulations which are not inconsistent with the provisions of this Agreement … [for] the protection of patents … and the prevention of deceptive practices …' For a general discussion of Article XX's requirements, see Johnson, *supra* note 108, 66-73.

132 In *United States – Section 337 of the Tariff Act of 1930*, op. cit., a GATT panel reviewing Section 337 of the US Tariff Act under Article XX (d) in 1989 as a violation of NT required under Article III found that it violated that obligation since it entailed

the Article XX exemption norms of GATT 1947 will provide some insight into the conventional understanding of how the 'statutory' Article XX defence operates. This is an important inquiry in light of the scope and field of numbered exceptions provided under it. Some of the key sub-provisions relevant to the patenting life debate are Article XX(a) on the party's adoption of measures *necessary to protect public morals*[133] and Article XX(b) *necessary to protect human, animal or plant life or health* (repeated in the language of TRIPS discretionary norms), subject to the proviso in the 'chapeau' of Article XX that the measure must not be applied 'in a manner which would constitute a means of arbitrary or unjustifiable discrimination between countries where the same conditions prevail, or a disguised restriction on trade'.[134] These provisions have been explored by a number of panels and AB in their reports. While not binding on states outside of the dispute, panel and AB decisions provide useful insight into the workings of the affirmative defence and are essential guidelines on issues of interpretation.[135]

---

domestic enforcement measures procedurally weaker for foreigners than that offered to nationals violating American IPRs. The NT required 'treatment no less favourable' under Article III which was interpreted as requiring 'effective equality of opportunities' in domestic laws and regulations. Once the EU established its complaint under Article III, the burden shifted to the US as defending party to justify the measures under Article XX. The panel found that measures were not 'necessary' to the enforcement of domestic intellectual property laws as other countries enforced their domestic IPRs adequately and equally through measures that applied to both nationals and foreigners and provided the same legal processes.

133 See e.g. Steve Charnovitz (1998) 'The Moral Exception in Trade Policy' 38 Va. J. Int'l 689.

134 While there is debate about whether the opening language of Article XX means that Article III on NT applies regardless or whether a weaker standard of it applies only to the matters listed due to the words 'between countries where the same conditions prevail', The GATT panel decision of US-Section 337 of the Tariff Act, *supra* note 463 held that the Article XX(d) exception may be claimed for measures that would otherwise violate Article III, see discussion in Trebilcock and Howse, *supra* note 117.

135 *Japan – Taxes on Alcoholic Beverages, supra* note 124 at 16 provides: 'Adopted panel reports are an important part of the GATT *acquis* ... They create legitimate expectations among WTO members, and, therefore, should be taken into account where they are relevant to any dispute. However, they are not binding, except with respect to resolving the particular dispute between the parties to that dispute ....' In addition, the AB concurred with the panel that unadopted reports have no legal status in the GATT or WTO system but could be used as useful guidance to a panel. But, the AB in *Argentina-Measures Affecting Imports of Footwear, Textiles, Apparel and Other Items* (WT/DS121/AB/R), 27 March 1998, was critical of the panel for relying on unadopted panel rulings as legal authority and in the *Shrimp/Turtles* implementation ruling, *United States – Import Prohibitions of Certain Shrimp and Shrimp Products,* WT/DS58/AB/RW, the AB held that though not binding, future panels were to be guided by prior AB rulings.

The Article XX defence allows members to adopt certain measures notwithstanding GATT obligations. The complainant must first establish the inconsistency of the measure with GATT obligations. The burden then shifts to the defending party to establish that the measure falls within the scope of the exception and to do so that party needs to establish three criteria: that the policy objective of the measure is within the range of policies captured by the exception invoked; that the measure under attack is, in relation to Articles XX (a), (b) and (d), 'necessary' to meet the public policy objective and satisfies the remainder of the qualifying language internal to these paragraphs; and that the measure is applied in conformity with the introductory clause ('chapeau') of Article XX.

'Necessary' for Policy Objectives Permitted within Article XX

The question to be asked is whether the policy objective of the impugned measure is within the scope of Article XX and whether the contested measure is 'necessary' to meet this policy objective. In *US – Section 337 of the Tariff Act of 1930*,[136] a GATT panel held that even if a policy objective is within the scope of Article XX, a defending party cannot justify the measure as 'necessary' if there is an alternative measure available that is not inconsistent with other GATT provisions. If all measures directed at the objective are GATT inconsistent, the least inconsistent alternative should be adopted. The reasoning employed by the panel in this decision in deciding whether Article XX(d) applied was extended to Article XX(b) in *Thailand – Restrictions on Importation of and Internal Taxes on Cigarettes*.[137] Here, the panel accepted that smoking posed a serious health risk and that the Thai government's ban on foreign cigarettes (from the US) for this policy purpose, while allowing the sale of domestic cigarettes under the regulated industry, was within the scope of (rationally connected to) the exception. But, in determining whether it was 'necessary' the panel found that the measure was not the least restrictive of trade means available for achieving this objective and in applying a form of proportionality test, it determined that the ban could not be justified under Article XX(b) of the GATT. US cigarettes posed greater health risks by attracting young consumers and increasing the probability of their addiction through marketing, advertising and additives. Less restrictive measures could have targeted the control of advertising or content applied to all cigarettes on a non-discriminatory basis. But the panel's ruling reaches deep into the regulatory framework of a state with greater potential domestic costs for

136 *US – Section 337 of the Tariff Act of 1930*, op. cit.

137 *United States–Thailand – Restrictions on Importation of and Internal Taxes on Cigarettes*, DS10/R, adopted 7 November 1990, BISD 37S/200 [*Thai Cigarettes*].

monitoring and enforcement than a government may be able to afford and fails to give proper consideration to the defending party's interests.[138]

In *Tuna/Dolphins I*, the measures considered by the panel were American quantitative import restrictions (contrary to Article XI) on Mexican fish or fish products harvested, contrary to domestic standards used using a method that endangered the life of dolphins accompanying the schools of tuna in the Pacific Ocean. The panel determined the measures taken to protect dolphin life and preserve biodiversity were policies within the scope of the Article XX(b) and (g) exceptions, neither of which impose any limitation requiring that the life, health or natural resources being protected fall within the territorial jurisdiction of the defending party asserting sovereign jurisdiction under Article XX of the GATT. However, on the issue of proportionality required by 'necessary' in the Article XX paragraphs, the panel found that the US failed to show it exhausted all options less restrictive of trade and gave the Articles a narrow interpretation, finding that the exceptions did not extend to production or consumption outside of the defending party's jurisdiction.

In *Tuna/Dolphins II*,[139] the challenge was by the EU to a secondary embargo in US tuna legislation against tuna that compromised the life of dolphins in the *way* such tuna was harvested (through purse seine nets). The EU contested the GATT consistency of the US measure designed to prevent importation of tuna through third countries. The panel reconsidered its approach in the earlier tuna case and found that there was nothing requiring the policy objective under Article XX to be limited to protection within the defending party's territory and that global commons/environmental policies could be justified under this defence so long as the desired policy impact within the scope of the defence was not achieved solely through the inducement of policy changes in *other* countries; this is known as the 'extra-territoriality' doctrine and is not allowed as it would seriously impair GATT objectives (because one member could take trade action to enforce its domestic laws or standards in another member even if to protect animal health or exhaustable natural resources). The objective of the measure was determined to create positive policy changes in other jurisdictions outside the scope of Article XX (b) and (g).[140]

Conformity with the 'Chapeau'

The question that must next be addressed is whether the contested measure conforms to the 'chapeau' of Article XX. If a defending party is successful in establishing that the measure is directed at a policy within the scope of Article

138 See Trebilcock and Howse, *supra* note 117 at 516.

139 *United States – Restrictions on Imports of Tuna*, DS29/R, 16 June 1994, unadopted [*Tuna/Dolphins II*].

140 Ibid. at 5.26 regarding GATT Article XX(g) and at 5.38 regarding Article XX(b).

XX, and that it is 'necessary', or 'relating to' as the respective paragraphs warrant, the measure must of course be applied even-handedly in order to withstand allegations of discrimination and disguised protectionism.[141] The liberal use of the enumerated exceptions is firmly curtailed by reference to the 'chapeau' portion preceding it. In *Reformulated Gasoline*, the AB's review of the Panel's findings was significant in establishing a methodology for examining a contested measure under Article XX (b) and (g):

> [T]he AB in *Reformulated Gasoline* had stated that the application of Article XX is a two-step process, involving first of all a determination of whether the measures fall within a particular exception, and second, whether they meet the criteria in the chapeau … [T]he AB had distinguished between the first step, the determination of whether a measure falls within a particular exception, and the second step, that of ascertaining whether the *manner of application* of measures is reasonable or abusive.[142]

Equally important, is the AB's refusal to give the words 'necessary to' and 'relating to' anything other than an 'ordinary meaning' thereby rejecting the 'least-restrictive means' test from XX(g). The AB considered whether the measure as a whole, and not just the GATT-illegal element was aimed at conservation. The chapeau of Article XX provides:

---

141 *Canada – Measures Affecting Exports of Unprocessed Herring and Salmon*, L/6268, adopted 22 March 1988, BISD 35S/98, GATT Panel Report also discussed in Trebilcock and Howse, *supra* note 117 at 515. A Canadian regulatory measure requiring fish caught in Canadian waters be processed in Canada before export was held to be a violation of Article XI and failed to meet the Article XX(g) test set out by the GATT Panel who interpreted the wording that measures be 'related to' conservation of exhaustible natural resources 'as meaning 'primarily aimed at' such conservation, but it viewed this as weaker than the requirement of 'necessity' imposed by Article XX(b)'. Accurate statistical data, the purported objective of the measure, could be collected without the ban, and had been done so for other species. See also *In the Matter of Canada's Landing Requirement for Pacific Coast Salmon and Herring,* Final Report of the Panel, 16 October 1989 in which a seemingly neutral measure requiring that Canadian water fish be landed and unloaded in Canada was found to have a disparate impact on US destined fish and was struck down by an FTA panel as a violation of Article XI for disguised protectionism of the B.C fishing industry. The panel ruled that alternative means less trade restrictive were available to Canada to meet its objectives of monitoring and compliance with its domestic conservation scheme (i.e. co-operation with US authorities, on board inspections of catches and cargo, etc.). Article XX(g) did not apply as the measure's objective of conservation could not be considered the 'primary purpose'. Trebilcock and Howse argue that the 'the availability of a hypothetically less restrictive means of achieving the same goals should not necessarily mean that the measures under scrutiny fail the least-restrictive means test'. *Supra* note 117 at 517.

142 Emphasis in original, ibid. at 526–8 and their discussion of *Reformulated Gasoline*, *supra* note 124.

> [s]ubject to the requirement that such measures are not applied in a manner which would constitute a means of arbitrary or unjustifiable discrimination between countries where the same conditions prevail, or a disguised restriction on international trade, nothing in this Agreement shall be construed to prevent the adoption or enforcement by any contracting party of measures: …

The manner in which the measure is applied is tested against the chapeau and while the language has some ambiguity, the fundamental theme is to be found in the purpose and object of avoiding abuse or illegitimate use of the exceptions under Article XX.[143] The 'object and purpose' of a treaty is an important approach to interpreting trade obligations but it cannot supplant, according to the findings of the AB in *Beef Hormones*,[144] the need for a textual analysis to be made based on the wording of a provision.[145]

Does the Article XX Defence Apply?: Differentiating Article XX from the Equitable Conduct Defence

If the Article XX defence were to apply to TRIPS disputes, it would create another potential source of intertextual conflict (intra-system conflict) within the WTO. This next section considers whether the traditional Article XX defence applies and if not, the lessons that can be learned in interpreting TRIPS Article 8 as creating an affirmative defence.

---

143 *Reformulated Gasoline* AB, *supra* note 124. See also *Shrimp/Turtles*, *supra* note 135.

144 *EC Measures Concerning Meat and Meat Products (Hormones)* WT/DS58/AB/R Appellate Body Report adopted on 12 October 1998 [*Beef Hormones*]. The 'dioxin crisis' that hit Belgium in Spring of 1999 after it was discovered that the meat and eggs of chickens that had consumed tainted feed contained levels of the noxious chemical sufficient to harm its consumers, coupled with the hormone-beef concern, may have spurred the EU to solidify its position on the precautionary principle in a Communication [COM (2000) 1 final], see Natalie McNelis, 'EU Communication on the Precautionary Principle' (2000) J. of Int'l Eco. L. 545. Here the EU imposed an import ban on beef from cattle raised with growth hormone because of concerns that consumption of such beef may be carcinogenic. A violation of the WTO's Agreement on Sanitary and Phytosanitary Measures Article 1(2) and Annex A Nr 1.which limits national regulations to those aimed at protecting human, animal, or plant life or health from specifically defined 'food-borne' or 'pest-or disease-related' risks was found against the EU despite the fact that it was not shown that the ban had the effect of restricting international trade. The AB held that the EU was not entitled under the SPSA to maintain this ban.

145 See also the *Reformulated Gasoline*, *supra* note 124; *Japanese Alcoholic Beverages*, *supra* note 124.

It would appear that Article XX does not apply to TRIPS disputes. The VCLT (Article 30) provides that absent an express rule to the contrary, later specific agreements on the same subject by the same parties prevail over earlier treaties. While the introductory note to Annex 1A of the WTO Agreement provides that in the case of conflict between GATT 1994 and a provision of another Agreement in Annex 1A, the provision of the other Agreement prevails to the extent of the conflict, no WTO rule describes the relationship between agreements in different annexes. In *Canada – Certain Measures Concerning Periodicals*,[146] the issue of the relationship and the potential for inconsistencies between two agreements (GATT 1994 and GATS) were addressed. Both the panel and AB found that the obligations under the different agreements can co-exist without overriding each other and, therefore, state measures must conform to each agreement to which a state is party.[147] For example, a measure may be TRIPS consistent yet inconsistent with a different GATT provision but nevertheless justified within the exemption norm of the affirmative Article XX defence. For example, the basic GATT obligation relevant to domestic health regulation is non-discrimination. Robert Howse writes,

> health measures that neither treat products from some WTO Members better than others nor treat 'like' domestic products better than imports are consistent with WTO law, and do not require a justification under the health 'exception' in Article XX of the GATT because, in determining 'likeness', WTO rules permit health risks to be taken into account'.[148]

146 *Canadian Periodicals*, op. cit. Canada argued that its 80 per cent tax on advertising revenues from split-run editions of periodicals was a measure regulating access to the magazine ad market and covered by GATS under which Canada had not committed to the provision of advertising services and thus did not have to comply with the NT requirement of members (at 5–6). The panel had found that the measure (an import restriction on split-run periodicals) to secure compliance with tax laws did not fall within the scope of the policies excepted under Article XX; the defence failed without further analysis of the Article XX elements.

147 Ibid. Panel Report, at 73, para. 5.17 and AB Report at 17. This approach was adopted by the AB in *EEC – Regime for the Importation, Sale and Distribution of Bananas,* WT/DS27/AB/R (25 September 1997) at para. 221.

148 See Robert Howse 'The WHO/WTO STUDY on Trade and Public Health: A Critical Assessment', <http://faculty.law.umich.edu/rhowse/Drafts_and_Publications/WHOWTO.pdf#search=%22howse%20critical%20assessment%22> at 1. See also AB decision in *European Communities-Measures affecting Asbestos and Asbestos-containing Products*, Report of the Appellate Body, WT/DS135/AB/R (12 March 2001) where Canada challenged a ban by France on all asbestos and asbestos containing products and the AB, reversing the panel's finding, found that the health effects of the products could be considered in determining likeness under article III.4 but confirmed the panel's finding that the measure could be upheld under Article XX(b).

Let us imagine a state recognizes in its law the patentability of genetically modified or living modified organisms (GMOs and LMOs). This is completely consistent with its trade obligations under TRIPS. The state may have a related measure imposing regulatory standards for health risks related to trade in GMOs and LMOs and this measure may be contested under the GATT 1947. If the measure is non-discriminatory it should be found to be GATT consistent.[149] However, if the complainant proves that the measure is discriminatory or that the products are 'like' products so that the effect is discriminatory, then the burden would shift to the defending party to invoke GATT Article XX(b) proving that the measure was 'necessary' for the purpose of, *inter alia*, the protection of human life and health. A ban on a new variety hamster tobacco plant to be introduced into the environment, for example, would arguably fall within Article XX(b) as 'necessary to protect human, animal or plant life or health'.[150] Still, it would have to pass scrutiny under the *chapeau* to determine whether the restriction is an arbitrary, unjustifiable, or disguised protectionist measure. This analysis would occur outside of a TRIPS inquiry on the domestic patentability issue and is consistent with the AB's finding that measures must conform to each agreement to which a state is party. However, there is no prescriptive guidance on resolving intra-WTO conflict of norms as between, for example Article XX and a TRIPS obligation.

In *Japan – Taxes on Alcoholic Beverages,* the AB emphasized that a principal guide to treaty interpretation is the text of the treaty itself and this was echoed by the AB in *Beef Hormones.*[151]The affirmative defence of Article XX is, by the very language used in the text of the original GATT 1947, limited to 'this Agreement' as set out in its 'chapeau'. Its extension into the GATT 1994 was by direct incorporation of the earlier GATT; consequently the Article XX defence is limited to GATT 1994 obligations and does not apply to those arising under other agreements such as TRIPS. In two dispute-settlement reports on *Section 211 Omnibus Appropriations Act*,[152] whether the general

149 '[E]ven if imported product A and domestic product B are alike in many other respects, if imported product A creates health risks that domestic product B does not, it would be consistent with National Treatment under the GATT to ban or restrict A but not B'. Howse, WHO/WTO Study, *supra* note 148 at 1.

150 Although hamster gene tobacco to be smoked by consumers (as a commodity) may be found to be against the Cartegena Protocol for the protection of the Environment itself and if challenged under the GATT, the restricting Party may have to argue that the measure was necessary to protect human health under Article XX(b) relying more on the harm of tobacco than the fact that it is GM based on WTO jurisprudence to date.

151 *Japan – Taxes on Alcoholic Beverages*, *supra* note 124 at 13; *Beef Hormones*, *supra* note 144.

152 *United States – Section 211 Appropriations Act; United States – Section 322 Omnibus Appropriations Act of 1998*, complaint by EC (WT/DS176/AB/R).

exceptions of Article XX could be permitted under TRIPS was raised before the AB, which concluded that no new exception based on the GATT would be allowed if it violated substantive TRIPS provisions. This understanding is consistent with the 'latter rule' of the VCLT on treaty interpretation and can also be normatively justified on the basis that all attempts should be made to read out conflict from international treaty obligations.[153] Still, that is not to say that the Article XX defence (and jurisprudence) has no influential or instructive role; in fact, parties may be informed on a number of levels in formulating arguments consistent with human rights obligations but apparently inconsistent with TRIPS obligations.

First, there may be an argument to be made that certain provisions of Article XX have gained international recognition as part of customary law or *jus cogens* as within the sovereign jurisdiction of a state to regulate. This allocation to national governments of specific substantive though trade-related issues is a fundamental norm from which there can be no derogation and, to the extent of any inconsistency, may operate to invalidate any treating agreement. Article 31.3(c) of the VCLT provides under the general rule of interpretation that what shall be taken into account is '*any* relevant rules of international law *applicable in the relations between* the parties'. Evidence could be adduced from other international soft and hard law, from the existence of collateral treaties developed under the auspices of the UN, like the CBD, from environmental regulation, and from Article XX itself which is relevant to the relations between the parties, as well as from the principle now well entrenched since the *Lotus* case:[154] that the sovereignty of states is presumed to be unlimited in its own territory unless bound by specific rules of law and particularly so in these specific areas of law and public policy.[155] The merits

153 See Trebilcock and Howse, *supra* note 117 at 546, agreeing that 'as a general matter, the dispute-settlement organs ought not easily to assume a conflict of obligations exists, but instead begin by attempting to read the WTO rules in such a manner as to fulfill or not frustrate the other treaty rights and obligations'.

154 *S.S. 'Lotus' (France v. Turkey)*, 1927 PCIJ (Ser. A) No. 10 (7 September) at 18, the Permanent Court of International Justice finds: 'the first and foremost restriction imposed by international law upon a State is that – failing the existence of a permissive rule to the contrary – it may not exercise its power in any form in the territory of another State. In this sense jurisdiction is certainly territorial; it cannot be exercised by a State outside its own territory except by virtue of a permissive rule derived from international custom or from a convention'. See generally Pauwelyn, *supra* note 48 at 150 and footnote 199 describing how this decision stands for the 'residual negative principle' in international law which posits that '"everything which is not expressly prohibited is allowed"'.

155 This is contrary to the finding by the *Tuna/Dolphin II* panel, *supra* note 139, which held that other agreements were not relevant for the interpretation of GATT treaty obligations if they did not apply to the Contracting Parties, and suggesting that GATT was self-contained and impervious to rules of other international

and validity of such a claim are well beyond the scope of this text and will not be explored further. More important for present purposes, the approach to the affirmative defence may be helpful in embracing or rejecting a similar interpretive approach to any affirmative defence raised in relation to a *prima facie* inconsistent measure located within TRIPS.

The proposed Equitable Conduct Defence justifies deviations from trade obligations in terms of market access rules in the same manner that the Article XX defence does. Both are based on discretionary norms (permissions) created within the treaty text as a means of balancing state interests against obligations raised in other provisions of the same text. And just as Article XX could have been interpreted as a *right* but has instead been employed in the service of states as a *defence,* so too, I suggest, should Article 8 of TRIPS. The approach taken to formulating and asserting this defence before the WTO can well be enhanced and informed by the existing jurisprudence of Article XX.

The Equitable Conduct Defence: A Public Policy/Interest Affirmative Defence

The supporting text for founding an Equitable Conduct/Public Interest Defence (ECD) is in TRIPS Article 8 permitting states to adopt measures necessary to 'protect public health and nutrition' and 'to promote the public interest in sectors of vital importance to socio-economic and technical development'. To summarize the burdens analyzed so far, we can say:

The complaining party in a trade dispute must prove:

Claim 1: A violation of an Obligation as set out in a provision of the text or

Claim 2: The violation of a Right – that is, the defending party has exceeded the scope or jurisdiction of the discretion arising from a provision of the text and in exceeding (abusing) its right, has violated an obligation.

Within these two contexts, there may be shifting evidential burdens of proof but the legal burden remains with the complainant. Once discharged, the legal burden shifts to the defending party to establish an affirmative defence in order to avoid liability and, as discussed above, for Article XX this means proving that the measure falls within the language of Article XX but also,

regimes. Compare this with the AB decision in *Shrimp/Turtles*, *supra* note 135 where international environmental law (hard and soft) was applied for interpreting the scope of the expression 'exhaustible natural resources' in Article XX(g). See Trebilcock and Howse, *supra* note 117 and generally, Joost Pauwelyn, 'The Role of Public International Law in the WTO: How far can we go?' 95 AJIL 535 [Pauwelyn, 'PIL']. For a discussion on treaty obligations and *jus cogens*, Pauwelyn, *supra* note 48.

because of the AB's interpretation, that it does not *exceed* the statutory defence. Earlier, Article 8 was discussed as part of the rights creating norms of TRIPS, functioning similar to Article 31 on compulsory licensing, for example. However, Article 8 may need to be re-imagined as an affirmative defence akin in spirit and function to the GATT Article XX affirmative defence. Article 8's existence and dual identity as a *right* creating norm means that it may have a role in claim 1 or claim 2 in support of the defending party's shifting evidential burden but its greatest utility is as an affirmative defence.

The most compelling reason to consider Article 8 as creating a *defence* rather than a *right* is that the latter creates substantive protections for members to achieve particular public policy outcomes and the necessary institutional balance in their domestic IPRs by granting the discretion (through exemption/permission norms) to adopt diverse measures within their right (such as Article 30 and Article 31). In contrast, the concept of an affirmative 'equitable defence' imports two important principles and extends beyond simply a legal defence. First, the idea of a defence supports the need for the complainant to contest the consistency of an impugned measure against some *other* provision before the legal burden shifts to the defending party to establish legal justification under Article 8. This is different from the interpretive approach taken to the rights creating discretionary norms of TRIPS (i.e. Articles 27.2/3, 30, 31 etc.) considered in text above; there, the contestability of a measure was based on whether it fit into the scope of the rights creating norm as discerned from the language of the provision. If it did, then the complaint will have successfully been defended on the terms of the provision and if it did not, there would be liability without a defence. Second, there are important reasons for characterizing this defence as 'equitable' even though I argue it is derived from TRIPS' text and in that regard is a 'statutory' or in this case, a 'treaty based' defence. These are enumerated below:

1. 'equitable' is responsive to the idea of Article 8 as the overarching *Principles* governing the entire TRIPS agreement and locates these principles within broader understandings of national and international governance frameworks. It helps contextualize the defence as one necessary to overcome the formalism of the law in pursuit of just equitable outcomes within the broader relationships of disputing parties under TRIPS as required by the 'public interest' referred to in Article 8, however defined. It would enable governments to meet their other objectives such as 'equitable sharing' under the CBD and obligations such as the protection of human rights under the ICESCR.
2. Characterizing Article 8 as an equitable defence reflects the potential liability of a state to its respective publics under other instruments if strictly adhering to trade-related IPRs protection above all other competing domestic regulatory policy objectives. What is the responsibility of a representative government to its constituents and

should meeting that responsibility be fettered due to TRIPS mandated IPRs? For example, a state may be domestically liable for human rights violations under its own legislation, under its Constitution, and under international instruments. The Article 8 ECD is a procedural safeguard for minimizing state liability in one governance framework for upholding particular human rights obligations of a state (as per a competing governance framework) for which the state may be domestically liable. Article 8 is equitable because it helps minimize conflict into the reconcilable category discussed earlier. Otherwise a state will be placed in a very precarious position acting on the one hand in a manner to minimize its own potential liability to its citizenry domestically for HR violations but may nevertheless be exposed to foreign liability and resulting trade sanctions undermining the welfare interests and human rights of the very public whose HR were being protected. If the state takes a measure that is *prima facie* inconsistent with a TRIPS right or obligation, then that measure may be justified and should equitably be recognized as such if it falls within the language of Article 8, just as a public policy objective within the scope of Article XX is defensible if the measure taken is directed at achieving that public policy objective. This is 'equitable' in that it provides a trade-related safeguard against trade-related hazards of HR-promoting domestic measures. Successive governments should not be fiscally punished and subject to increased domestic liability to individual citizens for their respective individual rights and civil liberties on the basis of international trade and intellectual property obligations technocratically committed to and with current disutility. Robert Howse suggests greater domestic parliamentary control over the making of WTO rules and asks:

> But whether this cashes out into greater democratic legitimacy depends on how well informed parliamentarians are, to what extent they are independent rather than subject to the discipline of the Party whip and therefore no real check on the executive, and also the extent to which we believe that approval of today's government is enough to provide legitimacy for rules that will have significant impact long after that government is gone, and which it is costly for future government to reverse (as, in practical terms, it would either have to get the consent of all other members to change the rules or accept a waiver, or be faced with the very high-stakes choice of withdrawing from the WTO).[156]

Howse adds, 'Given the cost of reversibility by a future government,

156 Robert Howse, 'Human Rights in the WTO: Who's Rights, What Humanity? Comment on Petersmann', <http://www.jeanmonnetprogram.org/papers/02/021201-01.pdf> at 13.

my own view is that the people should be consulted directly by referendum on the results of the Doha round, and that all governments should undertake to translate the proposals into local languages, and distribute them to the entire populations …'. Trade theory and policy are designed for maximizing the public interest by reducing consumer costs through the reduction of trade barriers. The normative justification for trade and institutional IPRs is similarly focused on the public interest. If their internationally mandated protection results in welfare loss by impeding a state from fulfilling its HR obligations, then these instruments of private property are no longer instrumental as global public goods and result in incoherence of the international system for global (self) governance. After all, any gains to be had in giving human rights domestic priority would have to be set against the costs that non-compliance with international law might impose (such as carousel trade sanctions), costs that are ultimately borne *by the peoples* in the aggregate (primarily as consumers) in order to protect the rights of *peoples* (as citizens of a state) individuated and humanized.

3. The ECD is mindful of transaction costs and is therefore economically efficient. The public who is to benefit from the existence of IPRs has to pay in many ways indirectly and directly: first, as tax paying subsidizers of innovation facilitating institutions such as public education and health; second, as subsidizers of a publicly administered patent system and then again as rent payers on the patented inventions it generates; third, as subsidizers of government liability to constituents for violations of human rights obligations or conversely patent infringement; fourth, as subsidizers of the costs of defending complaints; and finally, as consumers bearing greater consumption costs upon an adverse finding of non-compliance by the WTO DSB for regulatory measures that are, I suggest, justifiable. If the right to health is at a minimum a negative right to have one's health free from government interference, it would impose on the government an obligation not to harm its individual citizens by creating statutory measures (IPRs) *contrary* to the protection of this right and the individual's liberty to self-realize it. If the claimant is able to establish that this interference is unconstitutional, a monetary award may be granted or patents invalidated. Assuming a government were held liable and forced to pay, the monies paid would come from public coffers and, effectively, the harm created to the benefit of private property rights holders in the patented invention would be further spread as a cost to society with the public having to pay for the patented invention *directly* in its grant and *indirectly* (through their taxes) for a trade-inconsistent measure that seeks to avoid *direct* royalty payment.
4. The ECD may restore some of the legitimacy of the WTO by enabling a fairer application of the rules and a medium for a political ethic without which the very existence of the WTO is threatened. In the absence

of a multilateral trade system, threats of unilateralism, illegitimate sanctioning, restrictive border and/or tariff barriers, and bilateralism between strong and weak countries may re-emerge with beggar-thy-neighbour policies of a state in pursuit of self-sufficiency. A 'race to the bottom' may take place as protectionist regimes send us back in time to a grim period of non-specialization, non-interdependence, non-cooperation, and a resulting poorer world. The ECD would recognize the importance of updating the WTO's mandate to meet current 'social' challenges' and the central role of the state in overcoming them by endorsing *free* and *fair* trade.[157] The paradoxes of the trade/human-rights/IPRs linkage should not be institutionally perpetuated and though the conundrum created is not easily reconcilable with the existing TRIPS jurisprudence, it may become so with the recognition of an Article 8 defence. What is needed as a guide to interpretation is a comprehensive understanding of the purpose, mandate, scope, jurisdiction, and unifying principle for the WTO and its place in global and local regulatory institutions. It will be recalled that Article 8 is introduced by the heading *Principles* and provides:

> 1. Members may, in formulating or amending their laws and regulations, adopt measures necessary to protect public health and nutrition, and to promote the public interest in sectors of vital importance to their socio-economic and technical development, provided that such measures are *consistent with the provisions of this Agreement.*
> 2. Appropriate measures, provided that they are consistent with the provisions of this Agreement, may be needed to prevent the abuse of intellectual property rights by right holders or the resort to practices which unreasonably restrain trade or adversely affect the international transfer of technology.

Using Article 8 as an affirmative defence rather than a right helps overcome criticisms by some commentators that Article 8 is tautological and, as an articulation of 'principles', has no direct bearing on the parties in terms of creating rights and obligations. The alleged tautology is that the discretion allowed under Article 8.1 is qualified by the language in the text 'provided that such measures are consistent with the provisions of this Agreement'. That ambiguous terminology finds its way into trade agreements is not

157 Trebilcock and Howse argue for rejuvenating the WTO: 'The perspective from which the policies of Members are examined needs revision. Currently, the only issue considered in these examinations is whether a Member's policies and practices support free trade. It is clear, however, that the perspective through which a Member's policies are reviewed should not be that of free trade alone but rather that of the "*functioning of the multilateral trading system*".' Trebilcock and Howse, *supra* note 117 at 583.

surprising and follows from the desire to make treaties more acceptable to different negotiating members. A member's adoption of a TRIPS-consistent measure can be for any purpose and so the inclusion of this qualification in Article 8 does seem a bit tautological. However, the interpretative principle of effectiveness (*ut res magis valeat quam pereat* [that the thing may rather have effect than fail]) provides that interpretation must give effect to all terms of the treaty. This principle was cited by the AB in *Japan – Taxes on Alcoholic Beverages* to support the finding that no clause or paragraph may be reduced to 'redundancy or inutility'.[158] The AB in *Reformulated Gasoline* referred to Article 31.3 of the VCLT to find that 'interpretation must give meaning and effect to all the terms of the treaty. An interpreter is not free to adopt a reading that would result in reducing whole clauses or paragraphs of a treaty to redundancy or inutility'.[159] This principle would support reading Article 8 as a defence (an argument could be made that it is based on a right such as self-determination in established sovereign areas) as opposed to a mere laudatory articulation of an overarching though unenforceable principle that would otherwise be superfluous; members inherently have a right to adopt *any* TRIPS-consistent measure for *any* purpose.

Jon Johnson argues that Article 8 does not provide 'true "exceptions" such as those in Article XX of the GATT 1994'.[160] To the contrary, there are good reasons for interpreting this permission norm as a full-fledged defence. In *Beef Hormones*, the AB found that the interpretive principle of *in dubio mitius* requires that where two plausible approaches to the interpretation of a treaty provision exist, the interpreter must adopt the interpretation that is less restrictive of the sovereignty of the state or states undertaking the obligation.[161] Certainly, interpreting Article 8 as an ECD is less restrictive of sovereignty over regulatory matters that are specifically exempted in the language of the provision's text. And while a purposive approach cannot

158 *Supra* note 124 at 14.

159 *Supra* note 124 at 22.

160 Johnson, *supra* note 108 at 199.

161 Compare this, however, with the panel decision in *Canada – Generic Medicines*, *supra* note 119 contesting the Canadian provision allowing competing generic manufacturers to test patented products before the required 20 year protection period had expired and to stockpile production of generics 6 months before the expiry. While the testing provision was upheld as GATT consistent, in a very strange decision of the panel, the stockpiling was not despite the fact that the nature of the rights is to protect the patentee from competition and thus rents during the patent period. The result was that the additional term it would take to produce generics post expiration would effectively be a windfall for the patentee constituting additional ex post rewards rather than ex ante incentives. In determining whether the measures fell within Canada's Article 30 rights that formed an exception to TRIPS, the panel's interpretation of 'limited' failed to consider Article 8 and the legitimate interests of the defending party to protect public health. Trebilcock and Howse, *supra* note 117 at 416.

compromise an interpretation on the textual wording of a provision in dispute settlement, 'textualism should not mean a neglect of inquiry into purpose and object, when considering the exact words of the text'.[162] The closing language of Article 8.1 was added at the last stages of the TRIPS negotiation.[163] On one level, it provides the rationale for measures taken under Articles 30, 31, and 40. However, our discussion illustrates that the utility of Article 8 must go beyond that if the principles of treaty interpretation are to be given any effect. Members need not provide the motivation for any measure until the measure is contested and need not provide any explanation where it is not. This legal practice is consistent with the textual language of Article 8. If a complainant is able to establish that a measure exceeds the defending party's rights under a different provision, like Articles 30, 31, and 40, then Article 8.1 will allow the defending party to try to provide legal authority and justification for that measure; this may be easier to do where the TRIPS is silent on a matter or has created an exemption or permission norm such as those related to the patenting of life in Article 27.2 and 27.3.

If it is a defence analogous to Article XX, the approach to interpreting this defence and the words 'necessary to protect' in Article 8 need not be subject to the burden imposed by the AB in *Reformulated Gasoline* which required the *defending* party to establish that the contested measure did not abuse the discretionary permission norms of Article XX, by proving that the contested measure was within the scope of Article XX *and* did not actually exceed it. Under Article XX jurisprudence, this led to the difficulty of proving that a measure 'necessary' was in fact the least restrictive means of achieving its objective. Given the fresh start under Article 8, we may adopt the Trebilcock and Howse recommendation of a rebuttable presumption. Once the defending party establishes that the measure is *prima facie* within the scope of the ECD within Article 8, the evidential burden will shift to the complainant to establish that the measure exceeds it. This shifts the formidable challenge of proof from the defending party to the complaining party where a measure is *prima facie* justified as defensible. Such a proposal to the Article 8 ECD thereby removes the bad faith implication created by the AB under Article XX.

Some readers may disagree with this characterization of Article 8 as a defence and yet insist that whatever the status of the provision, the language imposes the full burden on the defending party. In that case, the role of Article 8 would be confirmed as more of a defence than Article XX, since the latter artificially imposes the full burden on the defending party (by the AB's decision) but Article 8 would arguably accomplish this based on different textual language. A full burden of proof textually imposed for establishing a legal justification and authority is more consistent with our normative understanding of burdens of proof for defences in the law. At the same time, Article 8 has reduced the

162 Ibid. at 134.

163 Gervais, *supra* note 97 at 122.

scope of that burden by requiring proof of consistency not with the *Agreement* (required by Article XX) or *obligations under the Agreement* (the language in an earlier draft of Article 8), or some broader concept of *international trade* (as would be the case with non-violation complaints) but with the '*provisions* of the Agreement' (TRIPS).

The language of Article 8.2 does not provide any textual basis for limiting measures taken to prevent abuse to any particular conception of 'abuse' such as non-working or refusal to licence a patent. Rather, the anticompetitive behaviour of patent portfolios, bad patents and the unfair competition practices that result from the predatory uses of patents, would seem to be sufficient justification for invoking this defence. Such market practices may constitute unfair competition and result in the unreasonable restraint of trade domestically. There is nothing to suggest that the reference to 'trade' in Article 8.2 should be interpreted as *international* trade; in fact, the inclusion of the adjective 'international' before 'transfer of technology' seems to negate such an interpretation with respect to 'trade'.

On a common sense reading of this provision then, a complaining party would first have to establish that a measure was inconsistent with a provision of the TRIPS (a right was exceeded or obligation not met with possible shifting evidentiary burdens). If the complaining party is successful, the legal burden would shift to the defending party to invoke Article 8 as a defence with the full burden of proof of establishing, as with Article XX jurisprudence: 1) that the object and purpose of the measure falls within the scope of the Article 8 exceptions; 2) that it was 'necessary' or taken 'to promote' – different criteria just as in Article XX(b) and (g); and 3) is consistent with the provisions of the Agreement, which include NT and MFN obligations. On my formulation, for the complainant to succeed, it must discharge the burden to establish that the defending party's conduct somehow *exceeds* even the defence – that it abuses the degree of permitted state discretion under Article 8.

Trebilcock and Howse contend that the role and interpretation of Articles 7 and 8 remain uncertain:

> It remains an open question whether … a foreign patent-holder who refused to comply with policy measures aimed at facilitating technology transfer or preventing anti-competitive abuse of patent protection could be legally denied the level of protection specified in the TRIPS agreement. In other words, are domestic policy measures that condition the granting of rights under the TRIPS Agreement on compliance with the kinds of measures contemplated in Article 7 and 8 'consistent' with the TRIPS Agreement? A further interpretive issue is whether Articles 7 and 8 could be used as a 'shield' by developing or other countries against unilateral US action in response

> to policies in conformity with the Uruguay Round TRIPS Agreement but nonetheless considered 'unfair' by US trade authorities? [164]

I suggest that the appropriate role of Article 8 within TRIPS disputes is as an affirmative defence. Defences in law must *a priori* be principled. The argument for the ECD would suggest 'yes' to be the appropriate response to their interpretive questions and the language consistent with the 'provisions'. Even if the full legal burden was to be imposed on the defending party under Article 8, it is much narrower and less onerous than that under Article XX's chapeau where the AB has interpreted 'unjustifiable discrimination' and 'a disguised restriction on international trade' sometimes in terms of *the least restrictive means,* creating a high evidential hurdle for the defending party.

The need to adhere to the textual language of a provision is consistent with Articles 3.2 and 19.2 of the DSU that stipulate 'the panel and Appellate Body cannot add to or diminish the rights and obligations provided in the covered agreements'. Recognizing an ECD would not be asking for 'law making' on the part of the panel or AB in the sense of domestic judicial activism but rather would be the appropriate way of *interpreting* what is already provided for in the Agreement text and has been renewed as a commitment in the Doha Ministerial Declaration where Articles 7 and 8 were highlighted as having special importance.[165] The Objectives articulated in Article 7 inform all of the state undertakings and the Principles of Article 8 defend them. This is an interpretative approach well within the mandate and jurisdiction of WTO jurists. Dinwoodie and Dreyfuss urge that

> Panels must take seriously the autonomy interests implicit in the structure of the international intellectual property system, and they must allow sovereigns to respond to changes in science, to the structure of their patent industries, or to other social needs. Otherwise, a series of worldwide disutilities will result ... [V]arious systemic values ... are crucial to this approach to analyzing TRIPS obligations: the incentives likely to optimize social utility may vary widely from country to country; permitting some diversity of approach allows nation states to act as laboratories in the development of international rules; affording space for the self-determination of sovereign states encourages voluntary and ultimately more effective compliance with international norms and, universality may have costs, whether measured in economic or non-economic terms.[166]

164 Trebilcock and Howse, *supra* note 117 at 411.

165 Doha Ministerial Declaration, *International Intellectual Property Law*, op. cit. para. 19.

166 Dinwoodie and Dreyfuss, *supra* note 175 at 448.

Additional Recommendations for the Stewardship of the WTO

> Civil society groups worried over threats from industry on the basis of TRIPS Agreement rules may find a modicum of comfort in the appointed guardian of the TRIPS Agreement – the WTO Appellate Body. There is a difference between threatening countries on the basis of unwarranted interpretations and the reality of persuading the AB. However, just as the US Supreme Court does not intervene on each occasion an accused is improperly imprisoned, so the AB is not involved in the daily implementation of the TRIPS Agreement. Its influence is somewhat remote and symbolic.[167]

This chapter has tried to provide a human rights framework for defending against an allegation that a measure is TRIPS-consistent when it is, nevertheless, justified on the basis of an affirmative defence supported by the text of Article 8. In addition to the substantive considerations, certain institutional reforms can be taken to facilitate the WTO's role as steward of trade and *other* values through the dispute settlement process. Many scholars have criticized the trade-IPRs linkage.[168] It is up to the WTO to mediate the competing values. Dinwoodie and Dreyfuss remind WTO jurists that it is:

> ... important for panels to keep what might be called the 'neo-federalist' underpinnings of the TRIPS Agreement in mind ... Even when prescriptive power resides at the international level, that power is typically the authority to set boundaries within which the WTO member can act rather than to impose a specific rule of law ... In the TRIPS context, these allocations have the additional effect of giving WTO members an important role in striking the producer/user balance of intellectual property law.[169]

Today, state agency is critical for managing the gradual erosion, nationally and internationally, of regulatory policy-making space by the judiciary, and the technocrats engaged in TRIPS-Plus negotiations. But, it is important to remember, as Frederick Abbott notes, 'IPRS are not only trade-related. They are also education-related, health-related, nutrition-related, defense-related,

---

167 Abbott, 'Toward New Era of Objective Assessment in the Field of TRIPS and Variable Geometry for the Preservation of Multilateralism' 8 *Journal of International Economic Law* 77, http://www.frederickabbott.com/uploads/77_1___toward_a_new_era_.pdf at 523.

168 For a critical review, see generally R. Michael Gadbaw, 'Intellectual Property and International Trade: Merger or Marriage of Convenience? (1989) 22 Vand. J. Trans. L 223.

169 Dinwoodie and Dreyfuss, *supra* note 166 at 447.

environment-related, energy related and so on'.[170] Abbott observes how the importance of a collaborative and cooperative ethos in the WTO:

> [m]ultilateral organizations such as the Food and Agriculture Organization and the World Health Organization have important interests in the way rights granted with respect to inventions within the scope of their regulatory mission are used. It is ... not remarkable that these organizations have sought to play a more significant role in the implementation of the TRIPS ... and in the formulation of new rules.[171]

On the issue of consultation and collaboration, greater efforts should be made to encourage the participation of the Specialized UN bodies in the consultation that occurs *before* a panel is struck. In addition, under Article 13.1 of the DSU, each panel has the right to seek information and technical advice from 'any individual or body which it deems appropriate'. Under Article 13.2, the panel may seek information from 'any relevant source and may consult experts to obtain their opinion on certain aspects of the matter'. The WTO would be well served by drawing on resources from international UN based institutions like WIPO and the World Health Organization (WHO) through expert consultation in dispute proceedings and by providing a more formal role for *Amicus Curiae* briefs by NGOs than currently embraced.[172] Such consultation and collaboration may also provide essential assistance in the work of the TRIPS Council, will promote the sharing of inter-institutional regulatory responsibility over these global public goods, as well as accountability, and therefore legitimacy, for their outcomes.[173] TRIPS Article 68 allows the TRIPS Council to consult and seek information 'from any source'; civil society and NGOs can play an important role in this respect. Article 1 of the DSU requires that 'the rules and procedures set out in this Understanding should be used to the extent necessary to avoid conflict'. Industry may resist, but this is a common tension 'between the regulator and regulated'[174]

170 Abbott, *supra* note 167 at 85.

171 Ibid.

172 Dunoff, *supra* note 62.

173 'Distributed Governance at the WTO-WIPO: An Evolving Model for Open-Architecture Integrated Governance' 3 *JIEL* (2000) 63 and Abbott 'Coherence, Co-option, Flanking and Containment: Distributed Governance Revisited' ASIL, HR and Trade Conference, April 2004.

174 Industry's resistance was evinced in the Access to Essential Medicines debate predating Doha (see www.accessmed-msf.org) and also during the negotiation of the international treaty on Plant Genetic Resources for Food and Agriculture at the FAO, where there was substantial lobbying by biotech intensive seed/plant producers to eliminate provisions regulating patenting of inventions from materials withdrawn from the multilateral system. See Abbott, *supra* note 157 at 85.

Third party participation by NGOs with *Amicus Curiae* briefs would be a vital step towards improving public participation and a more transparent process for achieving just results at the WTO.[175] Simultaneously, a conscious attempt to achieve more diversely composed panels under the DSU Article 8 through selection of panel members with great inter-disciplinary and inter-institutional strengths from varied cultural backgrounds would also help. And, if the WTO is increasingly perceived as a rule based judicial body, it would make good sense to integrate some of the institutional supports commonly associated with such a system such as the provision of Legal Aid to developing and least developed countries through a defence fund;[176] such initiatives are occurring at last. The WTO Doha Development Agenda Global Trust Fund was created following the 2001 Doha Ministerial Conference and funds trade-related technical assistance. Finally, the current funding for the WTO is dolefully inadequate. The annual budget is less than $100 million. 'As a result', Bhagwati writes, 'the WTO essentially has to rely for trade analysis on the "foreign legion" at the World Bank, the OECD etc … [At] WTO meetings, the world's media typically focus not on the WTO's analysis and economists, but on those from these other institutions. This is a travesty. It needs to be put right'.[177]

Conclusion

> We are in a period in which tremendous pressure is exerted at the bilateral and regional level for new and more restrictive rules that eliminate policy space in developing and developed countries. A presumed objective is to force the multilateral system to accommodate to a new reality or face the prospect of irrelevance. Benign neglect does not appear to be a constructive option.[178]

Attempts to liberalize trade require a concurrent commitment to legitimize the system through conformity with broader conceptions of normative social ideals of individual rights and distributive justice as articulated in UN HR and HD instruments. These universal commitments further the original ethos of cooperation through trade and promote the mutual reciprocal

---

175 Robert Howse, 'Membership and Its Privileges: The WTO, Civil Society, and the Amicus Curiae? Brief Controversy' (2003) 9 Euro. L. J. 496. Currently, participation is limited to what parties allow.

176 Helleiner, *supra* note 18.

177 Jagdish Bhagwati, 'Reshaping the WTO' (January/February 2005) Far Eastern Economic Review at 7, <http://www.columbia.edu/~jb38/FEER%20Final%20Edited%20by%20Restall%20and%20Bhagwati.pdf>.

178 Abbott, *supra* note 157 at 100.

benefit of all members, while leaving implementation and enforcement to the culturally diverse regulatory mechanisms of each state for implementation and enforcement. Successive WTO Round negotiations should endeavour to remedy apparent institutional inequalities and bridge the growing gap between stated ideals and real practice in order to preserve the welfare maximization mandate of liberal trade without losing sight of its ability to realize individual rights in the aggregate.

The normative interpretive prescriptions offered for reconciling obligations should remain relevant regardless of the full debate on the substance of any individual human right or rights in their aggregate. The tradition in the trade literature has been to discuss the trade-HR linkage nonetheless, not in terms of defined understandings of human rights, but more abstractly as a matter of preferences or in terms of the constitutionalization of human rights within the WTO.[179] Along these lines, one may conclude for a variety of reasons

179 See Ernst-Ulrich Petersmann, 'Time for Integrating Human Rights into the Law of Worldwide Organizations: Lessons from European Integration Law for Global Integration Law' Jean Monnet Working Paper 7/01, <http://ejil.org/forum_tradehumanrights/> on the right of property and freedom of contract as worthy of ICESCR HR protection but lacking protection because of an 'anti-market bias'. On this basis, Petersmann argues for the constitutionalization and enforcement of human rights in the WTO as a means of achieving democratic legitimacy and social justice [at 624] but his constitution has the idea of property central to it and has been criticized by leading human rights scholar Philip Alston, 'Resisting the Merger and Acquisition of Human Rights by Trade Law: A Reply to Petersmann' Paper No. 12/02, Jean Monnet Working Paper Series, Symposium, Trade and Human Rights: An exchange, <http://ejil.org/forum_tradehumanrights/> for approaching HR from 'within the citadel of international economic law' (at 2) in privileging the economic HR (the right to property and the right to trade) for integration in WTO law over all other HR such as the important social rights, such as the right to health and food. Alston writes that 'the real focus of his concept of human rights is remarkably narrow … In his scheme of things, the WTO is never going to be called on to promote social rights, which means that despite the homage paid to them they remain entirely marginal to the essential thrust of his proposals' (at 18). For additional criticisms of Petersmann's 'free trade constitutionalism', see Robert Howse, 'Human Rights in the WTO: Whose Rights, What Humanity?', *supra* note 156. Howse is critical of Petersmann's proposition that WTO legitimacy as a juridical system 'depends on the transformation of what he calls 'market freedoms' into 'fundamental rights' (at 5): 'To the extent that the public policies in question themselves happen to be based on human rights (for example, social rights), we can see clearly the hierarchy of rights that Petersmann is proposing. Social and their positive human rights may only be pursued by governments to the extent to which they can be shown as 'necessary' limits on market freedoms. But why not the reverse?' For a reply in defence , see Ernst-Ulrich Petersmann, 'Taking Human Dignity, Poverty and Empowerment of Individuals more Seriously: Rejoinder to Alston' Jean Monnet Working paper No.12/02, <http://www.jeanmonnetprogram.org/papers/02/021201-03.pdf> at 2, Petersmann claims his arguments have been misconstrued to support a 'free

that as between trade and human-rights obligations (however defined), the latter should always be given priority, *de facto* or *de jure*. However, such a conclusion is not without its problems. The universal trumping of human rights domestically is not the argument sought to be sustained in this text, although it is an argument that certainly can be supported by our discussion. Rather, a more modest and conservative view subsumed within this assertion is advanced as my central thesis based on the fact that there is a (functional) dependence of human rights on national and international constitutional restraints: states have human rights obligations. If a state seeks to give domestic legal effect to its HR obligations owed to its citizens and to prioritize them as policy preferences (in this case where industrial patent policy over life forms may conflict with human rights), *so long as it does so in a non-discriminatory manner*, economic and trade commitments should *not* serve as international legal impediments. The argument for a HR's approach to dispute settlement, and the ECD to this end, constitutes the second branch of the bifurcated prescription of how to ensure that trade rules do not create undue restraints on HR. One could go further and ask, as Robert Howse has eloquently done, 'Why not subject *free trade rules* to strict scrutiny under a necessity test, where these rules make it more difficult for governments to engage in interventionist policies to protect *social rights*?'[180]

Serious threats to the WTO's well-being are said to emanate primarily from two directions: 'the escalating erosion of non-discrimination and the steady encroachment by rich-country lobbies seeking to impose their unrelated agendas on trade agreements'.[181] Institutional reform requires, according to Bhagwati, two main changes: 'relaxing the "tightness" of obligations that the WTO now incorporates and creates political waves [sic], and augmenting its minuscule resources'.[182] Recognizing the ECD is one way to address the 'tightness'. The founding principle of non-discrimination (based on the NT and MFN clauses) was the fundamental criterion of the multilateral regime the GATT incorporated into the WTO system and Agreements. The MFN ensured that all GATT members automatically were extended the benefits of lowest tariff rates and other benefits extended to other members with few explicit exceptions (Article 24 for Preferential Treatment Agreements (PTAs)) and some transitional arrangements for LD and DCs. But today, Bhagwati

standing human right to trade' when in actuality they merely extend constitutional liberty rights: 'The everyday experience of billions of people who can survive only by trading the fruits of their labour in exchange for goods and services … should be recognized as a human right problem rather than merely as a legislative or administrative task to be left to "benevolent governments".' Interestingly, strict IPR protections may impede the liberty to labour.

180 Ibid. at 6.

181 Bhagwati, *supra* note 117.

182 Ibid. at 4.

writes, 'the central principle of non-discrimination has been virtually destroyed. Thus, PTAs have proliferated to close to 300 and the number is growing by the week. The agreements which the architects of the GATT thought would be minor exceptions have now swallowed up the trading system'.[183] This leads to what he calls the 'spaghetti bowl'; the 'world trading system is characterized by a chaotic criss-crossing of preferences, with a plethora of different trade barriers applying to products depending on which countries they originate from'.[184] The implication has been to convert the MFN into the LFN (least favoured nation), more often than not. What the ECD tries to achieve is some equilibrium that will allow state sovereign discretion within the limited textual scope provided for in Article 8 and with regard to a state's full set of obligations owed without compromising the respective parties' preference for using and maintaining the WTO system over other bilateral and regional arrangements. At the same time, the defence recognizes that even in democracies built on embedded liberalism and social welfare policies, individual civil liberties must not be compromised. By legitimizing appropriate state agency through a defence, a state may pursue HR goals in a manner conducive to fostering *home-grown development* without threats of unilateralism or trade sanction.

183 Ibid.

184 Ibid. at 6, Bhagwatti offers an example: 'MFN tariffs now apply only to five countries, with all others enjoying politically driven lower-tariff access to the EU under multiple PTAs, differentiated GSP (General scheme of Preferences), EBA (Everything but Arms) and other Schemes. Evidently, the MFN tariff in the EU has now become the LFN, the least favored nation, tariff!' and later adds: '[t]his is a fool's way of doing trade – not only does it destroy the efficient allocation of resources, but it flies in the face of the fact that today it is becoming almost impossible to define which product is whose. It is hard to believe that sensible men in charge of trade policy today, including the USTR, the EU Trade Commissioner and other luminaries of trade are so unmindful of the fact that, in the name of free trade, they are damaging the world trading system through discriminatory PTAs ....'

Chapter 8
Conclusion

> It's easier to resist at the beginning than at the end.
>
> Leonardo Da Vinci

> The State is made for man, not man for the State.
>
> Albert Einstein, *The World as I See It*

The patenting of life raises a variety of important issues that challenge our normative understanding of intellectual property rights and draw attention to the need for institutional reform within domestic and international law. Extending patent protection to genetic information surrenders control to property holders, primarily large corporations, over access and use of this information, and restricts individual liberty rights. It jeopardizes the future direction of genomics research and application. The individual and opportunity costs will continue to rise as we transition out of the knowledge economy to the bio-economy. Biopatenting is not only contestable by traditional doctrines in domestic patent law but also under numerous international human rights instruments that reflect the special and jurisdictionally sovereign character of, and in some instances our communal claims to, life and its genetic building blocks. These issues have become arguably more significant in Canada given our Supreme Court's determination that non-human higher life are *not patentable*. Yet, any life containing a patented subcomponent, such as a gene, or a vector would be an infringing *use* of the underlying patented invention. The Canadian government, though familiar with patent controversy at the intersection of health policy and human rights, has responded to this tension in other jurisdictions, but has not failed to critically examine how patent law may be impacting health policies, including predictive genetic testing, at home. As a result, Canada is often hailed as a global leader in taking the initiative to amend patent law to supply essential medicines to foreign (DC and LDCs) governments with little or no manufacturing capacity, but lacks similar initiative to address the persuasively more inclusive HR issues arising from biopatenting. The health rights and interests of our own citizens must be as important as those of our human family internationally, which Canadian law now laudably serves to protect.[1] Whether Canada or any nation decides

1 See 'Government of Canada Reinstates Legislative Proposals to Enable Export of Low-Cost Pharmaceutical Products to Least Developed and Developing Countries',

to allow patents on life is an issue that needs to be locally determined through a public participatory process with an engaged representative government. Strident efforts are needed to resolve the legal quagmire of biopatenting given how fundamental this issue is to the cultural, ethical and social fabric of our society. Continued failure to undertake a political process of review has led to an incremental development of patent law by judicial fiat that lacks a broader social context and a coherent foundation in national IPRs policy. It has resulted in the progressive emasculation of the state by its agents. Domestic patent protections unnecessarily exceed what is prescribed under TRIPS and limit the scope for optimal cross-policy coordination, further compromising the realization of HR by forfeiting important policy space preserved in TRIPS.

All nation states should pause to reconsider the objectives, structure, and practical function of domestic patent regimes and to contextualize the lessons to be learned to their own geopolitical and cultural setting. In the interim, the TRIPS-compliant recommendations made in this text should offer some relief from the pressure for convergence towards higher and stricter standards of protection. These constitute the first branch of my bifurcated prescription for regulatory agency over biopatenting and aim to empower domestic re-evaluation of an institutionalist instrument so it may better accord with its social purpose, the nature of the scientific paradigm, our collective interest in genetic information and patent doctrine. If and when national governments decide to act, then the preceding discussions should inform that process in order to achieve the least inefficient and most equitable design of patent law. Governments may express reservations regarding national regulatory space due to TRIPS, but these are largely illusory and draw attention to partisan politics. The domestic recommendations made for state agency are necessary, but may be insufficient – they may not be equally viable for all states. Their adoption may depend as much on infrastructure, industry, and technology as on political will. Nations, like Canada, have already established sophisticated IP systems, mature and uncorrupt judiciaries, democratic institutions for public participation and access to greater resources, Canada is thus advantaged in her ability to adopt these solutions I offer. Other nations may be too poor to succeed in implementing domestic TRIPS-compliant measures or the measures may be contested for a host of other reasons, including as negotiating leverage in existing trade disputes between the parties. Accordingly, the prescription for state agency cannot end merely with the contestable TRIPS-compliant national measures. Where a trade-impacting measure is non-discriminatory and adopted in good faith and with the objective of promoting human rights, the international community should refrain from bringing a WTO complaint, respecting that state's exercise of sovereignty and self-determination over its genetic and natural resources. However, if a complaint is brought, the potential

---

Industry Canada, 12 February 2004 News Release, <http://ic.gc.ca/cmb/welcomeic.nsf>.

for conflict should be resolved by the panel or AB with due regard to the ECD based on Article 8 of TRIPS. This broadened HR approach to TRIPS disputes is the second branch of the bifurcated prescription for state agency and completes the normative conceptual framework for legally resolving conflict in biopatenting. IPRS are not only trade-related, but also HR-related. The ECD creates a safety net whereby governments can garner the confidence to take appropriate domestic measures without fear of liability under TRIPS.

The growing animosity towards the WTO and the acrimonious protests surrounding Ministerial meetings in Seattle, Cancun, Montreal, and Italy indicate the need for improvement in WTO operations. The crisis of legitimacy the WTO now faces is not only in terms of issues relating to social justice and democratic accountability, but to the delicate balance that must be achieved in meeting the objectives of both the UN and the WTO through the protection of human rights and the promotion of human development. Renewed commitments outside of the WTO reinforce the need to integrate a human-rights and development framework for resolving disputes. A development approach to IPRs is needed. The analyses offered here leads one to conclude that WTO disputes *should not* arise over patenting life because of the inherent flexibilities built into TRIPS that were negotiated as a means of reflecting the sovereignty of states over their genetic resources confirmed by the CBD. But should TRIPS disputes emerge, and we can expect that they will, the ECD helps members and WTO jurists to prepare to meet the challenges of the process and to provide more just outcomes. Effective global governance means the human and social dimensions of trade may no longer be ignored.[2] In the absence of a global sovereign government, the hegemony of trade has forged the conditions for its own regime of neoliberal governmentality and a powerful politics of instrumental regulation. But trade is not and should not be valued as its own moral imperative. The role of the WTO is broader than some might envisage. WTO jurists should see themselves as steward of trade and *other* values – values which further the objectives, and therefore the survival, of a multilateral trade regime. The institutional structure and juridical dispute-resolution mechanism of the WTO can facilitate the realization of those values better than any existing alternative and invariably, since trade

2 These considerations are '... of extreme systemic importance for the system of international law[;] ... they are heavily value-laden and go to the heart of much of the critique against globalisation: is globalisation only about the economy and making profit or is it counterbalanced also by other factors such as environmental protection, development of weaker regions, social protection and safety nets?. Pauwelyn, *Conflict of Norms in Public International Law*, op. cit. at 2. Mendes and Mehmet, *Global Governance, Economy and Law*, op. cit. at 1 define global governance to include 'not only the institutions set up to deal with issues of global scope, but also the situations that evolve in the absence of appropriate and effective institutions to deal with such global matters'.

is a *means for achieving* HR and HD ends, the two regimes will complement. The operations of the WTO must remain within the paradigmatic framework of an integrated international *system* of law informed by the objectives of the twin institutions for global governance (trade and HR) and renewed commitments to cooperation based on redistributive justice and equal access to global public goods. Today's society demands no less.[3] Where the measure is taken to further the domestic realization of one set of international norms (HR) over another, the impugned measure may be made subject to scrutiny. State measures must remain faithful to fundamental trade values to ensure they are not merely disguised protectionism but in fact genuinely justified, and non-discriminatory as between states. It is the role of governments around the world to offer palatable substantive solutions to biopatenting and to ensure that their 'publics' are agreeable with their trade-offs. Some deviation from trade objectives are permissible, as they serve the ultimate spirit of trade and human-rights objectives once reconciled into an integrated global governance regime. From this perspective, they are defensible. HR policy preferences are simply another means for the state, alternative to trade, for achieving the same welfare-promoting objectives of development (such as improved living standards). Making HR a priority will help ensure a more stable, healthy, operational, economically viable, developmentally mature, and independent WTO member and should be encouraged. Rosemary Coombe sagely counsels:

> [t]he minimal preconditions which make the bestowal of authorial rights a social good in the public interest – however narrowly that public may be defined – are not in place in the social contexts in which we are so anxious to extend authorial power and privilege. This speaks, of course, to the absurdity of dividing questions of trade from question of human rights, and to the growing anachronism of modernity's mappings of public and private international law.[4]

A global culture that celebrates diversity – biodiversity, cultural diversity, and regulatory diversity – centred on the universality of human rights, is a culture worth preserving and proselytizing. TRIPS convergence in national IPRs need not lead to a 'dead end'. Judicial erosion of important policy space, narrow conceptions of justiciable human rights and general political inertia in patent policy are of grave concern, however, in that they emasculate stage agency, in particular the various regulatory options available to allay so-called TRIPS frictions involving HR and HD. Failing to embrace these options is a

3 '[f]air net benefits for all are an important ingredient of *successful cooperation*. But it is one thing to share net costs fairly and another to share net benefits fairly'. Kaul, *Providing Global Public Goods*, op. cit. at 58, emphasis added.

4 Coombe, 'Authorial Cartographies', op. cit. at 1365.

political failure that neglects a very important fact: a *realistic utopia* requires more than monetary aid.

Selected Bibliographic Sources

Legislation

Criminal Code, R.S.C. 1985, c.C-46.
Food and Drugs Act, R.S.C. 1985, c. F-27.
North American Free Trade Agreement Implementation Act, S.C. 1993, c.44.
Patent Act, R.S.C. 1970, c. P-4; as am. by R.S.C. 1970, c. 10 (2nd Supp.).
Patent Act, R.S.C. 1985, c. P-4.
Patent Rules, S.O.R./1996-423.
Plant Breeders' Rights Act, S.C. 1990, c. 20.

International Materials

Agreement on Implementation of Article VI of the Agreement on Tariffs and Trade, GATT BISD 15S/24

Agreement on Trade-Related Aspects of Intellectual Property Rights, being Annex IC to the Final Act and Agreement Establishing the World Trade Organization, 15 December 1993 (1994), 33 I.L.M. 81 [TRIPS].

Articles of Agreement of the International Monetary Fund. International Monetary Fund, <http://www.imf.org/external/pubs/ft/aa/index.htm>.

Charter of the United Nations, as signed 1945 and Amended 1965, 1968 and 1973. Charter of the United Nations, <http://www.un.org/aboutun/charter>.

Commission on Trade and Investment Policy. 'Regional Trade Agreements and the Multilateral Trading System (November 2002), International Chamber of Commerce, <http://www.iccwbo.org/home/statements_rules/statements/2002/Regional%20trade%20agreements_multilateral%20trading%20system.asp>.

Constitution of the Republic of South Africa 1996, No. 108 of 1996.

Director General of the WTO, Panitchpakdi, Supachai. *Prospects for the Millennium Round*, address, <http://www.moc.go.th/thai/dbe/global10.html>.

Doha Ministerial Declaration, Paragraph 31(i), adopted on November 2001, WT/MIN(01)/Dec/1 (20 November 2001), WTO, <http://www.wto.org/English/thewto_e/minist_e/min01_e/mindecl_e.htm>.

Eichengreen B. and Mussa, M. 'Capital Account Liberalization: Theoretical and Practica Aspects', IMF Occasional Paper No. 172, 1998.

European Commission. *Development and Implications of Patent Law in the Field of Biotechnology and Genetic Engineering* (Brussels: The Commission,

2002), The European Union On-Line, <http://europa.eu.int/eur-lex/en/com/rpt/2002/com2002_0545en01.pdf>.

European Patent Convention, 5 October 1973, as amended by the act revising Article 63 EPC of 17 December 1991, and by decisions of the Administrative Council of the European Patent Organization of 21 December 1978, 13 December 1994, 20 October 1995, 5 December 1996 and 10 December 1998, European Patent Office, <http://www.european-patent-office.org/legal/epc/e/ma1.html>.

European Patent Office. Press Release No. 3/92, 'European Patent for Harvard's Mouse' (March 1992).

——. Press Release No. 10/89, 'EPO Refuses Patent Application for Oncogenic Mouse' (October 1989).

General Agreement on Tariffs and Trade 1947, 30 October1947, 58 U.N.T.S. 187, Can. T.S. 1947 No. 27 (entered into force 1 January 1948) GATT, <http://pacific.commerce.ubc.ca/trade/GATT.html>.

General Agreement on Tariffs and Trade (1994), World Trade Organization, <http://www.wto.org/english/docs_e/legal_e/06-gatt.doc>.

General Council's statement on the draft Decision contained in document IP/C/W/405 to implement paragraph 6 of the Doha Declaration on the TRIPS Agreement and Public Health (30 August 2003), World Trade Organization, <http://www.wto.org/english/news_e/news03_e/trips_stat_28aug03_e>.

Hafner, Gerhard. 'Risk Ensuing from Fragmentation of International Law', ILC, Annex to Report on the Work of the 52nd session, General Assembly Official Records, 56th session, Supplement No. 10 (A/55/10), 321-339, United Nations, <http://www.un.org/law/ilc/reports/2000/english/annexe.pdf>.

Helleiner, G. 'Markets, Politics and Globalization: Can the Global Economy be Civilized?' UNCTAD 10th Raúl Prebisch Lecture, Palais des Nations, Geneva (11 December 2000), <http://www.unctad.org/en/docs/prebisch10th.en.pdf>.

'Human Development Report 2002: Deepening Democracy in a Fragmented World' Human Development Reports, <http://hdr.undp.org/reports/global/2002/en/>.

International Court of Justice, [1947] I.C.J. Acts & Doc. 94.

International Covenant on Civil and Political Rights (19 December 1966), 999 U.N.T.S. 171, Can. T.S. 1976 NO. 47, 6 I.L.M. 368, International Covenant, <http://www1.umn.edu/humanrts/instree/b3ccpr.htm>.

International Covenant on Economic, Social and Cultural Rights (16 December 1966) GA Res. 2200A(XXI), 21 UNGAOR Supp. No. 16, UN Doc. A/6316 (1966), 993 U.N.T.S. 3, International Covenant, <http://www.hrweb.org/legal/escr.html>.

International Law Commission (ILC) Study Group on 'Fragmentation of International Law' chaired by Professor Bruno Simma (9 May 2002). See *Daily Bulletin*, 54th session of the ILC, United Nations, <http://www.un.org/law/ilc/sessopms/54/jourchr.htm>.

Kevles, Daniel J. 'A History of Patenting Life in the United States with Comparative Attention to Europe and Canada' (Report to the European Group on Ethics in Science and New Technologies) (12 January 2002), European Commission, <http://europa.eu.int/comm/european_group_ethics/docs/study_kevles.pdf>.

Machlup, F. *An Economic Review of the Patent System* (1958) Study No. 15, US Senate Committee on the Judiciary, Sub-Committee on Patents, Trademarks, and Copyrights.

Manifest Destiny, PBS, <http://www.pbs.org/kera/usmexicanwar/dialogues/prelude/manifest/d2aeng.html>.

OECD, Joint Group on Trade and Competition. *Trade and Competition Policies Options for a Greater Coherence* (Paris: 2001), OECD, <http://www.oecd.org/LongAbstract/0,2546,en_2649_36442957_1901341_119699_1_1_1,00.html>.

——. *Science, Technology and Industry Outlook 1996*, Science, Technology and Industry Outlook 229. OECD STI Outlook, <http://www.oecd.org/document/39/0,2340,en_2649_34273_1814439_1_1_1_1,00.html>.

Patent Co-operation Treaty, 19 June 1970 (WIPO Publication No. 274(E)). World Intellectual Property Organization, <http://www.wipo.int/pct/en/texts/index.htm>.

Quigg, D.J. (Commissioner of the United States Patent and Trademark Office). Policy Statement on Patentability of Animals (7 April 1987) 1077 Off. Gaz. Pat. Office 24).

Reciprocal Trade Agreements Act of 1934 ch. 474, 48 Stat. 943, 19 U.S.C. § 1351).

Report of the Working Party on Organizational and Functional Questions, adopted on 28 February, 5 and 7 March, 1955, GATT, BISD 3S/231 and the text of the proposed Agreement in GATT, BISD Volume 1.

The Crucible Group II, 'Seeding Solutions: Policy Options for Plant Genetic Resources' vol. 1 (International Development Research Centre, 2000), 40.

The Final Act Embodying the Results of the Uruguay Round of Multilateral Trade Negotiations, concluded in Marrakesh, Morocco, on 15 April 1994, published in WTO Secretariat, The Results of the Uruguay Round of Multilateral Trade Negotiations, The Legal Texts (Geneva, 1995). World Trade Organization, <http://www.wto.org/english/docs_e/legal_e/03-fa.doc>.

The Vienna Convention on the Law of Treaties (full text) United Nations, <http://www.un.org/law/ilc/texts/treaties.htm>.

The Understanding Regarding Notification, Consultation, Dispute Settlement and Surveillance, BISD 26S/210.

'The 128 countries that signed GATT by 1994' World Trade Organization, <http://www.wto.org/english/thewto_e/gattmem_e.htm>.

Tokyo Round Codes (full text), World Trade Organization, <http://www.wto.org/english/docs_e/legal_e/prewto_legal_e.htm>.

'Translatio Studii et Imperii', <http://cla.calpoly.edu/~dschwart/engl513/courtly/translat.htm>.

United Nations, General Assembly. 25th Session, 1849th Meeting, 24 September 1970, Official Records (A/PV.1848).

——, General Assembly. 'World Convention on Human Rights' 12 July 2003, Official Records (A/CONF.157/23) United Nations, <http://www.unhchr.ch/huridocda/huridoca.nsf/(Symbol)/A.CONF.157.23.En?OpenDocument>.

——, General Assembly. 21st Session, 47th Meeting, 26 November 1999, Official Records, United Nations, <http://www.unhchr.ch/tbs/doc.nsf/(Symbol)/E.C.12.1999.9.En?Opendocument>.

United Nations Conference on Trade and Development (UNCTAD) Report 1997, Least Developed Countries Report, United Nations, <http://www.unctad.org/Templates/WebFlyer.asp?intItemID=3079&lang=1>.

——. 'Linking International Trade with Poverty Reduction' Least Developed Countries Report 2004, United Nations United Nations, <http://www.unctad.org/en/docs/ldc2004_en.pdf>, last visited 10 June 2004.

United Nations Conference on Trade and Employment, 'Final Act and Related Documents' (Havana Cuba: 21 November 1947–24 March 1948). World Trade Organization, <http://www.wto.org/english/docs_e/legal_e/havana_e.pdf>.

United Nations Development Program (UNDP). 1999 Report Globalization with a Human Face, United Nations, <http://hdr.undp.org/reports/global/1999/en/>.

——. Press Release (4 April 2001), United Nations Press Releases, <http://www.undp.org/dpa/pressrelease/releases/2001/april/4apr01.html>.

United Nations Educational, Scientific and Cultural Organization (UNESCO). Universal Declaration on the Human Genome and Human Rights, Adopted 29th Session of the General Conference of UNESCO, 1997, UNESCO, <http://portal.unesco.org>.

——. *Manual of the General Conference*, texts adopted by the General Conference at its 31st session (2002). UNESCO, <http://portal.unesco.org>.

——. News Release, 'Sustainable Development' (3 September 2002), UNESCO, <http://portal.unesco.org>.

United Nations, Global Compact. United Nations, <http://www.unglobalcompact.org/Portal/>.

United Nations, Research Institute for Social Development (UNRISD, 2000). United Nations, <http://www.unrisd.org>.

——. 'UNRISD 2000+: A Vision for the Future of the Institute' (April 2000), <http://www.unrisd.org/unrisd/website/document.nsf/0/0F54921FBCE191FB80256B67005C49DE?OpenDocument>.

——. *Universal Declaration on Cultural Diversity*. 31st Sess. of the General Conference of UNESCO, Paris, 2 November 2001, UNESCO, <http://unesdoc.unesco.org/images/0012/001271/127160m.pdf>.

——. *Universal Declaration of Human Rights*. GA Res. 217(III), UN GAOR, 3rd Sess, Supp. No. 13, UN Doc. A/810 (1948), <http://www.unhchr.ch/udhr/lang/eng.htm>.

United States Patents Quarterly 443 (Board of Patent Appeals and Interferences 1985).

World Bank 2000–2003 Annual Reports, <http://web.worldbank.org>.

UNCTAD. *World Investment Report: Transnational Corporations, Market Structure and Competition Policy* (Geneva: 1997).

United States Code § 35 (1926). U.S. Code, <http://www.access.gpo.gov/uscode/title35/title35.html>.

United States, Office of Technology Assessment. *Commercial Biotechnology, An International Analysis* (OTA-BA-218) (Washington, DC: US Congress, Office of Technology Assessment, January 1984).

'What are the Millennium Development Goals' UN Development Millennium Goals, <http://www.un.org/millenniumgoals/>.

Wolfenshon, James D. 'A Proposal For a Comprehensive Development Framework' (Discussion Draft), January 1999, The World Bank, <http://www.worldbank.org/cdf/cdf-text.htm>.

World Bank Development Report 2000/2001: 'Attacking Poverty', World Bank, <http://web.worldbank.org>.

World Trade Organization (WTO). *Differential and More Favourable Treatment, Reciprocity and Fuller Participation of Developing Countries*, WTO Dec. L/4903, B.I.S.D. (1979) 26S/203.

——. 'WTO in Brief' (28 July 2001) 5:28 ICTSD Bridges Weekly Trade News Digest, ICTSD, <http://www.ictsd.org/html/weekly/24-07-01/wtoinbrief.htm>.

——. 'Special Study on Trade and Competition Policy', in Annual Report for 1997 (Geneva: 1997), ch. IV, UNCTAD, World Investment Report: Transnational Corporations, Market Structure and Competition Policy (Geneva: 1997).

Jurisprudence

Amfac Foods Inc. v. Irving Pulp and Paper Ltd. (1984), 80 C.P.R. (2nd) 59 (F.C.T.D.).

Auton v. British Columbia, [2004] 3 S.C.R. 657.

Cadbury-Schweppes v. FBI Foods, [1999] 1 S.C.R. 142.

Catnic Components Ltd. v. Hill and Smith Ltd. [1982] R.P.C. 183; [1981] F.S.R. 60.

Comissioner of Patents v. Farbwerke Hoechst Aktiengesellschaft Vormals Meister Lucius and Bruning, [1964] S.C.R. 49.

Gosselin v. Quebec, [2005] 1 S.C.R. 238.

T19/90 Harvard/Onco-mouse OJ EPO 1990, 476 (Decision of the Technical Board of Appeal 3.3.2 of 3 October 1990).
Diamond v. Chakrabarty, 447 U.S. 303, 309 (1980) (citing S.Rep.No. 1979, 82nd Cong, 2nd Sess, 5 (1952); H.R.Rep.No. 1923, 82nd Cong, 2nd Sess, 6 (1952)).
European Patent Application No. 86 304490.7.
Ex Parte Allen 2 U.S.P.Q. 2nd 1425 (P.T.O. Bd. App. & Int. 1987) ; (1987), 1077 Official Gazette of the U.S.
Ex Parte Latimer, March 12, 1889, C.D, 46 O.G. 1638, U.S. Patent Office, *Decisions of the Commissioner of Patents and of the United States Courts in Patent Cases … 1889* (Washington, DC: Government Printing Office, 1890).
Kirin Amgen Inc v. Hoechst Marion Roussel Ltd. E.W.J. No. 3792 (QL), [2002] EWCA Civ. 1096 (C.A.).
Free World Trust v. Electro Santé Inc, [2000] 2 S.C.R. 1024, 2000 SC 66.
Harvard College v. Canada (Commissioner of Patents), [2002] S.C.J. No. 77, 2002 SCC 76.
In Re Howard Florey Institute-Relaxin, [1995] E.P.O.R. 541 [EPO Opposition Division].
Micro Chemicals Ltd. v. Smith Kline and French Inter-American Corp. (1971), [1972] S.C.R. 506.
Monsanto Canada Inc. v. Schmeiser (2001), 12 C.P.R. (4th) 204 (F.C.T.D), aff'd [2003] F.C. 165 (F.C.A.).
Moore v. Regents of the University of California, 51 Cal. 3rd 120, 125; 793 P.2nd 479 (Supreme Court Cal, 1990).
Parker v. Flook, 437 U.S. 584.
President & Fellows of Harvard College v. Canada (Commissioner of Patents), [2000] 4 F.C. 528.
Risi Stone Ltd. v. Group Permaco In. (1990), 29 C.P.R. (3rd) 243 (F.C.T.D.) (interlocutory); (1995), 65 C.P.R. (3rd) 2 (F.C.T.D.) (trial).
Tennessee Eastman Co. et al. *v. Commissioner of Patents*, [1974] S.C.R. 111, 8 C.P.R. (2nd) 201.
Xerox of Canada Ltd. v. I.B.M. Canada Ltd. (1977), 33 C.P.R. 24.

Secondary Material: Monographs

Aoki, Keith. *Seed Wars: Cases and Materials on Intellectual Property and Plant Genetic Resources* (Durham, NC: Carolina Academic Press, 2006).
Aristotle. *Rhetoric* (Montana: Kessinger Publishing, 2004).
Baldwin, R. *Non-tariff Distortions of International Trade* (Washington, DC: The Brookings Institution, 1970).
Claude E. Barfield, *Free Trade, Sovereignty, Democracy: The Future of the World Trade Organization* (Washington, DC: AEI Press, 2001).

Beale, Howard K. *Theodore Roosevelt and the Rise of America to World Power* (Baltimore, MD: Johns Hopkins University Press, 1956).
Bercovitch, Sacvan. *The American Jeremiad* (Madison, WI: University of Wisconsin Press, 1978).
Blakeney, Michael. 'Stimulating Agricultural Innovation in K.E. Maskus and J.H. Reichman, eds, *International Public Goods and Transfer of Technology under a Globalized Intellectual Property Regime* (Cambridge, MA: Cambridge University Press, 2005).
Bottomley, Stephen and Kinley, David, eds. *Commercial Law and Human Rights* (Brookfield, VT: Dartmouth, 2002).
Brandt, Richard B. *A Theory of the Good and the Right* (New York: Prometheus Books, 1998).
Charnovitz, Steve. *Trade Law and Global Governance* (London: Cameron May, 2002).
Cheyfitz, Eric. *The Poetics of Imperialism: Translation and Colonization from the Tempest to Tarzan* (Philadelphia: University of Pennsylvania Press, 1997).
Coicaud, Jean-Marc and Heiskanen, Veijo, eds. *The Legitimacy of International Organizations* (New York: United Nations University Press, 2001).
Cook, Curtis. *Patents, Profits and Power* (London: Kogan Page, 2002).
Cornish, William. *Intellectual Property: Omnipresent, Distracting, Irrelevant* (New York: Oxford University Press, 2004).
Cornish, William R. and Llewelyn, M. and Adcock, M. *Intellectual Property Rights (IPRs) and Genetics: A Study into the Impact ad Management of Intellectual Property Rights within the healthcare Sector* (Cambridge: Public Health Genetic Unit, July 2003), Public Health Genetics Unit Homepage, <http://www.phgu.org.uk/about_phgu/s-ipr1.doc>.
Croome, John. *Reshaping the World Trading System: A History of the Uruguay Round*, 2nd and rev. edn (Boston: Kluwer Law International, 1999).
Cutler, A. Claire. *Private Power and Global Authority: Transnational Merchant Law in the Global Political Economy* (Cambridge: Cambridge University Press, 2003).
D'Amato, Anthony and Long, Doris Estelle, eds, *International Intellectual Property Law* (Boston: Kluwer Law International, 1997).
Davies, Kevin. *Cracking the Human Genome: Inside the Race to Unlock Human DNA* (New York: The Free Press, 2001).
de Carvalho, Nuno Pires. *The TRIPS Regime of Patent Rights* (The Hague: Kluwer Law International, 2002).
de Montesquieu, Charles et al. *Montesquieu: The Spirit of the Laws* (Cambridge: Cambridge University Press, 1989).
Donnelly, J. *Universal Human Rights in Theory and Practice*, 2nd edn (Ithaca, NY: Cornell University Press, 2003).
Drahos, Peter. *A Philosophy of Intellectual Property* (Brookfield, VT: Dartmouth, 1996).

—— and John Braithwaite. *Information Feudalism* (London: Earthscan Publications, 2002).

—— and Mayne, Ruth. *Global Intellectual Property Rights: Knowledge, Access and Development* (New York: Palgrave, 2002).

Dutfield, Graham. *Intellectual Property Rights, Trade and Biodiversity: Seeds and Plant Varieties* (London: Earthscan Publications, 2000).

Erbisisch, E.H. and Maredia, K.M., eds. *Intellectual Property Rights in Agricultural Biotechnology* (Cambridge: CABI Publishing, 2004).

Evans, R.G, Barer, M.L. and Marmor, T.R. *Why Are Some People Healthy and Others Not? The Determinants of Health of Populations* (New York: Aldine Transaction, 1994).

Fecenko, Mark J. *Biotechnology Law: Corporate-Commercial Practice* (Markham, ON: Butterworths, 2002).

Fletcher, Ian et al., eds. *Foundations and Perspectives of International Trade Law* (London: Sweet and Maxwell, 2001).

Fox, H.G. *The Canadian Law and Practice Relating to Letters Patent for Inventions*, 4th edn (Toronto: Carswell, 1969).

Freeman, Mark and Van Ert, Gibran. *International Human Rights Law* (Toronto: Irwin Law, 2004).

Gardner, J.P, ed. *Human Rights as General Norms and A State's Right to Opt Out: Reservations and Objections to Human Rights Conventions* (London: British Institute of International and Comparative Law, 1997).

Garth, Bryant G. and Dezalay, Yves, eds. *Global Prescriptions: The Production, Exportation, and Importation of a New Legal Orthodoxy* (Ann Arbor, MI: University of Michigan, 2005).

Gervais, Daniel. *The TRIPS Agreement: Drafting History and Analysis*, 2nd edn (London: Sweet and Maxwell, 2003).

Gibbons, Michael et al. *The New Production of Knowledge* (London: Sage, 1994).

Glasbeek, Harry. *Wealth by Stealth: Corporate Crime, Corporate Law, and the Perversion of Democracy* (Toronto: Between the Lines, 2002).

Gold, E.R. *Body Parts: Ownership of Human Biological Materials* (Washington, DC: Georgetown University Press, 1996).

Gordon, Wendy. *Intellectual Property Theory Intensive Course Reader* (Toronto: University of Toronto Press, 2000).

Graham, Edward. *Fighting the Wrong Enemy: Anti-Global Activists and Multinational Enterprises* (Washington, DC: Institute for International Economics, 2000).

Habermas, Jürgen. *The Future of Human Nature* (Cambridge: Polity Press, 2003).

——. *Between Facts and Norms* (Cambridge: Polity Press, 1966).

Hart, H.L.A. *The Concept of Law*, 2nd edn (New York: Oxford University Press, 1997).

Hastrup, Kirsten, ed. *Legal Cultures and Human Rights: The Challenge of Diversity* (New York: Kluwer Law International, 2001).
Henkin, Louis. *How Nations Behave*, 2nd edn (New York: Columbia University Press, 1979).
——. *The Rights of Man Today* (Boulder, CO: Westview Press, 1978).
Hietala, Thomas R. *Manifest Design: Anxious Aggrandizement in Late Jacksonian America* (Ithaca, NY: Cornell University Press, 1985).
Higgins, R. *Problems and Process: International Law and How we Use It* (New York: Oxford University Press, 1995).
Hobbelink, Henk. *Biotechnology and the Future of World Agriculture* (London: Zed Books, 1991).
Horsman, Reginald. *Race and Manifest Destiny: The Origins of American Racial Anglo-Saxonism* (Cambridge, MA: Harvard University Press, 1981).
Hudec, Robert E. *Essays on the Nature of International Trade Law* (London: Cameron May International Law and Policy, 1999).
Hutchinson, Allan and Patrick Monahan, eds. *The Rule of Law: Ideal or Ideology* (Toronto: Carswell, 1987).
Iriye, Akira. *The Cambridge History of American Foreign Relations, Volume III: The Globalizing of America, 1912–1945* (Cambridge: Cambridge University Press, 1993).
Jackson, John H. *World Trade and the Law of GATT* (Indianapolis, IN: Bobbs-Merrill, 1969).
——, Davey, William J. and Sykes, Alan O. Jr. *Legal Problems of International Economic Relations*, 3rd edn (St Paul, MD.: West Publishing, 1995).
Jessup, Philip. *Transnational Law* (New Haven, CT: Yale University Press, 1956).
Johnson, Jon R. *International Trade Law* (Toronto: Irwin Law, 1998).
Judson, Horace. *The Eighth Day of Creation: Makers of the Revolution in Biology* (New York: Cold Spring Harbor Laboratory Press, 1996).
Kanavox, Panos and Christina Golna. *WTO and Patents: The Impact on the Pharmaceutical Industry* (London: Informa Pharmaceuticals, 2000).
Kant, Immanuel. *The Metaphysics of Morals*, trans. by Gregor, M. (Cambridge: Cambridge University Press, 1991).
Kaul, Inge et al., eds. *Providing Global Public Goods: Managing Globalization* (Toronto: Oxford University Press, 2003).
Kindred, Hugh M. et al. *International Law Chiefly as Interpreted and Applied in Canada*, 6th edn (Toronto: Emond Montgomery Publications, 2000).
Kuhn, Thomas. *The Structure of Scientific Revolutions* 2nd edn (Chicago: University of Chicago Press, 1970).
LaFeber, Walter. *The Cambridge History of American Foreign Relations, Volume II: The American Search for Opportunity, 1865–1913* (Cambridge: Cambridge University Press, 1995).

Lauren, Paul Gordon. *The Evolution of Human Rights: Visions Seen* (Philadelphia: University of Pennsylvania Press, 1998).
Leeming, David Adams. *The World of Myth: An Anthology* (New York: Oxford University Press, 1990).
Lloyd, Dennis. *The Idea of Law: A Repressive Evil or Social Necessity?* (London: Penguin Books, 1983).
Locke, John. *The Second Treatise of Government*, ed. by Thomas P. Peardon. (New York: Bobbs-Merrill, 1952).
Maine, Sir Henry. *International Law* (London: John Murray, 1890).
Matsushita et al. *World Trade Organization: Law, Practice, and Policy* (Oxford: Oxford University Press, 2003).
Maskus, Keith. *Intellectual Property Rights in the Global Economy* (Washington, DC: Institute for International Economics, 2000).
Mason, J.K. and McCall Smith, R.A. *Law and Medical Ethics* (London: Butterworths, 1999).
McCoy, Drew R. *The Elusive Republic: Political Economy in Jeffersonian America* (Chapel Hill, NC: University of North Carolina Press, 1988).
McElheny, Victor. *Watson and DNA: Making a Scientific Revolution* (New York: Basic Books, 2004).
McKibben, Bill. *Enough: Staying Human in an Engineered Age* (New York: Audio Renaissance Audiobook, 2003).
Meldrum, A.N. *The Eighteenth-Century Revolution in Science – The First Phase* (Calcutta: Longmans, Green and Co., 1930).
Mendes, Errol and Mehmet, Ozay. *Global Governance, Economy and Law: Waiting for Justice* (New York: Routledge, 2003).
Merton, R. *The Sociology of Science: Theoretical and Empirical Investigations* (Chicago: University of Chicago Press, 1973).
Mgbeogi, Ikechi. *Global Biopiracy: Patents, Plants, and Indigenous Knowledge* (Ithaca, NY: UBC Press, 2006).
Norhaus, W. *Invention, Growth and Welfare: A Theoretical Treatment of Technological Change* (Cambridge, MA: MIT Press, 1969).
Ortino, Federico. *Basic Legal Instruments for the Liberalization of Trade: A Comparative Analysis of EC and WTO Law* (Portland, OR: Hart Publishing, 2003).
Pauwelyn, Joost. *Conflict of Norms in Public International Law: How WTO Law Relates to Other Rules of International Law, Cambridge Studies in International and Comparative Law* (New York: Cambridge University Press, 2003).
Penrose Edith, *The Economics of the International Patent System* (Baltimore, MD: Johns Hopkins University Press, 1951).
Plato. *The Laws*, trans. by Thomas L. Pangle (Chicago: University of Chicago Press, 1998).
Pocock, John. *Virtue, Commerce and History* (Cambridge: Cambridge University Press, 1985).

Purves, William K. et al. *Life: The Science of Biology, Volume 1: The Cell and Heredity*, 5th edn (Sunderland, MA: Sinauer Associates Inc., 1998).
Rawls, John. *The Law of Peoples* (Cambridge, MA: Harvard University Press, 1999).
Rivette, K.G. and Kline, David. *Rembrandts in the Attic: Unlocking the Hidden Value of Patents* (Boston, MA: Harvard Business School Press, 2000).
Schacter, Oscar. *International Law in Theory and in Practice* (Netherlands: Martinus Nijhoff Publishers, 1991).
Sell, Susan K. *Private Power, Public Law: The Globalization of Intellectual Property Rights* (New York: Cambridge University Press, 2003).
——. *Power and Ideas: The North-South Politics of Intellectual Property and Antitrust* (Albany, NY: State University of New York Press, 1998).
Sen, Amartya. *Poverty and Famines* (Oxford: Clarendon Press, 1981).
——. *Development as Freedom* (New York: Anchor, 2000).
—— and Dreze, Jean. *Hunger and Public Action* (Oxford: Clarendon Press, 1989).
Shaw, Malcolm N. *International Law*, 5th edn (New York: Cambridge University Press, 2003).
Shiva, Vandana. *Biopiracy: The Plunder of Nature and Knowledge* (Toronto: Between the Lines, 1997).
Stiglitz, Joseph E. *Globalization and Its Discontents* (New York: W.W. Norton and Company, 2002).
Thomas, Jeffery S. and Michael A. Meyer, eds. *The New Rules of Global Trade: A Guide to the World Trade Organization* (Toronto: Carswell, 1997).
Townson, Monica. *Health and Wealth: How Social and Economic Factors Affect our Well Being* (Toronto: Lorimer, 1999).
Trebilcock, Michael and Robert Howse. *The Regulation of International Trade* 2nd edn (London and New York: Routledge, 1999).
——. *International Trade Regulation*, 3rd edn (London and New York: Routledge, 2005).
Vaver, David. *Intellectual Property Law: Copyright, Patents, Trade-Marks* (Toronto: Irwin Law, 1997).
Weinrib, Ernest. 'The Intelligibility of the Rule of Law' in Hutchinson, Allan C. and Monahan, Patrick J., eds, *The Rule of Law: Ideal or Ideology* (Toronto: Carswell, 1987).
Weiss, Linda. *The Myth of the Powerless State* (New York: Cornell University Press, 1998).
Wolfe, James H. *Modern International Law: An Introduction to the Law of Nations* (Upper Saddle River, NJ: Pearson Education, 2002).
Wollstonecraft, Mary. *A Vindication of the Rights of Woman* (London: Penguin Classics, 2004).
Woods, Gord. *The Creation of the American Republic, 1776–1787* (Chapel Hill, NC: University of North Carolina Press, 1998).

Secondary Material: Articles

Abbott, Frederick. 'The TRIPS Agreement, Access to Medicines and the WTO Doha Ministerial Conference' (September 2001).

Abraham, Carolyn. 'Scientists look at Creating a Human-Mouse Embryo' *Globe and Mail* (28 October 2002), *Globe and Mail*, <http://rtnews.globetechnology.com/servlet/ARticleNews/tech/RTGA/20021128/wxmous11>.

Adcock, Mike. 'Myriad Breast Cancer Patent Revoked after Public Hearing' (24 May 2004), University of Sheffield, <http://www.shef.ac.uk/bioethics-today/archives/files/Patentscomm.html>.

Adelman, Jeremy and Centeno, Miguel Angel. 'Between Liberalism and Neoliberalism: Law's Dilemma in Latin America' in Dezalay, Yves and Garth, Bryant G. eds, *Global Prescriptions: The Production, Exportation, and Importation of a New Legal Orthodoxy* (Ann Arbor, MI: University of Michigan, 2005).

Allison, John R. and Lemley, Mark A. 'Empirical Evidence on the Validity of Litigated Patents' (1998), Social Science Research Network, <http://papers.ssrn.com/sol3/papers.cfm?abstract_id=118149>.

Alston, J.P., Pardey, G. and Rosenboom, J. 'Financing Agricultural Research: International Investment Patterns and Policy Perspectives' (1998) 26 *World Development* 1045.

Amani, Bita. 'Has Patenting Gone Too Far?' *Innovate* (Spring 2004) 15, University of Toronto – Faculty of Law, <http://www.law.utoronto.ca/documents/publications/innovate_spring04.pdf>.

——. 'The Promise and Perfidy of Patents: Biotechnology, the Genetic Revolution, and the Invention of "Invention"' [on file with the author].

——. 'Patents and Public Health' (2002) 22:3 *Health L. Can.* 76.

——. 'A Penchant for Persian Rugs Over Palatable Products: The Use of Geographic Appellations as Trade-Marks' (2000) 14.2 *I.P.J.* 185; 14.3 *I.P.J.* 313.

—— and Coombe, Rosemary. 'The Human Genome Diversity Project: The Politics of Patents at the Intersection of Race, Religion, and Research Ethics' 27:1 *Law and Pol'y* 152.

Anderson, Robert D. and Holmes, Peter. 'Competition Policy and the Future of the Multilateral Trading System' (2002) *J. Int'l Econ. L.* 531–63.

Andrews, Lori B. and Paradise, Jordan. 'Gene Patents: The Need for Bioethics Scrutiny and Legal Change' (2005) 5 *Yale J. Health Pol'y L. and Ethics* 403–12.

Aoki, Keith. 'Weeds, Seeds and Deeds: Recent Skirmishes in the Seed Wars' (2003) 11 *Cardozo J. Int'l and Contemp. L.* 247.

Audley, John and Florini, Ann M. 'Overhauling the WTO: Opportunity at Doha and Beyond' (Policy Brief) (October 2001), Carnegie Endowment for International Peace, <http://www.ceip.org/files/pdf/pb6-AudleyFlorini.pdf>.

Austin, Lisa and Amani, Bita. 'Patents on Genes: Identifying Issues and Responses' (presented to the Ontario Provincial Advisory Committee on New Genetic Technologies, October 2001), annex to 'Legal and Ethical Challenges of New Predictive Genetic Testing Report of the Legal and Ethical Subcommittee of the Provincial Advisory Committee on New Predictive Genetic Technologies (2003) (Co-Chairs of the Committee: T. Lemmens and R. Mykitiuk, authors/contributors: Mireille Lacroix, Lisa Austin, Bita Amani), <http://www.health.gov.on.ca/english/public/pub/ministry_reports/geneticsrep01/genetic_report.pdf>; reproduced in Lemmens, Trudo et al., *Reading the Future?: Legal and Ethical Challenges of New Predictive Genetic Testing* (Montreal: Editions Thémis, 2007).

Babula, Jared. 'Transgenic Crops: A Modern Trojan Horse' (1999) 3:1 *J.L. and Soc. Challenges* 131.

Baghwati, Jagdish. 'Borders Beyond Control', (January/February 2003) 82:1 *Foreign Affairs* 98.

Bartels, L. 'Article XX of GATT and the Problem of Extraterritorial Jurisdiction: The Case of Trade Measures for the Protection of HR' (2002) 36 *J. World Trade* 353.

Barton, J.H. 'Non-Obviousness' (2003) 43 *IDEA* 475.

——. 'Reforming the Patent System' (2000) 287 *Science* 1933.

Beales, Howard J. III. 'Modification and Consumer Information: Modern Biotechnology and the Regulation of Information' (2000) 55 *Food and Drug L.J.* 105.

Becker, David. 'Apple Patented by Microsoft: Apparently, Intellectual Property Does Grow on Trees' (4 May 2004), CNET News.com, <http://news.com.com/2100-1008-5205574.html>.

Bell D. and Renner, M. 'The War on Terrorism Needs a Marshall Plan',, <http://www.net-about-town.com.au/imaginepeace/articles/marshallplan.htm>.

Bertini, Catherine. 'Triple North Korea Food Acid U.N. Urges' *Toronto Star* (30 November 2000), A16.

Black, Julia. 'Critical Reflection on Regulation' (2002) 27 *Australian Journal of Legal Philosophy* 1.

Blackett, Adelle. 'Whither Social Clause? Human Rights, Trade Theory and Treaty Interpretation' (1999) 31 *Colum. H.R.L. Rev.* 1.

Blichner, Lars C. and Molander, Anders. 'What is Juridification?' Arena – Centre for European Studies, <http://www.arena.uio.no/events/documents/PAPER_002.pdf>.

Bohman, James. 'Constitution Making and Democratic Innovation: The European Union and Transnational Governance' (2004) 3 *Eur. J. Pol. Theory* 315.

Boyle, James. 'The Second Enclosure Movement and the Construction of the Public Domain' (Winter/Spring 2003) 66 *SPG Law and Contemp. Probs.* 33.

Branstetter, Lee G. 'Do Stronger Patents Induce More Local Innovation' (2004) 7:2 *J. Int'l Econ. L.* 359.

Brashear, Andrea D. 'Evolving Biotechnology Patent laws in the United States and Europe: Are they Inhibiting Disease Research?' (2001) 12 *Ind. Int'l and Comp. L. Rev.* 183.

Bratic, W., McLane, P. and Sterne, R. 'Business Discovers the Value of Patents' *Managing Intellectual Property* (September 1998).

Brown, William Y. 'Promise and Peril' (2001) 18:5 *The Environmental Forum* 20.

Burk, Dan L. and Lemley, Mark A. 'Is Patent Law Technology-Specific?' Center for Innovation Law and Policy, <http://www.innovationlaw.org/pages/DanBurk.doc>.

Calabresi, Guido and A. Douglas Malamed, 'Property Rules, Liability Rules, and Inalienability: One view of the Cathedral' (1972) 85 *Harv. L. Rev.* 1089.

Campbell, Eric and Blumenthal, David. 'The Selfish Gene: Data Sharing and Withholding in Academic Genetics' (31 May 2002) *Journal of the American Medical Association*, ScienceCareers.org, <http://sciencecareers.sciencemag.org/career_development/previous_issues/articles/1680/the_selfish_gene_data_sharing_and_withholding_in_academic_genetics/(parent)/12098>.

'Canadian Biotechnology Advisory Committee Non-Governmental Organization Hearing on the Intellectual Property/Patenting of Higher Life Forms Project Steering Committee', Canadian Biotechnology Advisory Committee, <http://cbac-cccb.ca/epic/internet/incbac-cccb.nsf/vwapj/ngo-reportv4_e.pdf/$FILE/ngo-reportv4_e.pdf>.

Cantor, Alison E. 'Using the Written Description and Enablement Requirements to Limit Biotechnology Patents' (2000) 14.1 *Harvard Journal of L. and Tech.* 267.

Cho, Sungjoon. 'Linkage of Free Trade and Social Regulation: Moving Beyond the Entropic Dilemma' (2005) 5 *Chi. J. Int'l L.* 625.

Clement, Tony. 'Just What the Doctor Ordered?' *Innovate* (Spring 2004) 22, University of Toronto – Faculty of Law, <http://www.law.utoronto.ca/documents/publications/innovate_spring04.pdf>.

Cohen, Morris R. 'Property and Sovereignty' (1927) 13 *Cornell L.Q.* 8.

Coombe, Rosemary J. 'Intellectual Property, Human Rights and Sovereignty: New Dilemmas in International Law Posed by the Recognition of Indigenous Knowledge and the Conservation of Biodiversity' (1998) 6 *Ind. J. Global Legal Stud.* 59.

——. 'The Recognition of Indigenous Peoples' and Community Traditional Knowledge in International Law' (2001) 14 *St Thomas L. Rev.* 275.

——. 'Authorial Cartographies: Mapping Proprietary Borders in a Less-Than Brave New World' (1996) 48.5 *Standford L. Rev.* 1357.

——. 'Legal Claims to Culture in and against the Market: Neoliberalism and the Global Proliferation of Meaningful Difference' (2005) 1 *Law, Culture and Humanities* 32, York University, <http://www.yorku.ca/rcoombc/publications/Coombe_Legal_Claims_to_Culture.pdf>.

Crook, Clive. 'Globalization and Its Critics: A Survey of Globalization' *The Economist* (29 September 2001) 3-30.

Crooks, Ed. 'Prizes for Saving Lives: Drug Companies Will not develop New Vaccines unless they are Awarded for It' *Financial Times* (25 April 2001), CID at Harvard University, <http://www.cid.harvard.edu/cidinthenews/articles/FT_042701.html>.

Curtis, John M. 'Trade and Civil Society: Towards Greater Transparency in the Policy Process', <http://www.international.gc.ca/eet/pdf/12-en.pdf#search=%22john%20curtis%20trade%20and%20civil%20society%22> at 302.

Daar, A.S., Thorsteindóttir H., Martin, D.K. et al. 'Top 10 Biotechnologies for Improving Health in Developing Countries' (2002) 32 *Nature Genetics*, 229–32, University of Toronto<http://www.utoronto.ca/jcb/research/documents/top10ng.pdf>.

Daily Yomiuri. 'Hepatitis Antibody made from GM Rice' (1 November 2000), Crop Biotechnology News, <http://www.agbios.com/static/news/NEWSID_1763.php>.

Das, Bal Gopal. 'IP Dispute, GATT, WIPO: Of Playing by the Game Rules and the Rules of the Game (1994) 35 *J.L. and Tech* 149.

Davis, John Renfro. '"The World Turned Upside Down": Contemplations from the Marianas Trench', <http://www.contemplator.com/england/worldtur.html>.

Demaine, Linda J. and Fellmeth, Aaron Xavier. 'Reinventing the Double Helix: Novel and Nonobvious Reconceptualization of the Biotechnology Patent' (2002) 55 *Stanford Law Review* 304.

de Sousa Santos, Boaventura. *Toward a New Common Sense: Law. Science and Politics in the Paradigmatic Transition* (London and New York: Routledge, 1995), cited in Cutler, Claire, *Private Power and Global Authority: Transnational Merchant Law in the Global Political Economy* (Cambridge: Cambridge University Press, 2003) 19.

Dezalay, Yves and Garth, Bryant G.. 'Legitimating the New Legal Orthodoxy' in Dezalay, Yves. and Garth, Bryant G., eds, *Global Prescriptions: The Production, Exportation, and Importation of a New Legal Orthodoxy* (Ann Arbor, MI: University of Michigan, 2005).

Dinwoodie, Graeme B. and Dreyfuss, Rochelle Cooper. 'International Intellectual Property Law and the Public Domain of Science' (2004) 7(2) *JIEL* 431.

Drahos, Peter. 'Expanding Intellectual Property's Empire: the Role of FTAs' GRAIN, <http://www.grain.org/rights/tripsplus.cfm?id=28>.

Dresser, Rebecca. 'Ethical and Legal Issues in Patenting New Animal Life' (Summer 1988) 28 *Jurimetrics Journal* 399.

Dreyfuss, R.C. 'Examining State Street Bank: Developments in Business Method Patenting' (2001) *Computer und Recht International* 1.

Driesen, D. "What is Free Trade?' The Real Issue Lurking Behind the Trade and Environment Debate' (2001) 41 *Va. J. Int'l L.* 279.

Durrell, Karen Lynn. 'Intellectual Property Protection for Plant Derived Vaccine Technology: Here they Come are we Ready or Not?' 10:3 *Lex Electronica (*Winter 2006), <http://www.lex-electronica.org/articles/v10-3/durell.pdf>.

Dutfield, Graham. 'The Public and Private Domains: Intellectual Property Rights in Traditional Knowledge' (2000) 21 *Science Communication* 274.

——. 'Intellectual Property Rights and Biodiversity: Conflict or Synergy?' ch. 5 in *Intellectual Property Rights, Trade, and Biodiversity* (London: Earthscan Publications, 2000).

——. 'Sharing the Benefits of Biodiversity: Is there a Role for the Patent System?' (2002) 5:6 *Journal of World Intellectual Property* 899.

Dworkin, Ronald. 'Objectivity and Truth: You'd Better Believe It' (1996) 25 *Philosophy and Public Affairs* 87.

Eisenberg, Rebecca S. and Merges, Robert P. 'Opinion Letter as to the Patentability of Certain Inventions Associated with the Identification of Partial DNA Sequences' (1995) 23 *AIPLA Q.J.* 1.

Flood, Colleen and Hinds, Greg. 'Future of Health Care in Canada: The Who, What and How Much of Publicly Funded Health Care' (Spring 2004) 1 *Innovate* 23, University of Toronto – Faculty of Law, <http://www.law.utoronto.ca/documents/publications/innovate_spring04.pdf>.

Forcese, Craig. 'Globalizing Decency: Responsible Engagement in an Era of Economic Integration' (2002) 5 *Yale Human Rts and Dev. L.J.* 1.

Fried, Ina. 'Gates Wants Patent Power' (29 July 2004), CNET News.com, <http://news.com.com/Gates+wants+patent+power/2100-1014_3-5288722.html?tag+st.rc.targ_mb>.

Gallini, Nancy and Scotchmer, Suzanne. 'Intellectual Property: When is it the Best Incentive System?' in Jaffe, Adam and Lerner, Joshua and Stern, Scott, eds, *Innovation Policy and the Economy* (Cambridge, MA: MIT Press, 2002), David Levine's Economic and Game Theory Page, <http://www.dklevine.com/archive/scotchmer-when-is-ip-best.pdf>.

Gambril, David. 'Court Allows Patent on Harvard Mouse: Decision Paves the Way for Patenting all Life Forms except Humans' *Law Times*, <http://www.canadalawbook.ca/headlines/headlines52_arc.html>.

Gates, Bill. 'Will Frankenfood Feed the World? Genetically Modified Food has Met Fierce Opposition among Well Fed Europeans, but It's the Poor and the Hungry who Need it Most' *Vision 21* 1555:25 (19 June 2000), TIME.com, <http://www.time.com/time/magazine>.

Geller, Paul Edward. 'An International Patent Utopia' (2003) 85 *J. Pat. and Trademark Off. Soc'y* 582.

'Genetics and Patenting' Human Genome Project Information, <http://www.ornl.gov/sci/techresources/Human_Genome/elsi/patents.shtml#6>.

Glashausser, Alex. 'Difference and Deference in Treaty Interpretation' (2005) 50 *Vill. L. Rev.* 25.

Gold, E.R. 'Biomedical Patents and Ethics: A Canadian Solution' (2000), 45 *McGill L. J.* 413.

Gold, E.R. and Adams, W.A. 'The Monsanto Decision: The Edge or the Wedge' (2001) 19 *Nat. Biotechnol.* 587.

Goodman, Ryan and Jinks, Derek. 'Measuring the Effects of Human Rights Treaties'. (2003) 14 *European Journal of International Law* 1.

Gordon, Wendy. 'Of Harms and Benefits: Torts, Restitution, and Intellectual Property' (June 1992) XXI *J. Legal Stud.* 449.

Gore, Charles. 'The Least Developed Countries 2000 Report, "Aid, Private Capital Flows and External Debt: The Challenge of Financing Developments in the LDCs"' (Geneva: UNCTAD 2000) United Nations Conference on Trade and Development, <http://www.unctad.org/en/pub/ps1ldc00.en.htm>.

Gostin, Lawrence O. 'World Health Law: Toward a New Conception of Global Health Governance for the 21st Century' (2005) 5 *Yale J. Health Pol'y L. and Ethics* 413.

Gulbrandsen, Carl. 'Stem Cell Patent Holder's View of the California Challenge' (16 November 2004), Wisconsin Technology Network, <http://wistechnology.com/article.php?id=1352>.

Hall, Bronywn H. and Ziedonis, R.H. 'The Patent Paradox Revisited: An Empirical Study of Patenting in the U.S. Semiconductor Industry, 1979–1995' (2001) 32 *Rand J. Econ.* 102.

Hand, Graham. 'Policy Aspects of Reservations and Objections to Human Rights Conventions' in Gardner, J.P, ed., *Human Rights as General Norms and A State's Right to Opt Out: Reservations and Objections to Human Rights Conventions* (London: British Institute of International and Comparative Law, 1997). 117.

Hannum, Hurst. 'The UDHR in National and International Law' (1989/90) 12:1 Interights Bulletin, Interights, <http://www.interights.org/pubs/Old%20Bulletin%20PDFs/Bulletin%2012.1.pdf>.

Hardin, Garret. 'The Tragedy of the Commons' (13 December 1968) 162 *Science* 1243. Die Off, <http://dieoff.org/page95.htm>.

Hatfield, Mark O. 'From Microbe to Man' (1995) 1 *Animal L.* 5.

Heller, Michael A. and Eisenberg, Rebecca S. 'Can Patents Deter Innovation: The Anticommons in Biomedical Research' (1 May 1998) *Science* 280, Science/AAAS, <http://www.sciencemag.org>.

Hess, Charlotte and Ostrom, Elinor. 'Ideas, Artifacts, and Facilities: Information as a Common-Pool Resource' (2003) 66 *Law and Contemp.*

Probs. 111, Duke Law, <http://www.law.duke.edu/journals/lcp/articles/lcp66dWinterSpring2003p111.htm>.

Ho, C.M. 'Patent Law and Policy Symposium: Re-engineering Patent Law: The Challenge of New Technologies: Part III: International and Comparative Law Issues: Splicing Morality and Patent Law: Issues Arising from Mixing Mice and Men' (2000) 2 *Wash. U. J. L. and Pol'y* 247.

Howse, Robert. 'India's WTO Challenge to Drug Enforcement Conditions in the European Community Generalized System of Preferences: A Little Known Case with Major Repercussions for "Political" Conditionality in US Trade Policy' (2003) 4 *Chicago J. Int'l L.* 385.

——. 'The WHO/WTO Study on Trade and Public Health: A Critical Assessment.' (2004) 24:2 *Risk Analysis* 501, <http://faculty.law.umich.edu/rhowse/Drafts_and_Publications/WHOWTO.pdf#search=%22%22The%20WHO%2FWTO%20Study%20on%20Trade%20and%20Public%20Health%3A%20A%20Critical%20Assessment.%22%22>.

'Human Rights Approaches to Sustainable Development' (May 2002), UN Non-Governmental Liaison Service, <http://www.un-ngls.org/documents/pdf/roundup/ru90hrsd.pdf>.

Hunter, Jordana. 'Broken Promises: Trade, Agriculture and Development in the WTO' (May 2003) 4 *Melbourne J. Int'l L.* 299.

Ignatieff, Michael. 'Human Rights as Politics and Idolatry' in Gutman, A. ed, *Human Rights as Politics and Idolatry* (Princeton, NJ: Princeton University Press, 2001) at 55.

Janis, Mark D. 'Sustainable Agriculture, Patent Rights, and Plant Innovation' (2002) 9:1 *Ind. J. Global Legal Stud.* 91.

Jenkins, Neil. 'In Brief: The Impact of the EU Biotechnology Directive on the Patenting of Biotechnology' (2001) 96 *Patent World* 128.

Jonas, Hans. 'Lasst uns einen Menschen klonieren' in Jonas, Hans, *Technik, Medizin und Eugenik. Zur Praxis des Prinzips Verantwortung* (Frankfurt am Main: Suhrkamp, 1985).

Jones, B.A. *New York Times* (10 July 1968), cited in Baldwin, R. *Nontariff Distortions of International Trade* (Washington, DC: The Brookings Institution, 1970).

Jones, Phillip B.C. 'Will the Oncomouse Squeak Through the Supreme Court of Canada?' (July 2002) *Information Systems for Biotechnology News Report* 9, Information Systems for Biotechnology, <http://www.isb.vt.edu/articles/jul0206.htm>.

Joyner, Daniel H. 'Bridging the Gap between International Law and Foreign Policymaking' (2003) 31 *Denv. J. L. and Pol'y* 437, University of Denver Sturm College of Law, <http://www.law.du.edu/ilj/online_issues_folder/joyner.pdf>.

Kennedy, David. 'The Forgotten Politics of International Governance' (2001) 2 *E.H.R.L.R.* 117.

Kennedy, Kevin. 'A Review of Globalization and its Discontents' (2003) 35 *Geo. Wash. J. Int'l L. and Econ.* 251.

Kesan, Jay P. 'Carrots and Sticks to Create a Better Patent System' (2002) 12:2 *Berkeley Tech. L. J.* 763.

Korff, Baron S.A. 'An Introduction to the History of International Law' (1924) 18 *Am. J. Int'l L.* 249.

Koskenniemi, Martii. 'International Law in Europe: Between Tradition and Renewal' (2005) 16 *European Journal of International Law* 1.

Kremer, Michael. 'Creating New Markets for New Vaccines' (13 April 2000), International AIDS Economic Network, <http://www.iaen.org/files.cgi/57_kremervacc1.pdf>.

Laidlaw, Stuart. 'Reckoning with Technology Law in Court: Monsanto Decision Hurt Equity, Innovation: Expert' *Toronto Star* (21 June 2004), D1.

LaMonica, Martin. 'IBM retains patent crown' (12 January 2004), CNET News.com, <http://news.com.co/2100-1008_3-5138578.html>.

Leaffer, Marshal A. 'Protecting United States IP Abroad' (1991) 76 *Iowa L. Rev.* 272.

Leahy, Stephen. 'Monsanto Victory Plants Seed of Privatization' *Inter Press Service* (5 October 2004), Global Policy Forum, <http://www.globalpolicy.org/socecon/trade/gmos/2004/1005seeds.htm>.

Leckow, Ross. 'Bringing the Disenfranchised to the Table- Lessons of Conditionality' in *The Measure of International Law: Effectiveness, Fairness and Validity*, Canadian Council on International Law 31st Annual Conference 2002 (New York: Kluwer Law International, 2004).

Lemley, Mark. 'Rational Ignorance at the Patent Office' (2001) 95:4 *Nw. U. L. Rev.* 1495.

LePage, Michael. 'They Came, They Glowed…They Could Conquer the Pet Trade' *New Scientist* (20 December 2003–9 January 2004), 24, <http://www.newscientist.com>.

Lerner, J. 'Patent Policy Shifts and Innovation over 150 Years' (2002) 92 *Am. Econ. Rev. (P. and P.)* 221.

Lindert, Peter H. and Williamson, Jeffery G.. 'Does Globalization Make the World More Unequal?' National Bureau of Economic Research, Working Paper 8228 (April 2001), National Bureau of Economic Research, <http://www.nber.org/papers/w8228>.

Lippman, Abby. 'Prenatal Genetic testing and screening: Constructing needs and reinforcing inequities' (1991) 17 *Am. J. L. and Med.* 15.

——. 'Worrying – and worrying About – the Geneticization of Reproduction and Health' in Basen, G., Eichler, M. and Lippman, A., eds, *Misconceptions: The Social Construction of Choice and the New Reproductive Technologies*, vol. 1 (Ottawa: Voyageur Press, 1993).

Lowensteyn, Peter. 'Grotius and the Socioeconomic Development of the United Provinces Around 1600', Hugo Grotius <http://www.lowensteyn.com/grotius/>.

Lowi, Theodore J. 'The Welfare State, The New Regulation, and the Rule of Law' in Hutchinson, Allan C. and Monahan, Patrick J., *The Rule of Law: Ideal or Ideology* (Toronto: Carswell, 1987).

Marden, Emily. 'The Neem Tree Patent: International Conflict over the Commodification of Life' (Spring 1999) 22 *B.C. Envtl. Aff. L. Rev.* 279.

Macdonald, Stuart. 'When Means Become Ends: Considering the Impact of Patent Strategy on Innovation' (2004) *Information Economics and Policy* 16:1, *Information Economics and Policy*, <http://else.hebis.de/cgi-bin/sciserv.pl?collection=journals&journal=01676245&issue=v16i0001>.

McBride, Jeremy. 'Reservations and Capacity of States to Implement Human Rights Treaties' in Gardner, J.P, ed., *Human Rights as General Norms and A State's Right to Opt Out: Reservations and Objections to Human Rights Conventions* (London: British Institute of International and Comparative Law, 1997).

McDonald, Jan. 'It's Not Easy Being Green': Trade and Environment Linkages beyond Doha' in Buckley, R., ed., *The WTO and The Doha Round: The Changing Face of the World Trade Organization*, Global Trade and Finance Series, vol. 4 (New York: Kluwer Law International, 2003).

Messer, Ellen. 'Food Systems and Dietary Perspective: Are Genetically Modified Organisms the Best Way to Ensure Nutritionally Adequate Food?' (Fall 2001) 9:1 *Ind. J. Global Legal Stud.* 65.

Meurer, Michael J. 'Business Method Patents and Patent Floods' (2002) 8 *J. L. and Pol'y* 309.

Mgbeogi, Ikechi. 'Patents and Traditional Knowledge of the Uses of Plants: Is a Communal Patent Regime Part of the Solution to the Scourge of Bio Piracy?' (Fall 2001) 9:1 *Ind. J. Global Legal Stud.* 163.

Milanovic, Branko. 'True World Income Distribution, 1988 and 1993: First Calculations Based on Household Surveys Alone' (2002) 112 *Econ. J.* 51.

Moffat, Viva R. 'Mutant Copyrights and Backdoor Patents: The Problem of Overlapping Intellectual Property Protection' (2004) 19 *Berkeley Tech. L. J.* 1473.

Mossoff, Adam. 'Rethinking the Development of Patents: An Intellectual History, 1550–1800' (2001) *Hastings L.J.* 1255.

Nelson, Richard R. 'The Market Economy and the Scientific Commons' (2004) 33:3 *Research Policy*, 455.

'Nine Things Farmers Need to Know about the Seed Sector Review' (13 May 2004), National Farmers Union, <http://www.nfu.ca/seedsector.pdf>.

Oczek, Jeremy P. 'In the Aftermath of the 'Terminator' Technology Controversy: Intellectual Property Protections for Genetically Engineered Seeds and the Right to Save and Replant Seed' (2000) 41 *B.C.L. Rev.* 627.

Ogus, A.I. 'Regulatory Law: Some Lessons from the Past', (1999) 12:1 *Legal Studies* 1.

'Origins of the Human Rights Regime' (2000) 54 *Int. L. Org.* 217.

Pahuja, Sundhya. 'This is the World: Have Faith' (2004) 15 *European Journal of International Law* 2 also in Wright, Shelley. *International Human Rights, Decolonisation and Globalisation: Becoming Human.* (London and New York: Routledge, 2001).

Petersmann, Ernst-Ulrich. 'Time for Integrating Human Rights into the Law of Worldwide Organizations: Lessons from European Integration Law for Global Integration Law' Jean Monnet Working Paper 7/01, The Jean Monnet Program, <http://www.jeanmonnetprogram.org/papers/01/012301.html>.

——. 'The Transformation of the World Trade System through the 1994 Agreement Establishing the World Trade Organization' (1995) 6 *European Journal of International Law* 161, European Journal of International Law, <http://www.ejil.org/journal/Vol6/No2/art1.html>.

Putnam, Jonathan D. 'The Price We Pay for Drug Research' *Innovate* (Spring 2004) 26, University of Toronto – Faculty of Law, <http://www.law.utoronto.ca/documents/publications/innovate_spring04.pdf>.

Quigley, William. 'Catholic Social Thought and the Amorality of Large Corporations: Time to Abolish Corporate Personhood', (2004) 5 *Loy. J. Pub. Interest*, Loyola University, <http://www.loyno.edu/~quigley/catholicsocialthoughtamoralityoflargecorp.pdf>.

Rai, Arti, 'Addressing The Patent Gold Rush: The Role of Deference to PTO Patent Denials' (2000) 2 *Wash. U.J. L. and Pol'y* 199.

Rebak, Gary L. 'Patently Absurd: Corporations are Increasingly Converting the Shield of Patent Protection into the Sword of Unfair Competition' *FORBES ASAP* 169:14 (24 June 2002), 44.

Redgwell, Catherine. 'The Law of Reservations in respect of Multilateral Conventions' in Gardner, J.P, ed., *Human Rights as General Norms and A State's Right to Opt Out: Reservations and Objections to Human Rights Conventions* (London: British Institute of International and Comparative Law, 1997).

Reichman, Jerome H. 'From Free Riders to Fair Followers: Global Competition Under the TRIPS Agreement' (1997) 29:1–2 *J. Int'l. L. and Pol.* 12.

——. 'The TRIPS Agreement Comes of Age: Conflict or Cooperation with the Developing Countries' (Summer 2000) 23:3 *Case W. Res. J. Int'l L.* 441.

——. 'Minimum Standards of Intellectual Property Protection' in Correa, Carlos and Yusuf, Abdulqawi A., eds, *Intellectual Property and International Trade* (Boston: Kluwer Law International, 1998).

——. 'Of Green Tulips and Legal Kudzu: Repackaging Rights in Subpatenable Innovation' (2000) 53 *Vand. L. Rev.* 1743.

—— and Franklin, Jonathan A. 'Privately Legislated Intellectual Property Rights: Reconciling Freedom of Contract with Public Good Uses of Information' (1999) 147 *U. Pa. L. Rev.* 875.

—— and Lange, David. 'Bargaining Around the TRIPS Agreement: The Case for Ongoing Public-Private Initiatives to Facilitate Worldwide Intellectual Property Transactions' (Fall 1998) 9:1 *Duke J. of Comp. and Int'l L.* 11.

Ricardo, David. 'The Principles of Political Economy' (1817) full text McMaster University, <http://www.socsci.mcmaster.ca/~econ/ugcm/3ll3/ricardo/prin/prin1.txt>; and Fordham University, <http://www.fordham.edu/halsall/mod/ricardo-summary.html>.

Robertson, John A. 'Procreative Liberty in the Era of Genomics' (2003) 29 *Am. J. L. and Med.* 439.

Rock, Melanie. 'Genetic Norms, Eugenic Logic and UNESCO's International Bioethics Committee' (1997) 7 *Eubios Journal of Asian and International Bioethics* 108, <http://www.biol.tsukuba.ac.jp/~macer/EJ74/ej74g.html>.

Roy, Jeffrey. 'Canada's Technology Triangle' in de la Mothe, John and Paquet, Gilles, eds, *Local and Regional Systems of Innovation* (The Hague: Kluwer Academic Publishers, 1998).

Sagoff, Mark. 'Biotechnology and Agriculture: The Common Wisdom and its Critics' (Fall 2001) 9:1 *Ind. J. Global Legal Stud.* 13.

Sampford, Charles et al. 'Living Up to the Promises of Global Trade' in Buckely, Ross, ed, *The WTO and The Doha Round: The Changing Face of the World Trade Organization*, Global Trade and Finance Series, vol. 4 (New York: Kluwer Law International, 2003).

Schmidt, Markus. 'Reservations to United Nations Human Rights Treaties-the Case of the Two Covenants' in Gardiner, J.P., *Human Rights as General Norms And A State's Right To Opt Out* (London: British Institute of International and Compartive Law, 1997).

Sell, Susan K.'Post-TRIPS Developments: The Tension Between Commercial and Social Agendas in the Context of Intellectual Property' (2002) 14 *Fla. J. Int'l L.* 193.

Shand, Hope. 'Gene Giants: Understanding the "Life Industry"' in Tokar, Brian, ed., *Redesigning Life? The Worldwide Challenge to Genetic Engineering* (Montreal: McGill-Queen's University Press, 2001).

Shapiro, Carl. 'Navigating the Patent Thicket: Cross Licences, Patent Pools, and Standard-Setting' (March 2001), University of California Berkeley, <http://faculty.haas.berkeley.edu/shapiro/thicket.pdf>.

Shiva, Vandana. 'Genetically Engineered Vitamin A Rice": A Blind Approach to Blindness Prevention' in Tokar, Brian, ed, *Redesigning Life? The Worldwide Challenge to Genetic Engineering* (Montreal: McGill-Queen's University Press, 2001).

——. 'Biopiracy: The Theft of Knowledge and Resources' in Tokar, Brian, ed, *Redesigning Life? The Worldwide Challenge to Genetic Engineering* (Montreal: McGill-Queen's University Press, 2001),.

——. 'Biopiracy: The Plunder of Nature and Knowledge' (Spring 1998) 7:4 *Global Biodiversity* 38.

Singer, Joe. 'No Right to Exclude: Accommodation and Private Property' (1995–1996) 90 *Nw. U. L. Rev.* 1299. Harvard Law School, <http://www.law.harvard.edu/faculty/jsinger/bibliography/pdfs/right.pdf>.

Singh, Someshwar. 'Trips Regime At Odds With Human Rights Law, Says UN Body' (2000) Third World Network, <http://www.twnside.org.sg/title/odds.htm>.

Slaughter, Anne-Marie. 'Breaking Out: The Proliferation of Actors in the International System' in Dezalay, Yves. and Garth, Bryant G. eds, *Global Prescriptions: The Production, Exportation, and Importation of a New Legal Orthodoxy* (Ann Arbor, MI: University of Michigan, 2005),12.

'Special Issue on the Legalization of World Politics of International Organization' (2000) 54 *Int'L Org.* 3.

Stumberg, Robert and Warren, William. 'The Boston Tea Party Revisited: Massachusetts Boycotts Burma', National Conference of State Legislatures, <http://www.ncsl.org/programs/pubs/599burma.htm>.

Sun, Haochen. 'The Road to Doha and Beyond: Some Reflections on the TRIPS Agreement and Public Health' (2004) 15 *European Journal of International Law* 1.

Sunstein, Cass R. 'Beyond the Republican Revival', (1988) 97 *Yale L.J.* 1539.

Tauli-Corpuz, Victoria. 'Biotechnology and Indigenous Peoples' in Tokar, Brian, ed., *Redesigning Life? The Worldwide Challenge to Genetic Engineering* (Montreal: McGill-Queen's University Press, 2001).

Tansey, Geoff. 'Food Security, Biotechnology and Intellectual Property: Unpacking some Issues around TRIPS' (July 2002), <http://www.geneva.quno.info/>.

'Too Many Patents? US Plans Hearings' *Lawyer's Weekly* (30 November 2001).

'Technologies for Genomic Mapping, Sequencing and Analysis' (March 1997) 26.8 NIH Guide, National Human Genome Research Institute, <http://www.genome.gov/10000995>.

Trebilcock, Michael. 'Post Seattle Reflections: A Qualified Defense of the WTO and an Unqualified Defense of the International Rule of Law', Session 6, *Corporate Power, National Sovereignty, and the Rule of Law in a Global Economy* 319.

——. 'Critiquing the Critics of Economic Globalization' (2005) v. 1(102) *J. of Int. Law and Relations* 213.

——. 'What Makes Poor Countries Poor?: The Role of Institutional Capital in Economic Development' in Buscaglia, E. et al., eds, *The Law and Economics of Development* (Greenwich, CT: JAI Press, 1997).

—— and Howse, Robert, 'A Cautious View of International Harmonization: Implications from Breton's Theory of Competitive Governments' [unpublished paper presented 1 March 1999].

——. 'Trade Liberalization and Regulatory Diversity: Reconciling Competitive Markets with Competitive Politics' (1996) 6 *Eur J.L. and Econ.* 5.

'Understanding Global Trade and Human Rights: Report and Resource Guide for National Human Rights NGOs in View of the 2005 WTO Ministerial Conference, Hong Kong (MC6)' (July 2005) No. 423/2, International Federation for Human Rights, <http://www.fidh.org/IMG/pdf/wto423a.pdf>.

Van Steveninck, De Ruyter. 'Import Substitution and the Debt Crisis in Latin America' (1991) 3 *Timbergen Institute Research Bulletin* 133.

Vazquez, Carlos Manuel. 'Trade Sanctions and Human Rights: Past, Present, and Future' 6(4) *J. Int'l Econ. L.* 797.

Walterscheid, Edward C. 'The Use and Abuse of History: The Supreme Court's Interpretation of Thomas Jefferson's Influence on the Patent Law' (1999) *IDEA: The Journal of Law and Technology* 195.

Warren, Lynda M. 'Agrarian Law and Sustainable Development: Fifth World Conference on Agrarian Law' (1998) 1 *Biosciences and the Law* 411.

Westphal, Sylvia Pagan. 'Growing Human Organs on the Farm' *New Scientist* (20 December 2003–9 January 2004), 4.

'World Inequality', *BBC News* (18 July 2001) British Broadcasting Corporation<www.news.bbc.co.uk/1/hi/business/1442073.stm>.

Wysong, Pippa. 'Breaking Ground' *Financial Post* (13 November 2000), E1.

Yasuaki, Onuma. 'International Law in and with International Politics: The Functions of International Law in International Society' (2003) 14 *European Journal of International Law* 1.

Yelpaala, Kojo. 'Owning the Secret of Life: Biotechnology and property Rights Revisited' (2000) 32 *McGeorge L. Rev.* 111.

Yusuf, Abdulqawi A. 'TRIPs: Background, Principles and General Provisions' in Carlos Correa and Abdulqawi A. Yusuf, eds, *Intellectual Property and International Trade* (USA and Canada: Kluwer Law International, 1998).

Zagel, Gudrun Monika. 'WTO and Human Rights: Examining Linkages and Suggesting Convergence' (2005) 2:2 International Development Law Organization Voices of Development Jurist Paper Series; Social Science Research Network, <http://papers.ssrn.com/sol3/papers.cfm?abstract_id=740265>.

Government Documents

Auditor General of Canada, *Report of the Auditor General of Canada to the House of Commons, 1999* (Ottawa; Minister of Public Works and Government Services Canada, 1999).

Canada, Parliament, House of Commons Standing Committee on Foreign Affairs and International Trade, 'HIV/AIDS and the Humanitarian Crisis in Sub-Saharan Africa' (June 2003), Parliament of Canada, <http://www.parl.gc.ca>.

Canada, 2003 First Ministers' Accord on Health Care Renewal (Ottawa, September 2003), Health Canada, <http://www.hc-sc.gc.ca/english/hca2003/accord.html>.

——. 'Patenting of Higher Life Forms and Related Issues' (Report to the Government of Canada Biotechnology Ministerial Coordinating Committee) (June 2002).

——. 'Patenting of Higher Life Forms and Related Issues: Interim Report' (November 2001).

——. 'Patenting of Higher Life Forms and Related Issues: Summary of Comments, Responses to CBAC Interim Report' (6 June 2002).

Canadian Institute for Health Information Annual Report 2003–2004, <http://secure.cihi.ca/cihiweb/dispPage.jsp?cw_pagc=GR_1131_E>.

Clayman, Bruce. *Addendum to Technology Transfer at Canadian Universities: Fiscal year 2001 Update* (Report for the Canada Foundation for Innovation) (31 July 2003), SFU Research Matters, VP Research Reports, <http://www.sfu.ca/vpresearch/vpreports.htm>.

Health Canada. *Achieving Health for All: A Framework for Health Promotion* by Epp, Jake (Ottawa: Minister of Health, 1986).

——. Report of the National Forum on Xenotransplantation: Clinical, Ethical and Regulatory Issues, Ottawa (6-8 November 1997) (Ottawa: Health Canada, 1998), Health Canada, <http://www.hc-sc.gc.ca/hpb-dgps/therapeut/zfiles/english/btox/reports/frmrptx_e.pdf>.

Industry Canada, Canadian Intellectual Property Office. *Manual of Patent Office Practice*, 1998 ed, Canadian Intellectual Property Office, <http://strategis.ic.gc.ca/sc_mrksv/cipo/patents/mopop-e.pdf>.

Martin, Paul. Budget Speech (28 February 2000), Department of Finance Canada, <http://www.fin.gc.ca/budget00/speech/speech1e.htm>.

National Forum on Health. Canada Health Action: Building on the Legacy – Final Report of the National Forum on Health (Ottawa: Minister of Public Works and Government Services, 1997).

Schrecker, Ted and Wellington, Alex. 'Patenting of Biotechnological Innovations Concerning Animals and Human Beings' (Report prepared for the Canadian Biotechnology Advisory Committee Project Steering Committee on Intellectual Property and the Patenting of Higher Life Forms, 31 March 1999).

Schrecker, T., Elliott, C., Hoffmaster, C.B, Keyserlingk, E.W, and Somerville, M.A. 'Ethical Issues Associated with the Patenting of Higher Life Forms' (17 May 1997), Westminster Institute for Ethics and Human Values, McGill Centre for Medicine, Ethics and Law, Strategis.gc.ca, <http://strategis.ic.gc.ca/epic/internet/inippd-dppi.nsf/en/ip00095e.html>.

Conferences and Conference Papers

Amani, Bita. 'Patents, the Charter, and a Healthy Dose of Rights in Wrongs: The Poison is the Elixir for Life, Liberty, and Security of the Person', *Public Wrongs, Private Duties: Rethinking Public Authority Liability in Canada* (University of New Brunswick, Faculty of Law: 18–19 May 2006).

Dorland, Michael. 'Postmodern Law and the Translation of Cultures', Panamerican Colloquium on Cultural Industries and Dialogue between Civilizations in the Americas, Montreal, 22–24 April 2002, <http://www.er.uqam.ca/nobel/gricis/actes/panam/Dorland.pdf>.

'Dual Controversies of the Double Helix: Challenges of Regulating the Information and Property Aspects of Genetic Technology', *Reaching Through the Human Genome,* Keynote Address, Centre for Innovation Law and Policy, <http://www.innovationlaw.org/tip/pages/genetic_technology.htm>.

Epp, Jake. 'Achieving Health for All: A Framework for Health Promotion', First International Conference on Health Promotion, 1986).

Focus Resource Centre, <http://www.frcentre.net/library/AchievingHealthForAll.pdf>.

International Public Goods and Transfer of Technology Under a Globalized Intellectual, Property Regime Conference, Duke Law School, 4–6 April 2003, TRIPS Webcast, <http://www.law.duke.edu/trips/webcast.html>.

Kalanje, Christopher M. *WIPO-WASME Special Program on Practical Intellectual Property Issues*, Geneva, 6–9 October 2003, WIPO, <http://www.wipo.int/sme/en/activities/meetings/wipo_wasme_03/presentation/wipo_wasme_ipr_ge_03_12.pdf>.

MacCormick, Neil. 'On the Very Idea of Intellectual Property: An Essay According to the Institutionalist Theory of Law', University of Edinburgh, SCRIPT Presidential Lecture 2001, The Old College, 11 February 2002, Edinburgh Law School: Arts and Humanities Research Board<http://www.law.ed.ac.uk/ahrb/publications/online/maccormick.htm>.

Morrow, David A. and Smordin, Sandee. 'Examination of Biotech Patent Issues: The Harvard Mouse Case', International Conference on Intellectual Property, University of British Columbia, 19 September 2003.

Working Papers

Braga, Carlos A. Primo, Fink, Carsten and Sepulveda, Claudia Paz. 'Intellectual Property Rights and Economic Development' (March 2000) Washington DC, World Bank Discussion Paper No. 412.

Hall, Bronywn H. 'Business Method Patents, Innovation, and Policy' NBER Working Paper Series, Working Paper 9717, National Bureau of Economic Research, <http://www.nber.org/papers/w9717>.

Heald, Paul J. 'Mowing the Playing Field: Addressing Information Distortion and Asymmetry in the TRIPS Game' (2002) Vanderbilt Law and Economics Research Paper No. 02-21: Vanderbilt Public Law Research Paper 02-15, SSRN, <http://papers.ssrn.com/sol13/papers.cfm?abstract_id=319301>.

Howlett, Melanie J. and Christie, Andrew F. 'An Analysis of the Approach of the European, Japanese and United States Patent Offices to Patenting Partial DNA Sequences (ESTs)' (2004), University of Melbourne *Legal Studies* Research Paper No. 82, <http://ssrn.com/abstract=573184.

Levin, Jonathan and Levin, Richard. 'Patent Oppositions' (1 August 2002) Yale Law School John M. Olin Centre for Studies in Law, Economics, and Public Policy Working Paper Series, Working Paper No. 283, <http://lsr.nellco.org/yale/lepp/papers/283>.

Moser, P. 'How do Patent Laws Influence Innovation? Evidence from Nineteenth Century World Fairs' (2001) UC Berkeley: Working Paper.

Nuffield Council on Bioethics Discussion Paper, 'The Ethics of Patenting DNA' (July 2002), Nuffield Council on Bioethics, <http://www.nuffieldbioethics.org/publications/pp_0000000014.asp>.

Usher, Dan. 'The Distributive Implications of Patents on Indivisible Goods' (April 2004) Queen's Economic Department, Institute for Economic Research Working Paper No. 1018, <http://qed.econ.queensu.ca/pub/papers/abstracts/download/2004/1018.pdf>.

Other Material

Amani, Bita. 'The Case of the Harvard Mouse' (June 2003) CBC Radio One, *IDEA*S interview. Archived Canadian Broadcasting Corporation, <http://www.cbc.ca/ideas/calendar/2003/10_October.html>.

Canadian Cancer Society, 'Background on the Patenting of BRCA1 and 2 Genes', Canadian Cancer Society, <http://www.cancer.ca/ccs/internet/standard/0,3182,3172_31282995_32749610_langId-en,00.html>.

Canadian Intellectual Property Office, <http://cipo.gc.ca/>.

CBC Radio One, *IDEA*S Program 'How Humans Invented Animals' June (2003) archived Canadian Broadcasting Corporation, <http://www.cbc.ca/ideas/features/invented-animals/index.html>.

Cave, Damien.'Who ya Gonna Call? Patent Busters', Salon.com, <http://dir.salon.com/tech/view/2000/10/23/cella/index.html>.

Donahue, Marilyn. 'Donahue DNA', Marilyn Donahue, <http://www.mudhaus.com/marilyn/donahue>.

ETC Group, Action Group on Erosion, Technology and Concentration, 'Maize Rage in Mexico: GM Maize Contamination in Mexico – 2 Years Later' (10 October 2003), ETC Group, <http://www.etcgroup.org>.

Federalist Papers, Declaration of Independence, US Constitution, Historical Documents, <http://memory.loc.gov/const/fed/fedpapers.html>.

Funk and Wagnalls Standard Dictionary (New York: Funk and Wagnalls Company, 1977).

'Online Exhibits', The National Archives, <http://www.mkgandi.org/journalist/index.htm>.

'The GATT Years: From Havana to Marrakesh' World Trade Organization, <http://www.wto.org/english/thewto_e/whatis_e/tif_e/fact4_e.htm>.

International Encyclopedia of Unified Science, vol. 2, no. 2 (London: University of Chicago Press Ltd, 1962).

'Introduction to the Great War' Public Broadcasting Service, <http://www.pbs.org/greatwar/chapters/index.html>.

Jefferson, Thomas. 'Preamble to a Bill for the more General Diffusion of Knowledge, 1779' FE 2:221, Papers 2:527, University of Chicago Press Books Division, <http://press-pubs.uchicago.edu/founders/documents/v1ch18s11.html>.

Kennedy, Paul. 'How Humans inventing Animals' (2003) *IDEA*S Program, CBC RADIO, <http://www.cbc.ca/ideas/features/invented-animals/>.

Management of Breast Diseases, TransMed, <http://www.breastdiseases.com/risksfac.htm>.

Marcotte, Paul. 'The Biotechnology Industry and Animal Experimentation in Canada' Biotechnology Bulletin, Fasken Martineau (April 2003), <http://www.fasken.com/web/fmdwebsite.nsf/0/E7F4zd29E0BDFCB1785256D0100720F1D/$File/BIOTECHNOLOGY_BULLETIN_APRIL_2003.PDF?OpenElement>.

Oldham, Paul. *Global Status and Trends in IP Claims: Microorganisms* (United Kingdom: Centre for Economic and Social Aspects in Genomics, 2004), <http://www.cesagen.lancs.ac.uk>.

Report of the Standing Committee on Foreign Affairs and International Trade: Canada and the Future of the World Trade Organization, Advancing a Millennium Agenda in the Public Interest (Cndn Millenium Report), Chair Bill Graham, MP (Ottawa: House of Commons, Canada, 1999).

Robinson, Mary. Symposium on Human Rights in the Asia-Pacific Region (January 1998).

Sen, Amartya.Globalization: Value and Ethics, Talk given at Falcone Foundation, May 23, 2001 <http://www.fondazionefalcone.it/sen.html>, last accessed June 20, 2003.

Stone, Richard, 'An Islamic Science Revolution?' (2005) 309 *Science* 1802–1804, <http://www.sciencemag.org>.

Study on Canada's Human Rights Obligations, Human Rights Reports, <http://www.sen.parl.gc.ca/vpoy/english/Special_Interests/speeches/human_rights_report_response_300402.htm>.

The New Lexicon Webster's Dictionary of the English Language (1987) Canadian edn.

Torys Newsletter, No. 2002-29T (6 December 2002), <http://www.torys.com>.

Index